Synthesis and Modification of Nanostructured Thin Films

Special Issue Editor

Ion N. Mihailescu

MDPI • Basel • Beijing • Wuhan • Barcelona • Belgrade • Manchester • Tokyo • Cluj • Tianjin

Special Issue Editor
Ion N. Mihailescu
Lasers Department,
Laser-Surface-Plasma
Interactions Laboratory,
National Institute for Lasers,
Plasma, and Radiation Physics
(INFLPR)
Romania

Editorial Office
MDPI
St. Alban-Anlage 66
4052 Basel, Switzerland

This is a reprint of articles from the Special Issue published online in the open access journal *Nanomaterials* (ISSN 2079-4991) (available at: https://www.mdpi.com/journal/nanomaterials/special_issues/thin_film).

For citation purposes, cite each article independently as indicated on the article page online and as indicated below:

LastName, A.A.; LastName, B.B.; LastName, C.C. Article Title. *Journal Name* **Year**, *Article Number*, Page Range.

ISBN 978-3-03928-454-2 (Pbk)
ISBN 978-3-03928-455-9 (PDF)

Cover image courtesy of Marco Bobinger.

Synthesis and Modification of Nanostructured Thin Films

About the Special Issue Editor

Ion N. Mihailescu graduated from the Physics faculty at the University of Bucharest, Romania, in 1969. Since then, he has worked at the Institute of Atomic Physics, and at the National Institute for Laser, Plasma and Radiation Physics since 1996. He is fellow of OSA (2019), Dr. Honoris Causa of University of Cergy Pontoise, France (2017), the first "Galileo Galilei" recipient "for outstanding results in the field of optics under unfavorable circumstances" (1994), receiver of the "Dragomir Hurmuzescu" Prize of the Romanian Academy (1975) and of the 2012 Medal of Honor from the Institute of Atomic Physics, Romania, for "contribution to the field of laser-matter interactions". He has published 604 scientific papers, with more than 6000 citations and has an h-index of 34, according to Web of Science.

Preface to "Synthesis and Modification of Nanostructured Thin Films"

This volume is a response to a challenge I received from Nanomaterials. I was very pleased to find out that we share the opinion that "Synthesis and Modification of Nanostructured Thin Films" can be still considered a hotspot of the interdisciplinary research between lasers, optics, materials, electronics, informatics, telecommunications, biology, medicine, and probably many others. It is nowadays generally accepted that thin films allow for the production of highly competitive structures after complex design by rather simple and reliable techniques, with minimum consumption of materials and very low production costs. On the other hand, it was demonstrated that thin films should not only just reproduce the composition of the base material, but present a preferential structure and even morphology, in any case, guaranteeing the desired functionality. This Special Issue reviews original methods of thin films fabrication/synthesis or modification but also complex complementary characterization using various dedicated techniques. One expects plenty of future developments for thin films applications in key technological areas. I consequently accepted, with pleasure, the invitation to serve as Guest Editor for this Special Issue. Our common conviction was confirmed by the large number of submissions. We received more than 30 contributions, but because of scientific, editorial, and technical standards we felt obliged to refuse numerous manuscripts. It was a difficult decision because many indeed contained new ideas, and we therefore took the liberty of encouraging the authors to return with improved manuscripts for future dedicated issues of Nanomaterials. Finally, we decided to confine this Special Issue to 18 contributions which we considered to be original and which open new research areas into the field of thin film synthesis, modifications, and applications. The resulting issue proved to be rather international, with authors from 12 countries (China, Czech Republic, France, Germany, Iraq, Israel, Italy, Poland, Portugal, Romania, Taiwan, and USA). All of the contributions represent prominent universities and research centers all over the world. The Guest Editor and the MDPI staff are pleased to offer this Special Issue to all interested readers, including graduate and PhD students as well as postdoctoral researchers, but also to the entire community interested in the field of nanomaterials We are also convinced that the Special Issue can serve as a useful reference for libraries. We hope that this new contribution to a very productive and prospective field will result in further developments with great significance from both scientific but also applicative/industrial points of view. However, our opinion is that there exists plenty of room for the next stage in the development of research in this and related fields. Our reasons are mainly based on the expected development of instruments and methods for both the synthesis and characterization of thin films and their utilization. New applications are therefore expected to emerge soon.

Ion N. Mihailescu
Special Issue Editor

 *nanomaterials*

Editorial

Synthesis and Modification of Nanostructured Thin Films

Ion N. Mihailescu

Lasers Department, Laser-Surface-Plasma Interactions Laboratory, National Institute for Lasers, Plasma, and Radiation Physics (INFLPR), Strada Atomistilor, nr. 409, P.O. Box MG-36, RO-077125 Magurele, Ilfov, Romania; ion.mihailescu@inflpr.ro

Received: 25 September 2019; Accepted: 4 October 2019; Published: 9 October 2019

The idea of nanomaterials, nanoscience, and nanotechnologies was formulated by Richard Feynman in 1959 in his famous lecture "There's Plenty of Room at the Bottom". He said that "The principles of physics, as far as I can see, do not speak against the possibility of maneuvering things atom by atom".

Since then, a nanomaterials "revolution" followed, confirming their superior properties, as e.g., toughness, strength, hardness, resistance to corrosion and wear, and various thermal, magnetic, and optical features. In this context, the fabrication and characterization of thin films remains the main cornerstone of nanotechnologies.

As a rule, films are deposited onto solid surfaces to obtain better properties. A thin film is basically defined as a low-dimension material fabricated by assembling atomic/molecular/ionic species with a final thickness in the nm range. Nano-thin films can be essentially synthesized from any kind of material, which opens the way to vast application domains.

This Special Issue on "Synthesis and Modification of Nanostructured Thin Films" contains contributions about thin film synthesis, modification, and characterization for potential applications in leading domains. This collection of 18 research papers represents 136 authors from 12 countries, and is devoted to advanced topics in both the synthesis (13) as well as the modification (5) of nanostructured thin films. In particular, the thickness of films ranges from a few up to 250 nm.

The major compounds of key interest were studied, including AlGaN [1], Cu NWs [2], photonic crystal fibers [3], $LiNbO_3$ [4], Au nanoparticles [5], Al_2O_3/Tm_2O_3 [6], Ge-DLC [7], Mo/Ti [8], ZnTe:Cu [9], Ge Sb Te [10], noble metal nanoparticles [11], collagen/Zn^{2+}-substituted calcium phosphates [12], TiO_2 [13], Si-DLC [14], SHG in ZnO nanofilms [15], $Cu_2Mg_xZn_{1-x}SnS_4$ [16], ethylene vinyl acetate (EVA) matrices [17], and LIPSS [18].

In the opinion of this Editor, the main characteristic of this selection is the quite large range of applications, which extends from nanobiomedicine to solar cells.

Finally, one should stress upon the original character of all contributions, which will serve as active vectors in the years to come for the further dynamic development of new nanostructured thin film systems.

Funding: This research received no external funding.

Acknowledgments: INM kindly thank Core Programme—Contract 16N/2019 for the permanent support of his work. A special thanks to all the authors for submitting their studies to the present Special Issue and for its successful completion. INM also acknowledges the *Nanomaterials* reviewers for enhancing the quality and impact of all submitted papers.

Conflicts of Interest: The authors declare no conflict of interests.

References

1. Tasi, C.-T.; Wang, W.-K.; Ou, S.-L.; Hunag, Y.-S.; Horng, R.-H.; Wuu, D.-S. Structural and Stress Properties of AlGaN Epilayers Grown on AlN-Nanopatterned Sapphire Templates by Hydride Vapor Phase Epitaxy. *Nanomaterials* **2018**, *8*, 704. [CrossRef] [PubMed]

2. Mock, J.; Bobinger, M.; Bogner, C.; Luigi, P.; Becherer, M. Aqueous Synthesis, Degradation, and Encapsulation of Copper Nanowires for Transparent Electrodes. *Nanomaterials* **2018**, *8*, 767. [CrossRef] [PubMed]

3. Malka, D.; Katz, G. An Eight-Channel C-Band Demux Based on Multicore Photonic Crystal Fiber. *Nanomaterials* **2018**, *8*, 845. [CrossRef] [PubMed]

4. Wu, R.; Wang, M.; Xu, J.; Qi, J.; Chu, W.; Fang, Z.; Zhang, J.; Zhou, J.; Qiao, L.; Chai, Z.; et al. Long Low-Loss-Litium Niobate on Insulator Waveguides with Sub-Nanometer Surface Roughness. *Nanomaterials* **2018**, *8*, 910. [CrossRef] [PubMed]

5. Rout, A.; Boltaev, G.S.; Ganeev, R.A.; Fu, Y.; Maurya, S.K.; Kim, V.V.; Srinivasa, K.R.; Guo, C. Nonlinear Optical Studies of Gold Nanoparticle Films. *Nanomaterials* **2019**, *9*, 291. [CrossRef] [PubMed]

6. Liu, Y.; Ouyang, Z.; Yang, L.; Yang, Y.; Sun, J. Blue Electroluminescent Al_2O_3/Tm_2O_3 Nanolaminate Films Fabricated by Atomic Layer Deposition on Silicon. *Nanomaterials* **2019**, *9*, 413. [CrossRef] [PubMed]

7. Jelinek, M.; Kocourek, T.; Jurek, K.; Jelinek, M.; Smolková, B.; Uzhytchak, M.; Lunov, O. Preliminary Study of Ge-DLC Nanocomposite Biomaterials Prepared by Laser Codeposition. *Nanomaterials* **2019**, *9*, 451. [CrossRef] [PubMed]

8. Shen, H.; Yao, B.; Zhang, J.; Zhu, X.; Xiang, X.; Zhou, X.; Zu, X. Effect of Thickness of Molybdenum Nano-Interlayer on Cohesion between Molybdenum/Titanium Multilayer Film and Silicon Substrate. *Nanomaterials* **2019**, *9*, 616. [CrossRef] [PubMed]

9. Chen, B.; Liu, J.; Cai, Z.; Xu, A.; Liu, X.; Rong, Z.; Qin, D.; Xu, W.; Hou, L.; Liang, Q. The Effects of ZnTe:Cu Back Contact on the Performance of CdTe Nanocrystal Solar Cells with Inverted Structure. *Nanomaterials* **2019**, *9*, 626. [CrossRef] [PubMed]

10. Bulai, G.; Pompilian, O.; Gurlui, S.; Nemec, P.; Nazabal, V.; Cimpoesu, N.; Chazallon, B.; Focsa, C. Ge-Sb-Te Chalcogenide Thin Films Deposited by Nanosecond, Picosecond, and Femtosecond Laser Ablation. *Nanomaterials* **2019**, *9*, 676. [CrossRef] [PubMed]

11. Tommasini, M.; Zanchi, C.; Lucotti, A.; Bombelli, A.; Villa, N.S.; Casazza, M.; Ciusani, E.; de Grazia, U.; Santoro, M.; Fazio, E.; et al. Laser-Synthesized SERS Substrates as Sensors toward Therapeutic Drug Monitoring. *Nanomaterials* **2019**, *9*, 677. [CrossRef] [PubMed]

12. Neacsu, I.A.; Arsenie, L.V.; Trusca, R.; Ardelean, I.L.; Mihailescu, N.; Mihailescu, I.N.; Ristoscu, C.; Bleotu, C.; Ficai, A.; Andronescu, E. Biomimetic Collagen/Zn^{2+}-Substituted Calcium Phosphate Composite Coatings on Titanium Substrates as Prospective Bioactive Layer for Implants: A Comparative Study Spin Coating vs. MAPLE. *Nanomaterials* **2019**, *9*, 692. [CrossRef] [PubMed]

13. Lungu, J.; Socol, G.; Stan, G.E.; Ştefan, N.; Luculescu, C.; Georgescu, A.; Popescu-Pelin, G.; Prodan, G.; Gîrţu, M.A.; Mihăilescu, I.N. Pulsed Laser Fabrication of TiO_2 Buffer Layers for Dye Sensitized Solar Cells. *Nanomaterials* **2019**, *9*, 746. [CrossRef] [PubMed]

14. Bociaga, D.; Sobczyk-Guzenda, A.; Komorowski, P.; Balcerzak, J.; Jastrzebski, K.; Przybyszewska, K.; Kaczmarek, A. Surface Characteristics and Biological Evaluation of Si-DLC Coatings Fabricated Using Magnetron Sputtering Method on Ti6Al7Nb Substrate. *Nanomaterials* **2019**, *9*, 812. [CrossRef] [PubMed]

15. Long, H.; Habeeb, A.A.; Kinyua, D.M.; Wang, K.; Wang, B.; Lu, P. Influences of Ga Doping on Crystal Structure and Polarimetric Pattern of SHG in ZnO Nanofilms. *Nanomaterials* **2019**, *9*, 905. [CrossRef] [PubMed]

16. Sui, Y.; Zhang, Y.; Jiang, D.; He, W.; Wang, Z.; Wang, F.; Yao, B.; Lili, Y. Investigation of Optimum Mg Doping Content and Annealing Parameters of $Cu_2Mg_xZn_{1-x}SnS_4$ Thin Films for Solar Cells. *Nanomaterials* **2019**, *9*, 955. [CrossRef] [PubMed]

17. Mariotti, G.; Vannozzi, L. Fabrication, Characterization, and Properties of Poly (Ethylene-Co-Vinyl Acetate) Composite Thin Films Doped with Piezoelectric Nanofillers. *Nanomaterials* **2019**, *9*, 1182. [CrossRef] [PubMed]

18. Alves-Lopes, I.; Almeida, A.; Oliveira, V.; Vilar, R. Influence of Femtosecond Laser Surface Nanotexturing on the Friction Behavior of Silicon Sliding Against PTFE. *Nanomaterials* **2019**, *9*, 1237. [CrossRef] [PubMed]

 *nanomaterials*

Article

Influence of Femtosecond Laser Surface Nanotexturing on the Friction Behavior of Silicon Sliding Against PTFE

Isabel Alves-Lopes [1,*], **Amélia Almeida** [1], **Vítor Oliveira** [1,2] **and Rui Vilar** [1]

[1] CeFEMA–Center of Physics and Engineering of Advanced Materials, Instituto Superior Técnico, Universidade de Lisboa, Avenida Rovisco Pais, 1049-001 Lisboa, Portugal

[2] Instituto Superior de Engenharia de Lisboa, Avenida Conselheiro Emídio Navarro No. 1, 1959-007 Lisboa, Portugal

* Correspondence: isabel.alves.lopes@tecnico.ulisboa.pt

Received: 29 July 2019; Accepted: 27 August 2019; Published: 30 August 2019

Abstract: The aim of the present work was to investigate the influence of laser-induced periodic surface structures (LIPSS) produced by femtosecond laser on the friction behavior of silicon sliding on polytetrafluoroethylene (PTFE) in unlubricated conditions. Tribological tests were performed on polished and textured samples in air using a ball-on-flat nanotribometer, in order to evaluate the friction coefficient of polished and textured silicon samples, parallel and perpendicularly to the LIPSS orientation. In the polished specimens, the friction coefficient decreases with testing time at 5 mN, while it increases slightly at 25 mN. It also decreases with increasing applied load. For the textured specimens, the friction coefficient tends to decrease with testing time in both sliding directions studied. In the parallel sliding direction, the friction coefficient decreases with increasing load, attaining values similar to those measured for the polished specimen, while it is independent of the applied load in the perpendicular sliding direction, exhibiting values lower than in the two other cases. These results can be explained by variations in the main contributions to friction and in the wear mechanisms. The influence of the temperature increase at the interface and the consequent changes in the crystalline phases of PTFE are also considered.

Keywords: laser surface texturing; laser-induced periodic surface structures; LIPSS; silicon; PTFE; friction

---

1. Introduction

Laser surface texturing is one of the most versatile methods for controlling the tribological properties of materials, since it allows creating a variety of surface textures at the micro and nanoscales in a wide range of materials, with excellent control of the surface features shape and size and negligible degradation of the bulk material [1]. Femtosecond lasers are particularly promising for this application because, due to their high intensity and extremely short interaction time, they significantly reduce undesirable thermal effects, leading to better control of the surface topography, improved accuracy, better reproducibility and less surface contamination [2–4]. Furthermore, due to the non-linear nature of the laser-material interaction mechanisms for these lasers, wide band gap transparent materials, as well as metals and semiconductors, can be easily processed [5].

When surfaces are irradiated with femtosecond laser pulses at fluences slightly above the material ablation threshold, parallel surface undulations known as laser-induced periodic surface structures (LIPSS) are formed on the surface of a wide variety of materials [6–9]. The surface topographies by femtosecond laser treatment were previously described by Bonse et al. [9] and Oliveira et al. [10–12]. In this paper, we consider LIPSS with a periodicity of the same order of magnitude of the radiation wavelength, known as low-spatial frequency LIPSS (LSFL). The formation of LSFL is due to a periodic modulation of the absorbed radiation intensity that arises as a consequence of the interference between the incident laser beam and electromagnetic waves propagating parallel to the surface [13,14]. This modulation of the laser absorbed energy generates a periodic variation of the material surface temperature and, consequently, of the transformations that occur in the material, leading to the imprinting of the surface topography by differential ablation [15]. LIPSS affect surface properties such as wettability [16], light reflectivity and transmittance [17–19], cell behavior and osseointegration capability of implant materials [16,20], bacteria adhesion and biofilm formation [21], as well as the materials tribological behavior, in particular its friction coefficient [6,22–28].

The studies available of the influence of LIPSS on the tribological behavior of materials reveal often a complex behavior [22–31]. The tribology of polytetrafluoroethylene (PTFE) has been extensively investigated (e.g., [32–38]), but the influence of surface nanotextures on this behavior remains poorly understood. He et al. [39] studied the effect of textures consisting of square section pillars and grooves of different dimensions on the friction of a poly(dimethylsiloxane) (PDMS) elastomer using nanoscratch testing. The tests were conducted at room temperature, using a 1.6 mm diameter 304 stainless-steel bearing ball, and a Rockwell indenter diamond tip with 25 μm radius, respectively, 5, 10 and 25 mN applied loads and 1 μm/s sliding speed. The authors found that the square pillar textures significantly reduce the macroscale friction coefficient, an effect explained by the reduction of the real contact area, and, consequently, of the adhesion contribution to friction, as compared to flat surfaces. A linear relationship between the friction coefficient and $(load)^{-1/3}$ was observed for the textured specimens, as predicted in classical friction theory for elastic contacts, but this relationship did not hold for the flat surfaces. Analysis of the experimental data using the Johnson–Kendall–Roberts theory [40], showed that this difference in behavior is due to a larger adhesion force in the case of the flat specimens, confirming the explanation provided for the effect of surface texturing on friction. At the microscale, the influence of texture on the friction coefficient is less noticeable, particularly at 25 mN load, because the reduction of adhesion is offset by the effect of mechanical interlocking due to the comparable size of the diamond tip and the surface features dimensions. In the case of the grooved textures, the friction coefficient is anisotropic, being largely controlled by the stick-slip behavior of the system.

The silicon-PTFE tribosystem is important for a wide range of applications, particularly in micro-electro-mechanical systems (MEMS). The use of polymers in microsystems has been increasing, as replacements of silicon in some parts (e.g., [41]), as coatings and lubricants (e.g., [42–44]) or in hybrid microdevices containing silicon electronics [45], and PTFE is one of the most attractive polymers for these applications due to its low dielectric constant, low surface adhesion, chemical inertness, electrical and thermal insulation properties and self-lubricating characteristics [46–49], so the study of the friction behavior of the silicon/PTFE system at the nanoscale is of utmost importance. In the present work, we studied the influence of surface textures produced by femtosecond laser on the friction behavior of single crystalline silicon sliding on PTFE, under unlubricated conditions.

2. Materials and Methods

<111> single crystal wafers of p-doped silicon with 525 ± 25 μm thickness were cut into approximately 1 cm × 1.5 cm samples and cleaned in an ultrasonic bath with acetone followed by isopropanol. Surface texturing was performed by the direct writing technique, using a Yb:KYW chirped-pulse regenerative amplification laser system (Amplitude Systèmes s-Pulse HP, Pessac, France) with a central wavelength of 1030 nm, pulse duration at full width at half maximum (FWHM) of 560 fs and a Gaussian intensity distribution in the cross-section of the beam, with a $(1/e^2)$ radius of $\omega_0 \sim 50$ μm at the sample plane, calculated using the method of Liu [50]. The laser beam was focused by a 10 mm focal length lens on a plane 12 mm above the surface of the specimen, leading to a spot size of about 100 μm. The polarization direction of the linearly polarized laser beam was controlled by a half-wave (λ/2) plate. The laser treatment was performed by moving the samples under the stationary beam using a computer-controlled XYZ stage. In order to cover all the surface with a uniform LIPSS

structure, scanning of the sample is performed by moving the stage in the XX direction under the stationary beam. After completing one track, the sample is displaced in the YY direction and then scanned in the XX direction again. The XX direction corresponds to the direction of the laser beam polarization, thus generating LIPSS perpendicular to the laser tracks. Consecutive tracks were partially overlapped, in the YY direction, by moving the specimen about 50% of the width of a single laser track in the XX direction. This allows texturing the entirety of the sample's surface with uniform LIPSS. The laser treatments were performed with an average pulse energy of 100 µJ, calculated from the average laser beam power measured with an Ophir Photonics 10A-SH-V1.1RoHS powermeter (Andover, MA, USA). Taking into consideration that the laser spot diameter at the specimens' surface was 200 µm, the average fluence was 0.32 J cm^{-2}. The scanning speed was 1 mm/s and the pulse frequency 200 Hz, leading to an average of 40 pulses per surface spot.

The textured surfaces were characterized using a LEICA DM 5500B binocular optical microscope (Wetzlar, Germany) and a JEOL JSM 7001F field emission gun scanning electron microscope (FEG-SEM, Akishima, Tokyo, Japan) operated in the secondary electrons imaging mode. The surface texture was quantitatively characterized by analyzing stereoscopic pairs of SEM images of the same areas using Alicona-MeX software (Graz, Austria).

The tribological tests were performed in a ball-on-flat linear reciprocating sliding configuration, using a CSM Instruments NTR1 nanotribometer and 3 mm diameter PTFE balls (Redhill, UK). The tests were performed at 5, 10 and 25 mN applied loads, with 1 cm/s maximum sliding velocity and 0.5 mm half amplitude, in dry conditions and at room temperature. The laboratory temperature and the relative humidity varied between 23 and 26 °C and 40% and 56%, respectively. To be able to perform the tests with the testing parameters ranges used, two medium load (ML) cantilevers with different stiffnesses were used. A ML1 cantilever was used for 5 mN applied load, while a ML2 cantilever was selected for the tests performed at 10 and 25 mN applied loads. The tests were performed with the sliding direction both parallel and perpendicular to the LIPSS direction. Their duration varied between 50 and 1000 sliding cycles (corresponding to about 15 to 450 s). At least three tests were performed for each set of conditions, in different regions of the specimen. Polished samples were tested under the same conditions for comparative purposes (reference samples). The evolution of the wear scars surface morphology was studied by scanning electron microscopy (SEM). Cross-sections of the specimens tested with 1000 cycles in the parallel sliding direction prepared by ion milling were also examined by SEM. Semi quantitative element analysis of the wear surfaces was performed by energy-dispersive X-ray spectroscopy (EDS,) using an EDS attachment to the FEG-SEM.

3. Results

3.1. Laser-Processed Surfaces

The SEM micrograph in Figure 1a,b illustrates the surface textures created by the femtosecond laser treatment. Figure 1c shows a 3D reconstruction of the surface and Figure 1d a SEM micrograph of the cross-section perpendicular to the LIPSS direction. Due to the Gaussian profile of the laser beam, the laser tracks are slightly deeper in the centre, creating a surface waviness with a period of approximately 50 µm and a depth of 200 nm when consecutive tracks are overlapped (Figure 1a). The LIPSS have approximately 730 nm average period and 230 nm height, as measured by analysis of stereoscopic pairs and in good agreement with literature [51]. The LIPSS form in the direction perpendicular to that of the laser beam polarization, indicated in Figure 1a by the double arrows. Their peaks are covered with ablation debris (Figure 1b), which can be removed by ultrasonic cleaning. The arithmetic roughness of the surface, Ra, is approximately 0.77 µm and its fractal dimension is approximately 2.1, obtained from the analysis of stereoscopic pairs. This topography corresponds to the formation of low-spatial frequency laser-induced periodic surface structures [9].

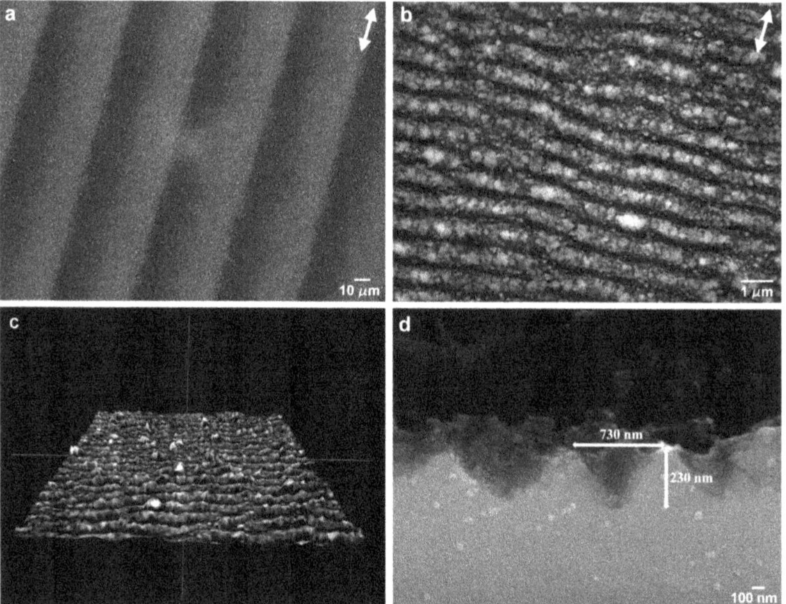

Figure 1. (**a**) SEM micrograph of a silicon surface textured with laser-induced periodic surface structures (LIPSS); (**b**) higher magnification image; (**c**) 3D surface reconstruction of the textured surface; (**d**) SEM micrograph of the cross-section of the LIPSS. The double arrows in (**a**,**b**) indicate the beam polarization direction.

3.2. Tribological Tests

Figure 2 illustrates the variation of the friction coefficient during three consecutive tests performed on different regions of polished Si specimens (reference samples) at 5 and 25 mN applied loads. Since the tribological tests were performed in linear reciprocating motion, the sliding distance is proportional to the number of cycles. The tests performed on polished specimens with 10 mN loads presented non-reproducible stick-slip behavior due to the cantilever, and the corresponding results were not included in the figure. Independently of the testing conditions, some dispersion of the friction coefficient values is observed, due to surface irregularities of the PTFE counterbodies. At 5 mN (Figure 2a), a short (<50 cycles) run-in period was observed during which the friction coefficient increased. After the run-in period, the friction coefficient decreased steadily from 0.60 to 0.54, 0.52 to 0.44 and 0.52 to 0.48 (between 100 and 1000 cycles for tests 1, 2 and 3, respectively). The average friction coefficient between 750 and 1000 cycles was 0.49 ± 0.05. At 25 mN (Figure 2b), after a run-in period lasting about 50 cycles during which the friction coefficient decreased, a slight but steady increase of the friction coefficient with the sliding distance was observed, eventually reaching a steady state after 750 cycles. The average value of the friction coefficient between 750 and 1000 cycles was 0.32 ± 0.05.

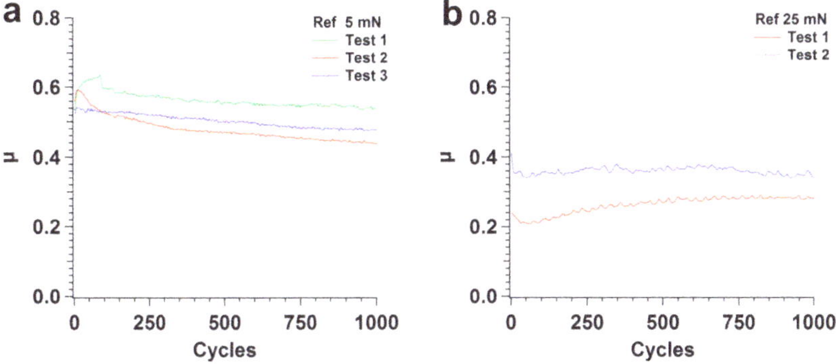

Figure 2. Variation of the friction coefficient with the number of cycles (proportional to the sliding distance) of consecutive tests for the polished Si specimen (Ref = Reference sample) at (**a**) 5 mN and (**b**) 25 mN.

The friction coefficient results obtained for the textured specimens are plotted in Figure 3. For tests at 5 mN in the parallel sliding direction there was poor reproducibility, for reasons that will be discussed later. For this applied load, the average value of the friction coefficient between 750 and 1000 cycles was 0.49 ± 0.08. For the tests at 10 and 25 mN in the parallel sliding direction, the reproducibility of the results was much better (Figure 3b,c). The friction coefficient decreased continuously since the beginning of the test, initially (0–50, 100 cycles) faster, and no run-in period could be distinguished. The average values of the friction coefficient between 750 and 1000 cycles, were 0.46 ± 0.04 and 0.35 ± 0.02 for 10 and 25 mN, respectively.

For the perpendicular sliding direction and an applied load of 5 mN (Figure 3d), after a short run-in period during which the friction coefficient decreased sharply, its value remained constant within the limits of the experimental error. The average value of the friction coefficient in the interval 750–1000 cycles was similar for in all tests, 0.25 ± 0.01. For 10 and 25 mN (Figure 3e,f), the friction coefficient decreased steadily from the beginning of the tests, reaching average values of 0.30 ± 0.03 and 0.25 ± 0.01 in the range 750–1000 cycles, respectively.

The variation of the average friction coefficient in the interval 750–1000 cycles with applied load is illustrated in Figure 4a. The friction coefficient decreased with the applied load for both polished and textured specimens when sliding in the parallel direction, but for the perpendicular sliding direction the variation was negligible. The lowest values of the friction coefficient were obtained for the perpendicular sliding direction, and the highest for the parallel sliding direction. The difference decreased with increasing load.

The three data sets were linearly fit by the least squares method for polished and textured specimens. The fitting was extremely good for the textured specimens in the parallel sliding direction ($R^2 = 1.00$), but less so for the polished specimens and the textured specimens tested in the perpendicular sliding direction ($R^2 = 0.37$ and 0.13, respectively). In case of elastic contact, the Hertz theory predicts that the real contact area was proportional to $(load)^{2/3}$ and thus that the friction coefficient would be proportional to $(load)^{-1/3}$ [52]. If the average friction coefficient was fitted with a $(load)^{-1/3}$ function the correlation was weaker for the textured specimens ($R^2 = 0.92$ and $R^2 = 0.01$ for the parallel and perpendicular sliding directions, respectively), as illustrated in Figure 4b, and better for the polished specimens ($R^2 = 0.65$).

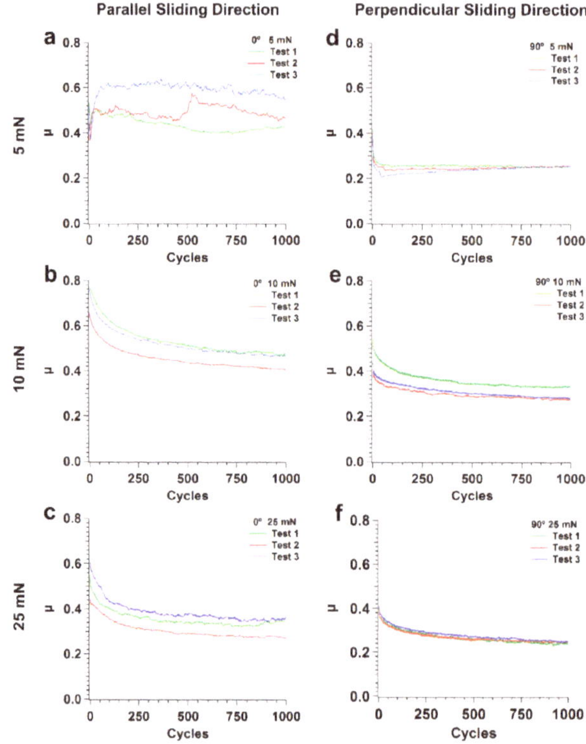

Figure 3. Variation of the friction coefficient with the number of cycles (proportional to the sliding distance) of three consecutive tests performed on textured specimen sliding in the parallel (0°) direction at (**a**) 5, (**b**) 10 and (**c**) 25 mN, and in the perpendicular (90°) direction at (**d**) 5, (**e**) 10 and (**f**) 25 mN.

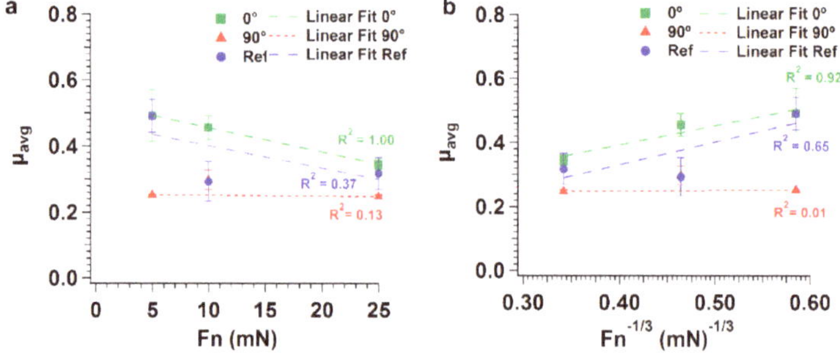

Figure 4. Variation of the average friction coefficient with (**a**) applied load and (**b**) with the applied load to the power (−1/3), for polished (Ref = reference sample) and textured specimens, in the parallel (0°) and perpendicular (90°) sliding directions.

3.3. Surface Morphology

SEM micrographs of the worn surfaces in the polished Si specimens are presented in Figures 5–8. It can be observed that a transfer layer of PTFE formed at the specimen surface for all applied loads. For 5 mN (Figure 5), the transferred material took the form of thin PTFE films with a fibrous structure (observed in the region indicated by the black arrow in Figure 5b). The dark regions observed in Figure 5c,d corresponded to thick lumps of PTFE. Ribbons a few microns wide (indicated by the black arrow) were drawn from these lumps of material, a process that was previously observed by Makinson and Tabor in similar testing conditions [35]. The surface morphology of wear tracks produced at 25 mN are presented in Figure 6. For these testing conditions, the transferred material was more irregularly distributed and presented a wider range of morphologies (Figure 6a). PTFE lumps and ribbons were observed, but thin films appeared in some areas as well (Figure 6b). Figure 6c shows PTFE ribbons drawn from a lump of transferred material and the fibers formed by the disaggregation of the ribbons. Ribbons usually form when PTFE lumps are rolled or twisted between the counterbody and the specimen surface [35]. They are formed by an arrangement of parallel fibers, which can be seen in the figure connecting the ribbons and the thin films. The lumps of PTFE consist of overlapped PTFE layers drawn at consecutive scans of the counterbody over the specimen surface. This is consistent with the mild stick-slip observed in the friction curves, independently of the load. On the other hand, the periodic adhesion between the transfer layer and the PTFE counterbody may also explain the large stick-slip oscillations observed for polished specimens and 10 mN applied load. In some areas the PTFE films detached from the surface, as illustrated in Figure 6d. In other regions, the film is smudged, indicating that it was dragged by the counterbody in its motion (Figure 6b). The relation between the PTFE fibers and the ribbons is illustrated in Figures 7 and 8, corresponding, respectively, to the zones indicated by A and B in Figure 6b. In Figure 7, fibers can be seen at the edges of a PTFE ribbon. Figure 8 depicts a SEM micrograph taken at a 45° tilt, that shows PTFE fibers drawn from the ribbons. The fibers were a few tens of nanometers wide and agglomerate to form films a few hundreds of nm wide. The area covered by the transferred PTFE film and the width of the wear tracks increased with increasing load.

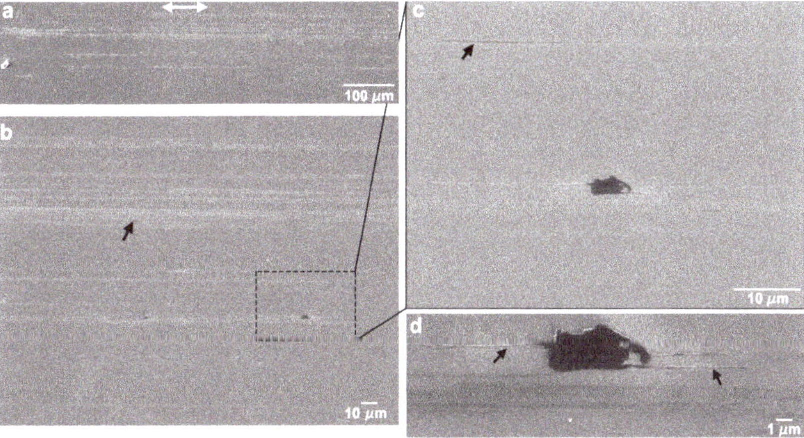

Figure 5. SEM micrographs of the centre of wear tracks on polished Si specimen performed at 5 mN. (**a**,**b**) General view of the wear track at different magnifications. (**c**) Transferred PTFE material. (**d**) Detailed view of the transferred PTFE material. The double arrow indicates the sliding direction.

The results of the EDS analysis performed on the wear tracks surfaces were only semiquantitative, because the electrons of the incident electron beam penetrated the material to a depth larger than the thickness of the transferred material, so characteristic X-rays were emitted simultaneously from the transferred material and from the substrate. Due to the irregular surface topography and the heterogeneity of the surface, the ZAF correction (atomic number (Z), absorption (A) and fluorescence (F)) necessary for quantification also could not be performed. However, since the Si characteristic X-rays were attenuated by the PTFE layer proportionally to the thickness of this layer, the percentage of Si measured at each point gave an indication of the thickness of the transfer material at that particular location. EDS point analysis was performed at the points indicated by 1 to 4 in Figure 6d, leading to the following results: Points 1 and 2, 100 wt.% Si (no transfer material film), Points 3 and 4, 94 wt.% Si, corresponding to regions covered with a transfer film. This value is typical of areas covered with transferred material, although in some regions where the transfer material thickness is larger, the Si content can be as low as 81 wt. %.

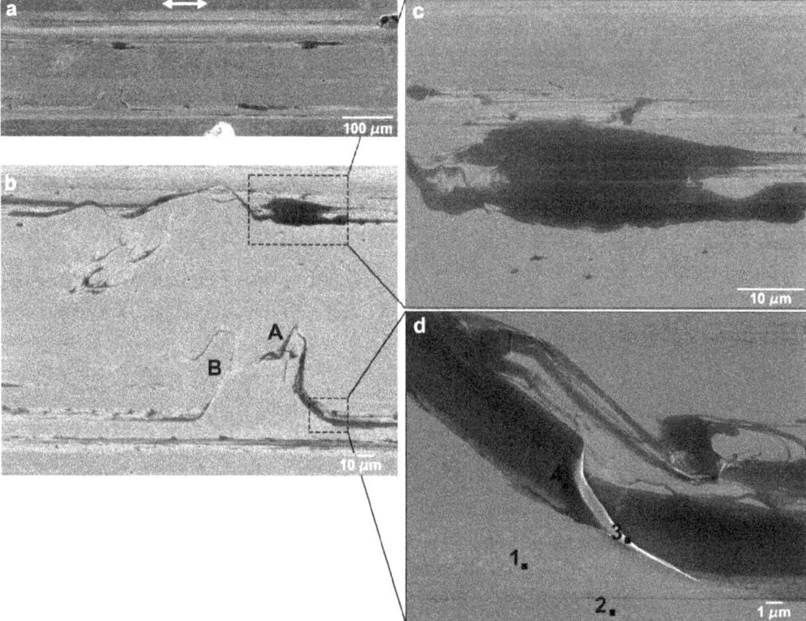

Figure 6. SEM micrographs of the centre of wear tracks on the polished Si specimen performed at 25 mN. (**a**,**b**) General views of the wear track at different magnifications. (**c**) Detail of the transferred PTFE lump. (**d**) Loosely adherent PTFE film. The double arrow indicates the sliding direction.

Figure 7. PTFE fibers drawn from PTFE ribbons in zone A of the wear track produced at 25 mN illustrated in Figure 6b.

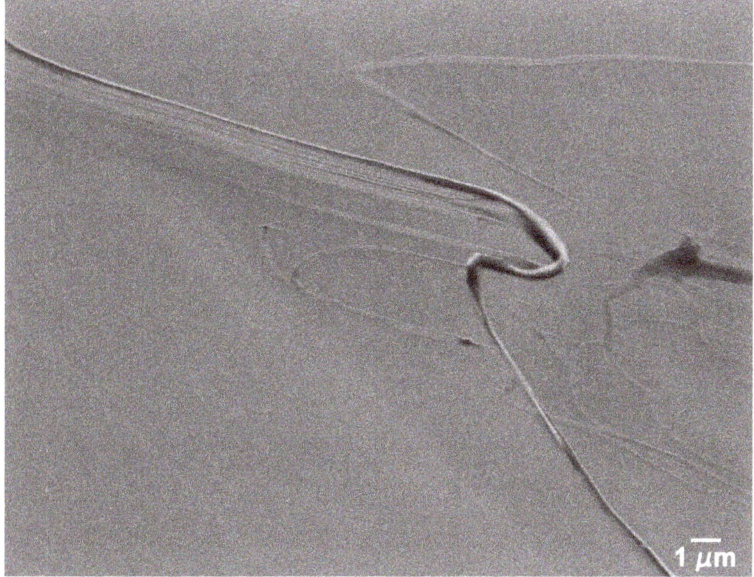

Figure 8. PTFE fibers drawn from a PTFE ribbon joining to form a thin film in zone B of the wear track produced at 25 mN illustrated in Figure 6b, taken at 45° tilt.

The morphology of the wear tracks created in the parallel sliding direction is depicted in the SEM micrographs of Figure 9. The wear tracks obtained with a load of 5 mN (Figure 9a,b, 50 and 1000 cycles, respectively) were covered with stripes of a transferred PTFE film, aligned with the sliding direction, which filled the valleys between the LIPSS crests, seen as darker regions in the SEM images. In some regions, thicker PTFE layers were observed (seen in more detail in Figure 10a), which sometimes were rolled between the two sliding surfaces to form twisted rolls (Figure 10b). The fibrous structure of the PTFE film was clearly observed in Figure 10c (indicated by the arrow). In the longer tests (1000 cycles), the LIPSS crests were worn out in the central region of the tracks (Figure 10b). For this applied load, the area covered by PTFE did not vary significantly with the number of cycles, as shown by comparing Figure 9a,b, but the thickness of the PTFE stripes increased, as demonstrated by the larger contrast of Figure 10b as compared to Figure 10a. For 10 mN (Figure 9c,d), the morphology of the transferred material was different. It consisted mainly of lumps, thick layers and ribbons, filling the spaces between the LIPSS, as shown in the region indicated by the black arrow in Figure 11. The transferred material did not form continuous stripes but concentrated at the elevations of the surface waviness (Figure 9c). The area covered with PTFE increased with the sliding distance (as shown by comparing Figure 9c,d), but the wear tracks were never uniformly covered with PTFE. The morphology of the transferred material for 25 mN was similar, but the area covered with transferred material was already important after 50 cycles (Figure 9e). This area increased with the duration of the tests (Figure 9f), but even after 1000 cycles PTFE-free regions were still observed in the depressions of the surface waviness (Figure 9f). Similarly, to what happens for 10 mN, the transferred material took the form of ribbons (shown by an arrow in Figure 12a), sheets and lumps of different thicknesses (Figures 9f and 12b). A SEM micrograph of a cross-section of a wear track (Figure 13) taken at 45° tilt showed that the transfer material filled the depressions between the LIPSS. This figure also shows how the PTFE lumps and ribbons attach to the surface through filaments of the same material.

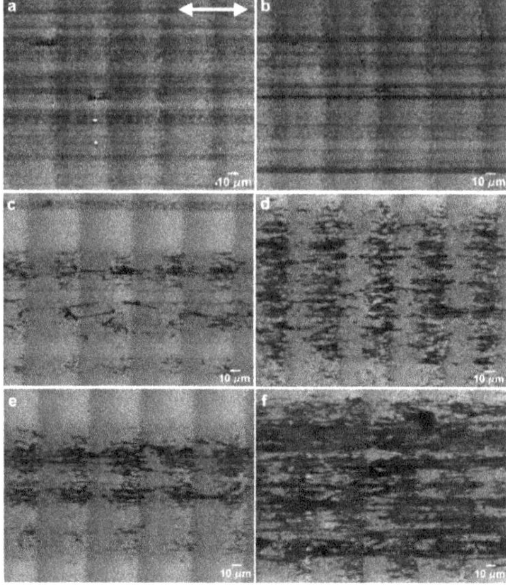

Figure 9. Wear tracks at 5 mN after (**a**) 50 cycles and (**b**) 1000 cycles; at 10 mN after (**c**) 50 cycles and (**d**) 1000 cycles; and at 25 mN after (**e**) 50 cycles and (**f**) 1000 cycles. The double arrow indicates the sliding direction, parallel to the LIPSS orientation.

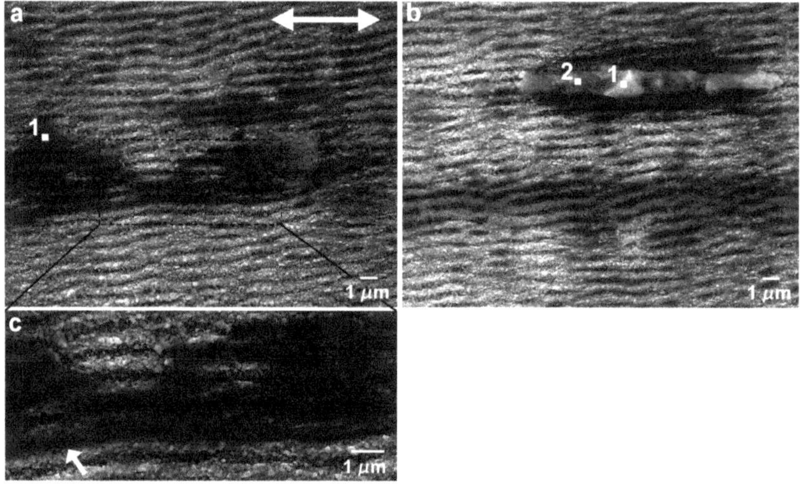

Figure 10. SEM micrographs of the wear tracks of the textured specimen in the parallel direction at 5 mN, after 1000 cycles, showing (**a**) PTFE transferred film and (**b**) worn LIPSS and PTFE roll. (**c**) Detailed view of (**a**), illustrating the fibrous nature of the PTFE film. The double arrow indicates the sliding direction.

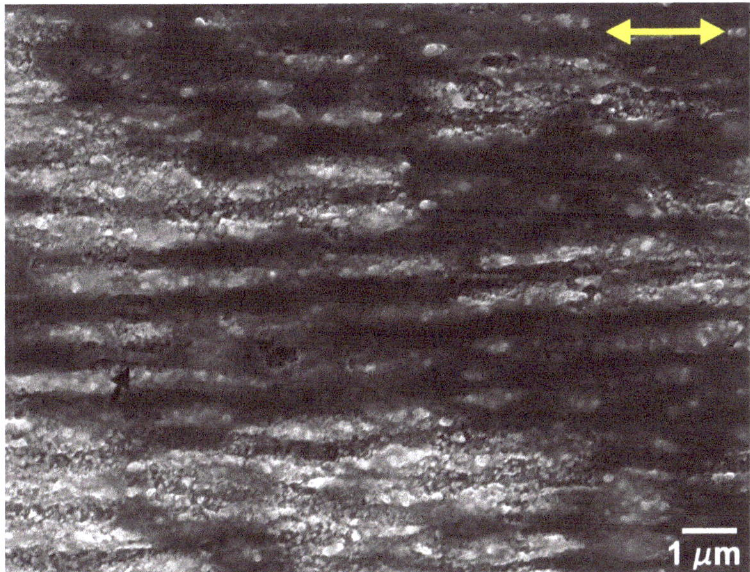

Figure 11. SEM micrograph of a wear track produced at 10 mN and 400 cycles. The double arrow indicates the sliding direction.

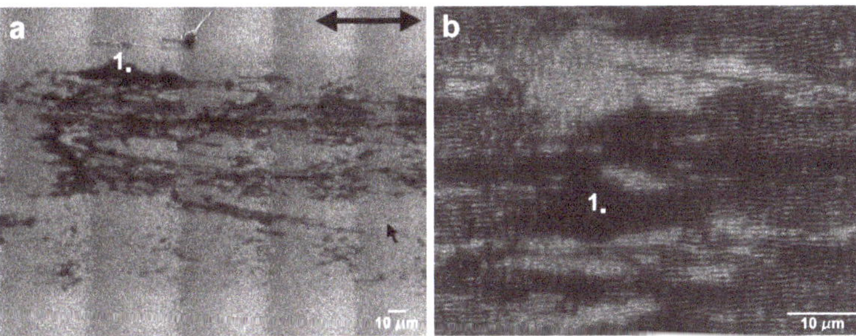

Figure 12. Energy-dispersive X-ray spectrometry (EDS) point analysis on the wear tracks produced with 25 mN load and (**a**) 50 cycles and (**b**) 1000 cycles. The double arrow indicates the sliding direction.

Figure 13. SEM micrograph of the cross-section of a wear track taken at 45° tilt. The double arrow indicates the sliding direction.

The differences in the thickness of the transferred layer from region to region were reflected in the results of the EDS analysis. The EDS analysis at point 1 in Figure 10a led to 73 wt. % Si, and at points 1 and 2 on the PTFE roll in Figure 9b to 78% Si and 72% Si, respectively, corresponding to a moderately thick layer of transferred material. For 25 mN at 50 and 1000 cycles the layers of transferred material were thicker: 49 wt. % Si was measured in point 1 of Figure 12a; while in point 1 of Figure 12b the Si content was 69 wt.%.

The evolution of the wear track morphology for the tests performed in the perpendicular sliding direction was analogous to the parallel sliding direction, with some differences (Figures 14 and 15). As for the parallel sliding direction, for 5 mN and 50 cycles (Figure 14a) the wear track presented stripes of a PTFE transfer film (darker in the SEM images) and a few areas where the PTFE transfer layer was sufficiently thick to cover the LIPSS. The area covered with PTFE did not change significantly with the number of cycles (compare Figure 14a,b). At higher loads (10 and 25 mN), the areas covered with thick PTFE layers were larger and in greater number, but they remained discontinuous, except after 1000 cycles at 25 mN. For 10 mN, the LIPSS peaks showed traces of wear in the central region of the track since 50 cycles, and the PTFE transfer film and lumps concentrated at the tracks periphery (Figure 15a). The area covered with PTFE increased with increasing sliding distance, but it did not cover entirely the wear tracks, even after 1000 cycles (Figure 14d). Contrarily to what happens in the parallel sliding direction, the transfer film did not fill the space between the LIPSS (Figure 16a) and was covered in some areas with PTFE rolls and ribbons. The evolution for 25 mN was similar, but faster. The transferred material took mainly the form of lumps and ribbons (Figure 16b). After 50 cycles (Figure 14e), a transfer film formed at the periphery of the wear tracks (Figure 15c), which extended to the centre with an increasing number of cycles (compare Figure 15c,d). After 1000 cycles, the transfer layer covered a significant proportion of the wear track surface, but not completely (Figure 14f). It extended to the entire length of the tracks in some regions, contrarily to what happened in the parallel sliding direction, where the transfer material layer was interrupted due to the surface waviness.

Figure 16a,b present the locations of EDS analysis, which led to the following results: (a) The points 1 to 4 (Figure 16a), corresponding to a PTFE ribbon, a PTFE film, the exposed specimen surface and PTFE fibers presented 43 wt.% Si, 77 wt.% Si, 93 wt.% Si and 64 wt.% Si, respectively. The points indicated by 1 to 3 in Figure 16b corresponded to different regions of the PTFE film, and led to 77 wt. % Si (Points 1 and 2) and 73 wt. % Si in Point 3. Point 4 corresponded to the exposed surface and led to 98 wt. % Si and 2 wt. % O.

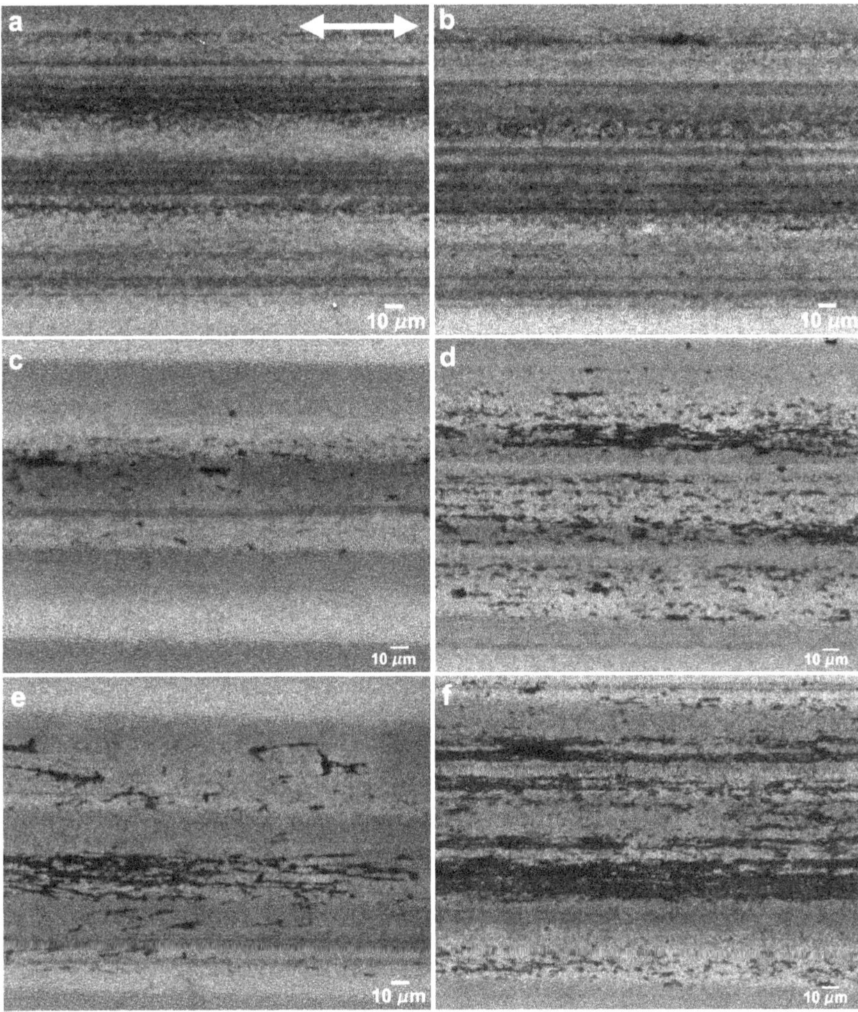

Figure 14. SEM micrographs of the central region of wear tracks produced in the perpendicular sliding direction at 5 mN after (**a**) 50 cycles and (**b**) 1000 cycles; at 10 mN after (**c**) 50 cycles and (**d**) 1000 cycles and at 25 mN after (**e**) 50 cycles and (**f**) 1000 cycles. The double arrow indicates the sliding direction.

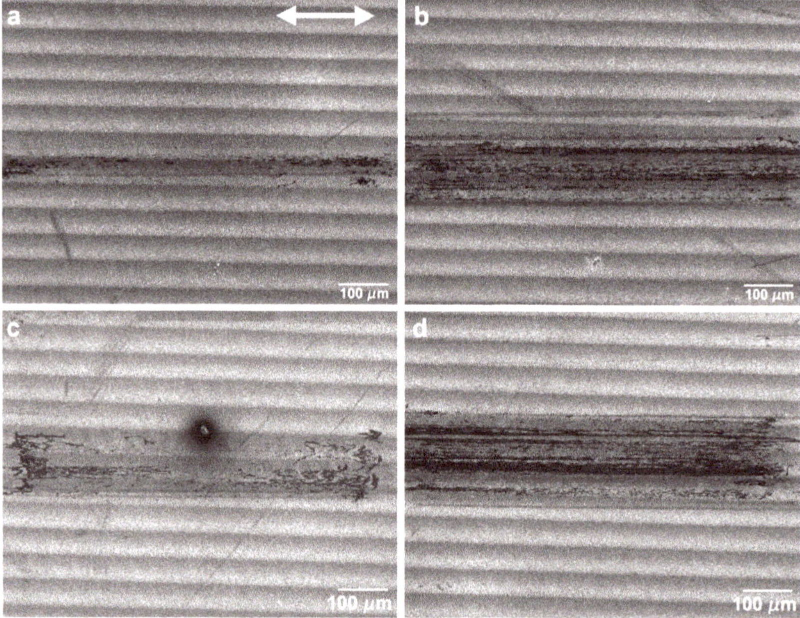

Figure 15. SEM micrographs of the wear tracks of the textured specimen in the perpendicular direction at 10 mN (**a**) after 50 cycles and (**b**) 1000 cycles and at 25 mN (**c**) after 50 cycles and (**d**) 1000 cycles. The double arrow indicates the sliding direction.

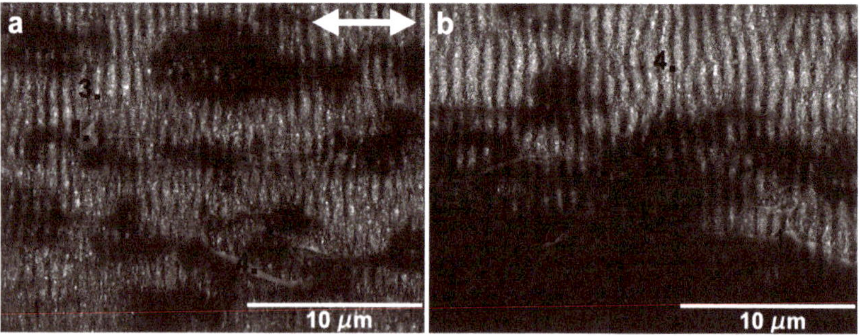

Figure 16. EDS point analysis on wear tracks performed with (**a**) 1000 cycles at 10 mN cycles and (**b**) 400 cycles at 25 mN. The double arrow indicates the sliding direction.

4. Discussion

In order to better understand the friction behavior of textured surfaces sliding against PTFE, it is necessary to take into consideration the morphology of the transferred material, and the deformation and fracture behavior of this polymer. PTFE is a semi-crystalline polymer, which undergoes several phase transformations up to its melting temperature [53]. The low temperature phase (usually designated phase II) presents a triclinic structure and is stable up to 19 °C. At this temperature, it transforms to phase IV, which presents a partially ordered hexagonal structure and is stable between 19 and 30 °C. The high temperature phase is less well-known but is characterized by a higher degree of disorder. Above 150 °C the material becomes amorphous [54]. As expected from these phase transitions, PTFE presents low ductility at low temperature, but the ductility increases considerably with temperature, while its tensile strength decreases, reaching a value of ~5 MPa at about 200 °C [46,54].

The fracture mechanism depends on temperature as well, a brittle behavior being observed below 19 °C and a ductile behavior, with considerable plasticity, above 30 °C [55]. The strength and ductility depend on the strain rate, that decreases with decreasing plasticity while the yield stress increases [56]. Between 25 and 150 °C, the true rupture stress decreases by one half [57] and the crack propagation is accompanied by the formation of fibrils, which provides an effective mechanism to dissipate energy. These mechanical properties have a direct impact on the friction behavior of PTFE. When sliding against highly polished surfaces of materials such as glass [35] and silicon oxide [58] at low sliding speeds (≤1 mm/s) and relatively high temperatures, a very thin film (10–40 nm) of PTFE is transferred to the countersurface, leading to an extremely low friction coefficient at room temperature (~0.04). These films are crystalline and highly ordered, even more than the bulk PTFE polymer, and form when PTFE strands consecutively attach to the substrate and are pulled off from the bulk material to yield an array of parallel fibers [59]. This film presents very low shear strength and, consequently, a very low friction coefficient, making thin PTFE films a highly effective solid-state lubricant. For higher sliding speeds, especially at low temperatures, the material presents a brittle behavior [60] and fractures randomly in irregular particles, which adhere to the countersurface [35]. These particles are then laminated between the two bodies to form films of PTFE with a fibrous morphology, typical of PTFE deformed at moderately high temperatures [60]. The friction coefficient is much higher than 0.3 [35].

The tests performed on polished silicon specimens in the present work lead to values of the friction coefficient of about 0.6, for an applied load of 5 mN, decreasing to ~0.5 when the number of cycles increases, and roughly constant values of 0.3–0.35 at 25 mN (Figure 2). This regime is preceded by a run-in period, during which the friction coefficient either increases (5 mN) or decreases (25 mN) and these variations can be explained by the initial reduction in surface roughness [61] as well as the growth of a low shear strength PTFE transfer film in some areas of the surface. The transfer of PTFE to the silicon surface occurs due to the large adhesion forces between the Si substrate and the PTFE (mainly van der Waals and electrostatic forces). Since, for the scanning speed used, deformation is occurring in the PTFE low plasticity regime, it fractures, forming particles that adhere to the Si surface and are deformed between the contact surfaces to form the transfer layers observed in Figures 4 and 5 [32,52,62,63]. After the run-in, the friction coefficient decreases progressively as the areas covered with PTFE expand gradually, preventing further pulling-off of PTFE particles. However, its value remains high (>0.4) because the transfer layer is thick and irregular due to the high sliding speed used (10 mm/s). This is confirmed by the SEM micrographs of Figure 5. On the other hand, 1000 cycles were not enough for the PTFE transfer film to cover completely the Si surface, so the friction coefficient did not reach a steady state (Figure 2a). For 25 mN, the overall value of the friction coefficient was lower than for 5 mN. The friction coefficient decreased slightly during the run-in period (Figure 2b), as the PTFE transfer layers begin forming on the Si surface [38]. This evolution was faster than for 5 mN and the higher interface shear stress might lead to a more perfect alignment of the PTFE chains parallel to the sliding direction, thus explaining the lower value of friction coefficient obtained as compared to that at 5 mN. Periodic oscillations of the friction force are observed indicating that stick-slip occurred during testing. Comparative tests performed with different cantilevers showed that the presence and amount of stick-slip depend on the cantilever and on the applied load, which explains the poor reproducibility of the results obtained for certain testing conditions (in particular, 10 mN). It is well known that the tangential contact stiffness determines the slope in the stick stage, while the stick-slip amplitude is determined by the cohesive strength and surface energy of the sliding interface [63–65]. The tests were performed with a ML1 cantilever (a soft cantilever), which revealed itself adequate for 5 mN, but too soft for 10 and 25 mN. On the other hand, the stiffer ML2 cantilever was adequate for 25 mN, but too stiff for 5 and 10 mN, and led to stick-slip, particularly for 5 mN applied load. No stick-slip was observed for the textured specimens, despite using the same cantilevers. It is well-known that the force required to overcome adhesion increases with increasing contact area. Since texturing reduces the contact area, adhesion is sufficiently reduced to avoid stiction. The average value of the friction coefficient in the steady state (after 750 cycles) is 0.32, similar to the value reported by Makinson and

Tabor [35] for PTFE sliding on glass in the high friction regime (>0.33). This value is expectable taking into consideration that, for the relatively high sliding velocity used in the present work (10 mm/s), PTFE must present a brittle behavior, with fracture and lump formation, accompanied by a large material displacement [35,63]. However, due to the stick-slip phenomenon, the transfer of PTFE to the Si surface is discontinuous. The regions where the transfer layer was thicker correspond to the stick (static) regions, where adhesion and, consequently, the friction coefficient were larger and the deformation of the polymer and the amount of transferred material were more important. In the slip region, PTFE was transferred to the counterbody more smoothly, in the form of a thinner fibrous film. In the slip stage, the interface shear strength was lower, causing the friction coefficient to decrease momentarily, until the moment when adhesion was re-established. The stick-slip regime was maintained during the entire test, so consecutive layers of PTFE were transferred to the same regions of the surface, building the thick multilayer transfer of polymer shown in Figure 6c. Since stick-slip was milder at 25 mN, this effect was less marked. The increase of the friction coefficient during the tests was due to the shearing and ripping out of the PTFE film from the Si surface, as shown in Figure 7 [34].

The friction coefficient of PTFE sliding on polished silicon decreased with increasing applied load (Figure 4), in good agreement with the literature [32,66]. This evolution is often observed for polymers, including PTFE [38,52,67]. According to the SEM observations (Figures 5 and 6), the area occupied by the transferred PTFE films increased with applied load, which in turn decreased the interfacial shear strength and, consequently, the friction force [66,68]. The friction behavior observed for polished silicon was similar to that previously observed by other authors for PTFE sliding on glass [35,38,63]. The present results also agreed with the conclusions of Blanchet and Kennedy [34] who found that the critical velocity for the mild-to-severe wear transition for PTFE sliding on stainless steel under an average contact pressure of 6.55 MPa was 0.8 cm/s, slightly lower than the sliding velocity used in our work.

The textures observed in Figure 1 consisted of low-spatial frequency laser-induced periodic surface structures (LSFL) [6]. The formation of these structures is believed to be due to the modulation of the absorbed radiation intensity created by the interference of the incoming laser beam with scattered radiation propagating parallel to the solid surface [14] or with surface plasmon polaritons excited by the laser beam [69]. The LIPSS peaks were covered with ablation debris. This could be explained by the fact that, when the predominant ablation mechanism at the absorbed laser intensity maxima is liquid spallation [15], the shallow liquid layer created by the laser beam is fragmented into a large number of extremely small droplets, which are expelled in the confined space between the LIPSS and partially redeposit on their peaks. As shown by these authors [15], the only phase transformation introduced in the surface layer of material by this surface treatment is melting of a layer of material with a thickness varying in the 65–130 nm range. The laser treatment also induces the formation of defects in a layer of material less than 1 μm deep, as shown by Sedao et al. [70], but these defects do not seem enough to change significantly the surface hardness of the material [71], so the observed effects are mainly to be accounted by the surface topography.

The presence of the LIPSS texture changes significantly the tribological behavior of the system. Firstly, the friction coefficient decreased during the run-in period, as the PTFE chains align with the sliding direction and the polymer starts being transferred to the Si surface, reducing adhesion and the surface roughness [62]. After the run-in period, the friction coefficient continued to decrease, in particular for the higher loads. The tests performed in the parallel sliding direction with 5 mN applied load presented low reproducibility. The PTFE transfer film was not uniform and regularly distributed over the wear track surface (Figure 9a,b). In some regions the transfer layer was sufficiently thick to cover the LIPSS and PTFE ribbons were drawn out from the PTFE layer (Figure 9a). On the other hand, the wear particles that were trapped between the two surfaces were rolled and took a cylindrical shape (Figure 9b). This complex morphology indicated that the PTFE counterbody was fracturing randomly and a range of complex interactions occurred during the tests, which explain the dispersion of the friction coefficient value and its abrupt variations observed in the tests. For 10 and

25 mN, instead of a thin PTFE film, a thick transfer layer formed, initially only at the elevations of the surface waviness caused by the overlap of consecutive laser tracks (Figure 9c,e). The relative area occupied by these layers increased progressively with the number of cycles, explaining the decrease in friction coefficient, but the wear track was never completely covered even after 1000 cycles (Figure 9f). As shown in Figure 13, the space between the LIPSS was occupied by the PTFE fibers, which anchor the PTFE transfer layers to the Si substrate.

When testing in the perpendicular sliding direction at 5 mN load, the friction coefficient remained almost constant after the run-in period, with an average value of 0.25, because the proportion of the wear track surface covered with PTFE did not increase significantly with the number of cycles. The transfer layer occupied a larger area, was more uniformly distributed and large PTFE particles were less frequent than for the parallel sliding direction (Figure 14a,b). Consequently, the friction curves were smoother and more reproducible. For 10 and 25 mN, the friction coefficient decreased continuously, approaching the steady state after 1000 cycles. At these loads, the PTFE film initially formed at the periphery of the wear track (Figure 15a,c), where the contact pressure was lower, then extended to the centre of the wear track, occupying preferentially the elevations of the surface waviness, which were now parallel to the sliding direction, but eventually covered the wear track almost completely (Figure 15b and d at 10 and 25 mN, respectively). Its distribution was more uniform than in the parallel sliding direction (Figure 15b,d). The fraction of the wear tracks covered with PTFE increased with the number of cycles, causing the slight decrease in the friction coefficient observed.

In order to explain the observed variation of PTFE/Si tribological behavior with the testing parameters, its mechanical properties must be taken into consideration. As the applied load increased, there was a monotonic increase of the interface temperature, which favored the transition of PTFE to less orderly crystalline phases (Phase I, above 30 °C) and led to a steady decrease in the strain to failure rate [57]. This decrease in PTFE ductility facilitated plastic deformation and, consequently, the formation of the PTFE transfer film at the Si surface.

In order to explain the observed behavior for polished and textured specimens, one must compare the contributions of adhesion and elastic deformation to friction in both cases. The relative contributions of adhesion and elastic deformation of asperities in the case of the textured specimens could be evaluated for the perpendicular sliding direction by estimating the elastic energy stored in the deformation field of the polymer, $U_{elastic}$, and the gain in adhesion energy, U_{adh}, by using the following equations [72]:

$$U_{elastic} = E\lambda h^2,\tag{1}$$

$$U_{adh} = \Delta\gamma\lambda^2,\tag{2}$$

where E is the elastic modulus of PTFE, h and λ are the height and width of the substrate's cavity occupied by the polymer, respectively and $\Delta\gamma$ is the change in surface free energy per unit area. Considering that the height and the width of the LIPSS were about 200 and 400 nm, respectively, then $U_{elastic}$ was ~ 3.68 pJ and $U_{adhesion}$ was ~ 0.04 pJ, indicating that the elastic contribution to friction was more significant than that of adhesion. For the parallel sliding direction, λ can be estimated as 6 µm, which is about 1/4 to 1/5 of the Hertzian radius of contact at 5 to 25 mN, and $U_{elastic}$ as ~ 110 pJ and $U_{adhesion}$ as ~ 36 pJ, which is more than two orders of magnitude larger than for the perpendicular sliding direction. The same estimates made for polished silicon lead to $U_{elastic}$ as ~ 46 pJ and $U_{adhesion}$ as ~ 100 pJ, assuming that for smooth surfaces $\frac{h}{\lambda}$ ~ 0.01 or smaller [72]. From these estimates alone, we could conclude that the contribution of adhesion is large for the polished specimens, but much smaller for the textured specimens. Thus, a lower friction coefficient is expected for the textured specimens. In fact, the friction coefficient of the textured specimens tested in the parallel sliding direction is very similar to the value obtained for polished specimens, while the value for the perpendicular direction, under similar testing parameters, is lower (0.46 and 0.35 for 10 and 25 mN as compared to 0.30 and 0.25; Figure 4). Texturing increases the surface roughness, favoring plastic instead of elastic contacts, and increasing the ploughing component of friction. This is supported by the fact that, while the average friction coefficient of the polished specimens shows good correlation with the applied load$^{-1/3}$, indicating a predominantly elastic contact, the average friction coefficient measured in the textured specimens shows a better correlation to the applied load, suggesting a transition to a predominantly plastic contact regime [64]. In the elastic contact regime, an increase of the surface roughness leads to a decrease of the real contact area and, consequently, of the adhesion contribution to friction. This increase of roughness may also lead to an increase of the plastic deformation and a shift in the wear regime from sliding wear to abrasive wear, affecting mainly the softer material. Thus, the presence of LIPSS may change the PTFE/Si wear mechanism, increasing the contribution of plastic deformation (ploughing and cutting) to the friction [64]. Due to this change in the wear mechanism, the proportion of PTFE transferred lumps is usually larger for the textured specimens, particularly at high loads.

5. Conclusions

1. Tests performed on polished Si specimens show that the friction coefficient decreased steadily with testing time for 5 mN, with an average value of 0.5, and it increased slightly with testing time for 25 mN, with an average value of 0.3. At 5 mN, the wear tracks were characterized by the formation of very thin PTFE films with a fibrous structure and of thicker layers of PTFE, from which ribbons, a few microns wide, were drawn. Due to the relatively high sliding speed used and the fact that PTFE was in the low plasticity regime, the PTFE film was generally thick and irregular, explaining the particularly high value of the friction coefficient at this load (0.5). The wear track area covered by PTFE increased with testing time, preventing further pulling-off of PTFE particles and causing the friction coefficient to decrease progressively. The wear track never completely covered the Si surface, and so the friction coefficient did not reach steady state. At 25 mN, the wear track showed similar elements (lumps, ribbons and thin film), but, despite these forming faster than at 5 mN, the wear track was never completely covered with PTFE due to the presence of mild stick-slip, which prevented the formation of a uniform film over the Si surface. Due to this discontinuous material transfer, the friction coefficient increased during the tests, as the PTFE film was sheared and ripped-out from the Si surface. The higher interface shear stress facilitated the alignment of the PTFE chains parallel to the sliding direction, explaining the lower friction coefficient values in comparison to the 5 mN load. Such a decrease of the friction coefficient with load could be explained by the increase in area covered with transferred PTFE with increasing load, which in turn decreased the interfacial shear strength and, consequently, the friction force.

2. Tests performed with similar parameters in textured specimens showed that LIPSS changed significantly the tribological behavior of this system. Overall, the friction coefficient decreased with testing time in both sliding directions, more significantly for higher applied loads, except at 5 mN in the parallel sliding direction where random fracture of the PTFE counterbody occurred. This decrease was due to the increase in the area occupied by PTFE, which was further facilitated at higher loads due to the increase in interface temperature and consequent decrease in PTFE ductility. For the parallel sliding direction, the transfer film tended to be thick and formed initially only at the elevations of the surface waviness caused by the overlap of consecutive laser tracks. It progressively expanded, but the wear track was never completely covered. For all applied loads, the PTFE fibers occupied the space between the LIPSS, anchoring the transfer layers to the Si substrate. For the perpendicular sliding direction, the PTFE film formed initially at the periphery of the wear track, then progressively extended to the centre, while still occupying preferentially the elevations of the surface waviness. After 1000 cycles and at higher loads, the wear tracks were almost completely covered, more uniformly and occupying a larger area than for the parallel sliding direction, explaining the lower friction coefficient obtained at a steady state.

3. Texturing increased the surface roughness, favoring plastic instead of elastic contacts, increasing the ploughing component of friction and shifting the predominant wear regime of PTFE from

sliding to abrasive wear. This change in wear mechanisms led to more PTFE transferred lumps in the textured specimens than in the polished ones, particularly at high loads. The orientation of the surface features, namely of the surface waviness, relative to the sliding direction also had an important effect on friction. Abrasion of PTFE was more pronounced when sliding parallel to the LIPSS because the surface waviness was transverse to the sliding direction, increasing further the ploughing contribution to friction in this case. The smaller abrasion and consequent more uniform distribution of the thin PTFE transfer film in the perpendicular sliding direction at all loads was responsible for the overall lowest friction coefficient values.

Author Contributions: Conceptualization, R.V.; Data curation, I.A.-L.; Formal analysis, I.A.-L.; Investigation, I.A.-L. and V.O.; Methodology, V.O.; Supervision, A.A. and R.V.; Validation, I.A.-L.; Visualization, I.A.-L.; Writing—original draft, I.A.-L. and R.V.; Writing—review & editing, I.A.-L., A.A., V.O. and R.V.

Funding: This research was funded by FCT–Fundação para a Ciência e Tecnologia, grant numbers UID/CTM/04540/2013 and PD/BD/143034/2018.

Acknowledgments: The authors would like to thank Rogério Colaço (IST) and Patrizia Paradiso (IST) for the helpful insights regarding the tribological tests.

References

1. Etsion, I. State of the Art in Laser Surface Texturing. *J. Tribol.* **2005**, *127*, 248. [CrossRef]
2. Chichkov, B.N.; Momma, C.; Nolte, S.; von Alvensleben, F.; Tünnermann, A. Femtosecond, picosecond and nanosecond laser ablation of solids. *Appl. Phys. A Mater. Sci. Process.* **1996**, *63*, 109–115. [CrossRef]
3. Momma, C.; Nolte, S.; Chichkov, B.N.; Alvensleben, F.V.; Tünnermann, A. Precise laser ablation with ultrashort pulses. *Appl. Surf. Sci.* **1997**, *109–110*, 15–19. [CrossRef]
4. Zeng, X.; Mao, X.L.; Greif, R.; Russo, R.E. Experimental investigation of ablation efficiency and plasma expansion during femtosecond and nanosecond laser ablation of silicon. *Appl. Phys. A* **2005**, *80*, 237–241. [CrossRef]
5. Balling, P.; Schou, J. Femtosecond-laser ablation dynamics of dielectrics: Basics and applications for thin films. *Rep. Prog. Phys.* **2013**, *76*, 036502. [CrossRef] [PubMed]
6. Vorobyev, A.Y.; Guo, C. Direct femtosecond laser surface nano/microstructuring and its applications. *Laser Photonics Rev.* **2013**, *7*, 385–407. [CrossRef]
7. Siegman, A.; Fauchet, P. Stimulated Wood's anomalies on laser-illuminated surfaces. *IEEE J. Quantum Electron.* **1986**, *22*, 1384–1403. [CrossRef]
8. Borowiec, A.; Haugen, H.K. Subwavelength ripple formation on the surfaces of compound semiconductors irradiated with femtosecond laser pulses. *Appl. Phys. Lett.* **2003**, *82*, 4462–4464. [CrossRef]
9. Bonse, J.; Kruger, J.; Hohm, S.; Rosenfeld, A. Femtosecond laser-induced periodic surface structures. *J. Laser Appl.* **2012**, *24*, 042006. [CrossRef]
10. Oliveira, V.; Ausset, S.; Vilar, R. Surface micro/nanostructuring of titanium under stationary and non-stationary femtosecond laser irradiation. *Appl. Surf. Sci.* **2009**, *255*, 7556–7560. [CrossRef]
11. Oliveira, V.; Polushkin, N.I.; Conde, O.; Vilar, R. Laser surface patterning using a Michelson interferometer and femtosecond laser radiation. *Opt. Laser Technol.* **2012**, *44*, 2072–2075. [CrossRef]
12. Oliveira, V.; Vilar, R.; Serra, R.; Oliveira, J.C.; Polushkin, N.I.; Conde, O. Sub-micron structuring of silicon using femtosecond laser interferometry. *Opt. Laser Technol.* **2013**, *54*, 428–431. [CrossRef]
13. Van Driel, H.M.; Sipe, J.E.; Young, J.F. Laser-induced periodic surface structure on solids: A universal phenomenon. *Phys. Rev. Lett.* **1982**, *49*, 1955–1958. [CrossRef]
14. Sipe, J.E.; Young, J.F.; Preston, J.S.; Van Driel, H.M. Laser-induced periodic surface structure. I. Theory. *Phys. Rev. B* **1983**, *27*, 1141–1154. [CrossRef]
15. Cangueiro, L.T.; Cavaleiro, A.J.; Morgiel, J.; Vilar, R. Mechanisms of the formation of low spatial frequency LIPSS on Ni/Ti reactive multilayers. *J. Phys. D Appl. Phys.* **2016**, *49*, 365103. [CrossRef]
16. Cunha, A.; Serro, A.P.; Oliveira, V.; Almeida, A.; Vilar, R.; Durrieu, M.-C. Wetting behaviour of femtosecond laser textured Ti–6Al–4V surfaces. *Appl. Surf. Sci.* **2013**, *265*, 688–696. [CrossRef]

17. Vorobyev, A.Y.; Guo, C. Antireflection effect of femtosecond laser-induced periodic surface structures on silicon. *Opt. Express* **2011**, *19*, A1031. [CrossRef] [PubMed]

18. Dusser, B.; Sagan, Z.; Soder, H.; Faure, N.; Colombier, J.P.; Jourlin, M.; Audouard, E. Controlled nanostructrures formation by ultra fast laser pulses for color marking. *Opt. Express* **2010**, *18*, 2913. [CrossRef]

19. Halbwax, M.; Sarnet, T.; Delaporte, P.; Sentis, M.; Etienne, H.; Torregrosa, F.; Vervisch, V.; Perichaud, I.; Martinuzzi, S. Micro and nano-structuration of silicon by femtosecond laser: Application to silicon photovoltaic cells fabrication. *Thin Solid Film.* **2008**, *516*, 6791–6795. [CrossRef]

20. Cunha, A.; Zouani, O.F.; Plawinski, L.; Botelho do Rego, A.M.; Almeida, A.; Vilar, R.; Durrieu, M.-C. Human mesenchymal stem cell behavior on femtosecond laser-textured Ti-6Al-4V surfaces. *Nanomedicine* **2015**, *10*, 725–739. [CrossRef]

21. Cunha, A.; Elie, A.-M.; Plawinski, L.; Serro, A.P.; Botelho do Rego, A.M.; Almeida, A.; Urdaci, M.C.; Durrieu, M.-C.; Vilar, R. Femtosecond laser surface texturing of titanium as a method to reduce the adhesion of Staphylococcus aureus and biofilm formation. *Appl. Surf. Sci.* **2016**, *360*, 485–493. [CrossRef]

22. Bonse, J.; Kirner, S.V.; Koter, R.; Pentzien, S.; Spaltmann, D.; Krüger, J. Femtosecond laser-induced periodic surface structures on titanium nitride coatings for tribological applications. *Appl. Surf. Sci.* **2017**, *418*, 572–579. [CrossRef]

23. Mizuno, A.; Honda, T.; Kikuchi, J.; Iwai, Y.; Yasumaru, N.; Miyazaki, K. Friction Properties of the DLC Film with Periodic Structures in Nano-scale. *Tribol. Online* **2006**, *1*, 44–48. [CrossRef]

24. Yasumaru, N.; Miyazaki, K.; Kiuchi, J. Control of tribological properties of diamond-like carbon films with femtosecond-laser-induced nanostructuring. *Appl. Surf. Sci.* **2008**, *254*, 2364–2368. [CrossRef]

25. Pfeiffer, M.; Engel, A.; Gruettner, H.; Guenther, K.; Marquardt, F.; Reisse, G.; Weissmantel, S. Ripple formation in various metals and super-hard tetrahedral amorphous carbon films in consequence of femtosecond laser irradiation. *Appl. Phys. A Mater. Sci. Process.* **2013**, *110*, 655–659. [CrossRef]

26. Chen, C.Y.; Chung, C.J.; Wu, B.H.; Li, W.L.; Chien, C.W.; Wu, P.H.; Cheng, C.W. Microstructure and lubricating property of ultra-fast laser pulse textured silicon carbide seals. *Appl. Phys. A Mater. Sci. Process.* **2012**, *107*, 345–350. [CrossRef]

27. Eichstädt, J.; Römer, G.R.B.E.; Huis in't Veld, A.J. Towards Friction Control using laser-induced periodic Surface Structures. *Phys. Procedia* **2011**, *12*, 7–15. [CrossRef]

28. Wang, Z.; Zhao, Q.; Wang, C. Reduction of friction of metals using laser-induced periodic surface nanostructures. *Micromachines* **2015**, *6*, 1606–1616. [CrossRef]

29. Bonse, J.; Koter, R.; Hartelt, M.; Spaltmann, D.; Pentzien, S.; Höhm, S.; Rosenfeld, A.; Krüger, J. Femtosecond laser-induced periodic surface structures on steel and titanium alloy for tribological applications. *Appl. Phys. A* **2014**, *117*, 103–110. [CrossRef]

30. Bonse, J.; Koter, R.; Hartelt, M.; Spaltmann, D.; Pentzien, S.; Hohm, S.; Rosenfeld, A.; Krüger, J. Tribological performance of femtosecond laser-induced periodic surface structures on titanium and a high toughness bearing steel. *Appl. Surf. Sci.* **2015**, *336*, 21–27. [CrossRef]

31. Bonse, J.; Höhm, S.; Koter, R.; Hartelt, M.; Spaltmann, D.; Pentzien, S.; Rosenfeld, A.; Krüger, J. Tribological performance of sub-100-nm femtosecond laser-induced periodic surface structures on titanium. *Appl. Surf. Sci.* **2016**, *374*, 190–196. [CrossRef]

32. Biswas, S.K.; Vijayan, K. Friction and wear of PTFE—A review. *Wear* **1992**, *158*, 193–211. [CrossRef]

33. Hornbogen, E.; Karsch, U.A. Frictional wear of polytetrafluoroethylene (PTFE). *J. Mater. Sci. Lett.* **1983**, *2*, 777–780. [CrossRef]

34. Blanchet, T.A.; Kennedy, F.E. Sliding wear mechanism of polytetrafluoroethylene (PTFE) and PTFE composites. *Wear* **1992**, *153*, 229–243. [CrossRef]

35. Mackinson, K.R.; Tabor, D. The friction and transfer of polytetrafluoroethylene. *Proc. R. Soc. London Ser. A Math. Phys. Sci.* **1964**, *281*, 49–61. [CrossRef]

36. Patil, P.B.; Deore, E.R. Friction and Wear Behaviour of PTFE & Its Composites: A Review. *Int. J. Res. Advent Technol.* **2015**, *3*, 2321–9637.

37. Steijn, R.P. The sliding surface of polytetrafluoroethylene: an investigation with the electron microscope. *Wear* **1968**, *12*, 193–212. [CrossRef]

38. Tanaka, K.; Uchiyama, Y.; Toyooka, S. Mechanism of Wear of Polytetrafluoroethylene. *Wear* **1973**, *23*, 153–172. [CrossRef]

39. He, B.; Chen, W.; Wang, Q.J. Surface texture effect on friction of a microtextured poly(dimethylsiloxane) (PDMS). *Tribol. Lett.* **2008**, *31*, 187–197. [CrossRef]

40. Johnson, K.L.; Kendall, K.; Roberts, A.D. Surface Energy and the Contact of Elastic Solids. *Proc. R. Soc. A Math. Phys. Eng. Sci.* **1971**, *324*, 301–313. [CrossRef]

41. Tambe, N.S.; Bhushan, B. Micro/nanotribological characterization of PDMS and PMMA used for BioMEMS/NEMS applications. *Ultramicroscopy* **2005**, *105*, 238–247. [CrossRef]

42. Satyanarayana, N.; Sinha, S.K.; Ong, B.H. Tribology of a novel UHMWPE/PFPE dual-film coated onto Si surface. *Sens. Actuators A Phys.* **2006**, *128*, 98–108. [CrossRef]

43. Wilk, S.J.; Goryll, M.; Laws, G.M.; Goodnick, S.M.; Thornton, T.J.; Saraniti, M.; Tang, J.; Eisenberg, R.S. Teflon^{TM}-coated silicon apertures for supported lipid bilayer membranes. *Appl. Phys. Lett.* **2004**, *85*, 3307–3309. [CrossRef]

44. Smith, B.K.; Sniegowski, J.J.; LaVigne, G. Thin Teflon-like films for eliminating adhesion in released polysilicon microstructures. *Sens. Actuators A Phys.* **1998**, *70*, 159–163. [CrossRef]

45. Quake, S.R. From Micro- to Nanofabrication with Soft Materials. *Science* **2000**, *290*, 1536–1540. [CrossRef] [PubMed]

46. Dupont Teflon PTFE. Available online: http://www.rjchase.com/ptfe_handbook.pdf (accessed on 1 October 2018).

47. Sperati, C.A.; Starkweather, H.W. Fluorine-Containing Polymers. II. Polytetrafluoroethylene. In *Fortschritte Der Hochpolymeren-Forschung*; Springer: Berlin/Heidelberg, Germany, 1961; pp. 465–495.

48. Rae, P.J.; Dattelbaum, D.M. The properties of poly(tetrafluoroethylene) (PTFE) in compression. *Polymer* **2004**, *45*, 7615–7625. [CrossRef]

49. Unal, H.; Mimaroglu, A.; Kadioglu, U.; Ekiz, H. Sliding friction and wear behaviour of polytetrafluoroethylene and its composites under dry conditions. *Mater. Des.* **2004**, *25*, 239–245. [CrossRef]

50. Liu, J.M. Simple technique for measurements of pulsed Gaussian-beam spot sizes. *Opt. Lett.* **1982**, *7*, 196–198. [CrossRef]

51. Bonse, J.; Baudach, S.; Krüger, J.; Kautek, W.; Lenzner, M. Femtosecond laser ablation of silicon-modification thresholds and morphology. *Appl. Phys. A Mater. Sci. Process.* **2002**, *74*, 19–25. [CrossRef]

52. Bhushan, B. *Introduction to Tribology*, 2nd ed.; Tribology Series; John Wiley & Sons: Hoboken, NJ, USA, 2013; ISBN 978-1-119-94453-9.

53. Clark, E.S. The molecular conformations of polytetrafluoroethylene: Forms II and IV. *Polymer* **1999**, *40*, 4659–4665. [CrossRef]

54. Blumm, J.; Lindemann, A.; Meyer, M.; Strasser, C. Characterization of PTFE using advanced thermal analysis techniques. *Int. J. Thermophys.* **2010**, *31*, 1919–1927. [CrossRef]

55. Brown, E.N.; Dattelbaum, D.M. The role of crystalline phase on fracture and microstructure evolution of polytetrafluoroethylene (PTFE). *Polymer* **2005**, *46*, 3056–3068. [CrossRef]

56. Nunes, L.C.S.; Dias, F.W.R.; Da Costa Mattos, H.S. Mechanical behavior of polytetrafluoroethylene in tensile loading under different strain rates. *Polym. Test.* **2011**, *30*, 791–796. [CrossRef]

57. Rae, P.; Brown, E. The properties of poly (tetrafluoroethylene)(PTFE) in tension. *Polymer* **2005**, *46*, 8128–8140. [CrossRef]

58. Uçar, A.; Çopuroǧlu, M.; Baykara, M.Z.; ArIkan, O.; Suzer, S. Tribological interaction between polytetrafluoroethylene and silicon oxide surfaces. *J. Chem. Phys.* **2014**, *141*, 164702. [CrossRef] [PubMed]

59. Breiby, D.W.; Sølling, T.I.; Bunk, O.; Nyberg, R.B.; Norrman, K.; Nielsen, M.M. Structural Surprises in Friction-Deposited Films of Poly(tetrafluoroethylene). *Macromolecules* **2005**, *38*, 2383–2390. [CrossRef]

60. Brown, E.N.; Rae, P.J.; Bruce Orler, E.; Gray, G.T.; Dattelbaum, D.M. The effect of crystallinity on the fracture of polytetrafluoroethylene (PTFE). *Mater. Sci. Eng. C* **2006**, *26*, 1338–1343. [CrossRef]

61. Blau, P.J. On the nature of running-in. *Tribol. Int.* **2005**, *38*, 1007–1012. [CrossRef]

62. Blau, P.J. Interpretations of the friction and wear break-in behavior of metals in sliding contact. *Wear* **1981**, *71*, 29–43. [CrossRef]

63. Pooley, C.M.; Tabor, D. Friction and Molecular Structure: The Behaviour of Some Thermoplastics. *Proc. R. Soc. A Math. Phys. Eng. Sci.* **1972**, *329*, 251–274. [CrossRef]

64. Hutchings, I.M. *Tribology: Friction and Wear of Engineering Materials*; Edward Arnold: London, UK, 1992; ISBN 9780340561843.

65. Lee, S.M.; Shin, M.W.; Lee, W.K.; Jang, H. The correlation between contact stiffness and stick-slip of brake friction materials. *Wear* **2013**, *302*, 1414–1420. [CrossRef]
66. Biswas, S.K.; Vijayan, K. Changes to near-surface region of PTFE during dry sliding against steel. *J. Mater. Sci.* **1988**, *23*, 1877–1885. [CrossRef]
67. Uchiyama, Y.; Tanaka, K. Wear laws for polytetrafluoroethylene. *Wear* **1980**, *58*, 223–235. [CrossRef]
68. Smurugov, V.A.; Senatrev, A.I.; Savkin, V.G.; Biran, V.V.; Sviridyonok, A.I. On PTFE transfer and thermoactivation mechanism of wear. *Wear* **1992**, *158*, 61–69. [CrossRef]
69. Bonse, J.; Rosenfeld, A.; Krüger, J. On the role of surface plasmon polaritons in the formation of laser-induced periodic surface structures upon irradiation of silicon by femtosecond-laser pulses. *J. Appl. Phys.* **2009**, *106*, 104910. [CrossRef]
70. Sedao, X.; Shugaev, M.V.; Wu, C.; Douillard, T.; Esnouf, C.; Maurice, C.; Reynaud, S.; Pigeon, F.; Garrelie, F.; Zhigilei, L.V.; et al. Growth Twinning and Generation of High-Frequency Surface Nanostructures in Ultrafast Laser-Induced Transient Melting and Resolidification. *ACS Nano* **2016**, *10*, 6995–7007. [CrossRef]
71. Maharjan, N.; Zhou, W.; Zhou, Y.; Guan, Y.; Wu, N. Comparative study of laser surface hardening of 50CrMo4 steel using continuous-wave laser and pulsed lasers with ms, ns, ps and fs pulse duration. *Surf. Coat. Technol.* **2019**, *366*, 311–320. [CrossRef]
72. Persson, B.N.J. Theory of rubber friction and contact mechanics. *J. Chem. Phys.* **2001**, *115*, 3840–3861. [CrossRef]

 *nanomaterials*

Article

Influences of Ga Doping on Crystal Structure and Polarimetric Pattern of SHG in ZnO Nanofilms

Hua Long [1],*, Ammar Ayesh Habeeb [2], Dickson Mwenda Kinyua [1], Kai Wang [1], Bing Wang [1] and Peixiang Lu [1,3],*

[1] Wuhan National Laboratory for Optoelectronics (WNLO) and School of Physics, Huazhong University of Science and Technology, Wuhan 430074, China; I201622178@hust.edu.cn (D.M.K.); kale_wong@hust.edu.cn (K.W.); wangbing@hust.edu.cn (B.W.)
[2] Physics Department of College Science, Diyala University, Baqubah 964, Iraq; ammarlaser72@yahoo.com
[3] Hubei Key Laboratory of Optical information and Pattern Recognition, Wuhan Institute of Technology Wuhan 430205, China
* Correspondence: longhua@hust.edu.cn (H.L.); lupeixiang@hust.edu.cn (P.L.); Tel.: +86-027-8754-3755 (H.L.)

Received: 30 April 2019; Accepted: 7 June 2019; Published: 21 June 2019

Abstract: The second-harmonic generation (SHG) in gallium doped ZnO (GZO) nanofilms was studied. The Ga doping in GZO nanofilms influenced the crystal structure of the films, which affected SHG characteristics of the nanofilms. In our experiments, a strong SHG response was obtained in GZO nanofilms, which was excited by 790 nm femtosecond laser. It was observed that the Ga doping concentrations affected, not only the intensity, but also the polarimetric pattern of SHG in GZO nanofilms. For 5.0% doped GZO films, the SHG intensity increased about 70%. The intensity ratio of SHG between the incident light polarization angle of 90° and 0°changed with the Ga doping concentrations. It showed the most significant increase for 7.3% doped GZO films, with an increased ratio of c/a crystal constants. This result was attributed to the differences of the ratios of d_{33}/d_{31} (the second-order nonlinear susceptibility components) induced by the crystal distortion. The results are helpful to investigate nanofilms doping levels and crystal distortion by SHG microscopy, which is a non-destructive and sensitive method.

Keywords: ZnO nanofilms; SHG; Ga doping; polarization angle

1. Introduction

Semi-conductor nanofilms are one of the most widely used nanomaterials due to their excellent properties. As one type of a range of widely used semiconductor materials, ZnO nanofilms are applied in the fields, including photo-voltaic devices, photo-catalysis, and bio-imaging for its characteristics, such as wide band gap and high transparency [1–3]. Doping can manipulate the optical and electrical components of the intrinsic ZnO materials. For instance, Ga-doped ZnO films have outstanding properties, such as the wider band gap and low reactivity. Because the radii of Ga and Zn atoms are similar, even at high doping concentrations, the Ga doping leads to a small lattice deformation in ZnO [4–7].

The non-linear optical properties of semi-conductor nanofilms have attracted a lot of attention due to their potential applications in nonlinear optical frequency converters, microscopic images, and all-optical communication. The developments in nonlinear optical techniques have opened a window into the morphological and structural characteristics for a variety of systems, even for biological systems. It can also be an alternative measurement scheme suitable for detecting dynamic processes. [8,9]. For instance, SHG is a very sensitive and non-destructive technique for several applications in various fields, such as crystal structural detection, cancer cell diagnostics, optical switches in nano-devices [10–15].

Previous studies have demonstrated highly efficient SHG in different semi-conductor nanomaterials, such as CdS, GaAs, ZnO [16–22]. Owing to its polarization sensitivity [23,24], the SHG method has been proven to be an optical method for the detection of crystal structures without damage or special environmental requirements. Furthermore, since the wavelength of pumping laser can be tuned conveniently, the SHG method is advantageous in identifying suitable wavelengths for different materials detection. Through tuning the pumping wavelength, this method can even be applied for depth-resolved detection within nanomaterials.

With the above advantages, the SHG method has been used to detect the crystal structure characteristics of the ZnO nanomaterials. Moreover, the efficient generation of SHG signal, from several ZnO nanostructures, has been reported. For instance, Neeman et al. reported crystallographic mapping of ZnO nanowires using the SHG method [25]. Han et al. also used the SHG microscopy method to detect the lattice distortion in a bent ZnO nanowire [26]. Much attention has been paid to the analysis of the crystal structure of nanomaterials using SHG signal. However, few studies have focused on the analysis of the crystal structure induced by doping, using polarimetric patterns of SHG in ZnO nanostructures. In this work, we investigated the Ga doping influences on the crystal structure and polarimetric pattern of SHG in ZnO nano-films. The deviation of crystal site symmetry in Ga-doped ZnO nano-films was also determined from X-ray diffraction (XRD) results, and it was found to be strongly correlated with the SHG polarity. We found that the SHG intensity increased by about 70% for 5.0% doped GZO films. The intensity ratio of SHG between incident light polarization angle of 90° and 0° changed with the Ga doping concentrations. This result was attributed to the differences of the ratios of d_{33}/d_{31} induced by the crystal distortion. The results are helpful to determine nanofilms doping levels and crystal distortion using SHG method, which is a sensitive and non-destructive method.

2. Materials and Methods

The GZO films were deposited on silica substrates using pulsed laser deposition (PLD). The GZO targets with different Ga concentrations (0, 2.93%, 5.0%, 7.3%, 9.9%, and 20.9%, respectively) were fabricated [27,28]. To obtain the GZO targets, the powders of Ga_2O_3 (99.99%) and ZnO (99.99%) were mixed and sintered at 1350 °C for 48h. To clean the silica substrates, a mixed solution of H_2SO_4 and H_2O_2 was firstly used. The substrates were soaked in the mixed solution for 1 h at 80 °C. Then the substrates were treated with ultrasonic cleaning in another solution of H_2O_2, NH_4OH, and H_2O for 1 h. Excimer laser (Lambda Physik, KrF, 248 nm, 5 Hz, Coherent, Santa Clara, CA, USA) was used in the PLD. Before film deposition, the vacuum chamber was evacuated. The background gas pressure was 4.5×10^{-3} Pa. The oxygen gas was then introduced in the vacuum chamber. During the deposition the oxygen pressure was kept at 0.2 Pa. The substrate temperature was kept at 80 °C in the deposition process. The deposition time was 40 min. After that, the films were annealed for 1 h at 500 °C in air.

The surface morphology of the GZO films were measured through field emission scanning electron microscopy (FESEM, Sirion 200, FEI Co., Hillsboro, OR, USA). The crystal structures of the GZO films were determined by X-ray diffraction (XRD, PANalytical B.V., Almelo, Horland, the precision of 2θ is 0.00001°). Optical transmittance spectra of the samples were measured by UV-visible spectrophotometer (HITACHI U3310, Tokyo; Japan).

The SHG characteristics of the GZO films were detected by a home-built optical microscopy setup, as illustrated in Figure 1a. To pump the GZO films, the femtosecond laser beam from a mode-locked Ti-sapphire laser system (Tsunami, Spectra-Physics, 50 fs, 76 MHz, @790 nm, Newport Co., Irvine, CA, USA) was focused on the GZO nanofilms by a 40× objective (NA = 0.55) [29,30]. The transmitted SHG signals were collected by another 40× objective. Then, the transmitted signals were detected by a CCD or through a fiber coupling by a spectrometer (Acton 2500i, Princeton Instruments, Trenton, NJ, USA). In front of the CCD and the spectrometer, a 750-nm short-pass filter was used to filter out the incident laser. The laser power was measured by a laser power meter with a precision of 0.1 mW. In order to adjust the intensity of the pumping laser, we used a half-wave plate (HWP1) and a polarizing beam splitter (BP). Moreover, another half-wave plate (HWP2) was used to change the polarization direction of the pump laser. Figure 1b illustrates the crystallographic frame and the geometry of the laboratory frame. The laser propagated along the z-axis in the laboratory frame, whereas the films lied in the xy plane. Angle β determined the polarization of the incoming laser. Angle φ was 0°. Figure 1c shows the optical microscope image of the SHG signal of ZnO nanofilm detected by the CCD.

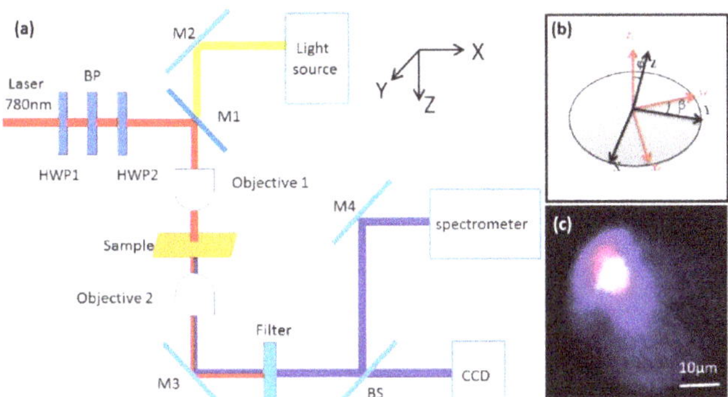

Figure 1. (**a**) Schematics of the experimental setup for measuring second-harmonic generation (SHG) signals; (**b**) geometry of the lab frame (X, Y, Z) and the crystal frame (x_c, y_c, z_c). While, ϕ is the angle between the z_c axis and the Z axis, β is the angle between the Y axis and the y_c axis on the xy plane; (**c**) optical microscope image of the SHG of the ZnO film.

3. Results and Discussion

The scanning electron microscopy (SEM) images of the GZO films surface (with doping concentrations of 0, 5.0%, 9.9% and 21.9%, respectively) were shown in Figure 2. The images show homogeneous and nano-crystalline films without laser-induced large particles on the surfaces. It can be observed that the grain sizes of the films were about several tens nanometers. And for the 21.9% doped ZnO film, the grain sizes show an obvious decrease.

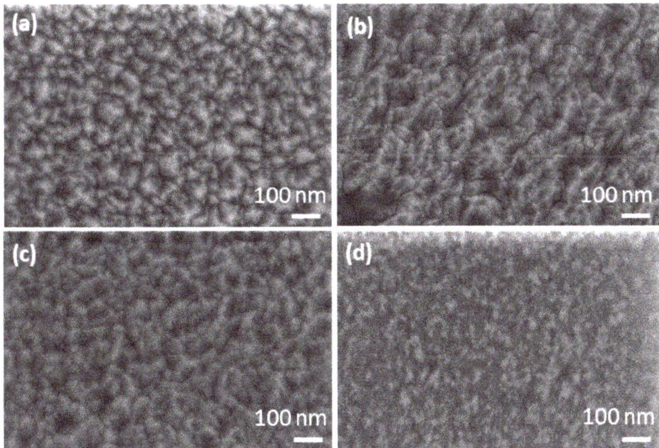

Figure 2. The SEM surface images of the gallium doped ZnO (GZO) films with different doping concentrations. (**a**) 0; (**b**) 5.0%; (**c**) 9.9%; (**d**) 21.9%.

To analyze the crystal distortion in ZnO nanofilms induced by doping, we compared the XRD patterns of the GZO films. As can be seen in Figure 3a, GZO films with hexagonal wurtzite structure were obtained. The XRD patterns show the characteristic diffraction peaks of the pure hexagonal wurtzite phase. For the undoped ZnO nanofilm, the diffraction peak of (002) at around $2\theta = 34.0°$ was strong. And the diffraction peak at around $2\theta = 30.5°$ corresponding to the wurtzite (100) peak was very weak. The strong (002) XRD peak indicates a high-quality crystal structure with c-axis preferred oriented. The preferential growth along c axis is due to the lower energy of growing along the [1] direction. Generally, the (0001) plane has the highest energy among all the facets. For instance, Zhan, et al., reported the (10$\bar{1}$0) plane has a surface energy of about 1.8 J m $^{-2}$. And the Zn-terminated (0001) surface has higher surface energy in the range from 2.5 to 5.8 J m^{-2}. So according to Gibbs-Wulff theory, the growth along [1] direction leads to a preference of exposing low energy (10$\bar{1}$0) planes [31–33]. Then, due to the lower energy of growth along the [1] direction, the film showed preferential growth along *c* axis. GZO films, with lower doping concentrations (2.9%, 5.0%, 7.3% and 9.9%), showed similar XRD patterns as ZnO. While, for 21.9% Ga doped GZO films, only a weak diffraction peak at $2\theta = 56.5°$ corresponding to (110) was observed. This was attributed to a significant decrease of crystal quality and a variation in the crystal structure.

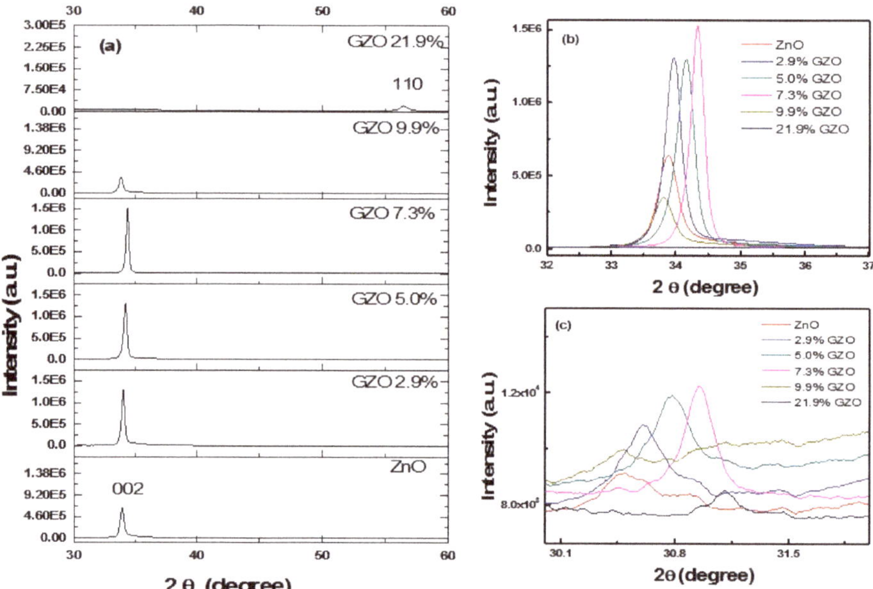

Figure 3. (**a**) XRD patterns of the GZO films; (**b**) the enlarged view of the diffraction peak of (002), showing the shifting of the diffraction peak position; (**c**) the enlarged view of the diffraction peak of (100), showing the shifting of the diffraction peak position.

The diffraction peaks around $2\theta = 34°$ and $2\theta = 30.5°$ of the GZO films were shown in details in Figure 2b,c respectively. We can see that the peaks gradually shifts to a level higher than 20 as Ga concentrations increased up to 7.3%. This shift indicates the reduction of the lattice spacing because the atomic radius of Ga (0.062 nm) is less than Zn (0.074 nm). When the Ga doping concentration was further increased to 9.9%, the two diffraction peaks shifted conversely towards lower 2θ. It might be induced by degradation of the crystal quality. Heavy doping Ga led to significant distortions and dislocations. For the GZO film with 21.9% Ga doped, no diffraction peak of (002) could be observed. A weak diffraction peak corresponding to (110) was observed. The heavily doped film were not grown along *c*-axis. The film grew with a preferred orientation along the (110) plane. These results were in agreement with the previous works [34–36].

From the XRD results, we are able to evaluate the grain size (D) using the following equation (Scherrer's formula) [37]:

$$D = k\lambda/(\beta_0\cos(\theta)) \tag{1}$$

where β_0 is the full width at half maximum (FWHM), k is a constant. The average grain sizes were calculated using this formula for GZO films. The calculated values are listed in Table 1. It can be noted that as the doping concentration increased to 7.3%, a slight increase in the grain size occurred. With further increase of the doping concentration, the grain size decreased.

Table 1. Grain size and lattice parameter of Ga-doped ZnO thin films.

Doping Concentration (at. %)	FWHM	2θ (Degree)	D (Grain Size) (nm)	c (Å)	a (Å)	c/a
0.0	0.326	33.91	25.47	5.285	3.381	1.563
2.9	0.197	34.02	42.17	5.273	3.370	1.564
5.0	0.211	34.18	39.37	5.245	3.350	1.565
7.3	0.154	34.33	53.71	5.219	3.332	1.566
9.9	0.253	33.81	32.78	5.296	3.383	1.565

The lattice constants (c and a) were also calculated for (002) and (100) plane from the XRD patterns. Table 1 show the calculated lattice parameters for GZO films. The calculated values of lattice spacing shows the reduction of the lattice constants in GZO nanofilms. The calculated values of c and a constants are close to the values of ZnO films previously reported [38–40]. The decrease of c and a constants with doping has been reported for GZO films fabricated by radio frequency magnetron sputtering method [41]. The observed reduction verifies the incorporation of Ga ions in ZnO lattice. The lattice reduction with the Ga doping concentration shows a significant reduction up to the 7.3% doped GZO nanofilms and then increases with further Ga doping. The incorporation of the Ga ions caused the reduction of the lattice, due to a smaller atomic radius of Ga than Zn. According to Vegard's law, the Ga element was successfully doped into the ZnO lattice because the lattice constants decreased linearly as the Ga concentration increased up to 7.3% [41]. The 7.3% Ga doping level may be close to the saturation of Ga doping in ZnO using PLD methods [42,43].

We then evaluated the crystal structure deformation caused by Ga doping by calculating the ratio of the lattice constant (c/a) for the GZO nanofilms. We then compared them with the standard c/a ratio of the undoped ZnO, with the hexagonal wurtzite structure. From Table 1, we can see that the GZO nanofilm with 7.3% Ga doping concentration have the largest c/a ratio, which means that the lattice distortion along *c* axis is the largest. Similar behavior has been observed for Eu doped ZnO nanowires [32,44]. Clearly the increase in the c/a ratio can be attributed to the introduction of Ga ions. The presence of O vacancies can also influence the system. Furthermore, since the Ga ions are smaller than the Zn ions, Ga ions and O vacancies may form defect complexes. As a result, we can expect the formation of several lattice distortions in the GZO nanofilms with further increasing doping level, and hence degradation of crystal structure quality.

Figure 4a show the optical transmission spectra of the GZO nanofilms. As can be seen in the figure, the transmittance in the visible range is in the range of 70–90%. In the SHG experiments, strong SHG signal can be observed around 390 nm. Figure 4b shows the spectra of SHG for the 2.9% doped GZO film under different excitation laser power. Figure 4c shows the corresponding plots of SHG intensity versus the incident laser power in logarithmic scale. The slope of the fitting line is about 1.9. It confirms a quadratic response of the SHG intensity as a function of the exciting laser power, verifying a second-order nonlinear process [45,46]. And the measurements indicate a SHG susceptibility of ~15 pm/V at 810 nm [47,48]. The value is close to the reported values [49].

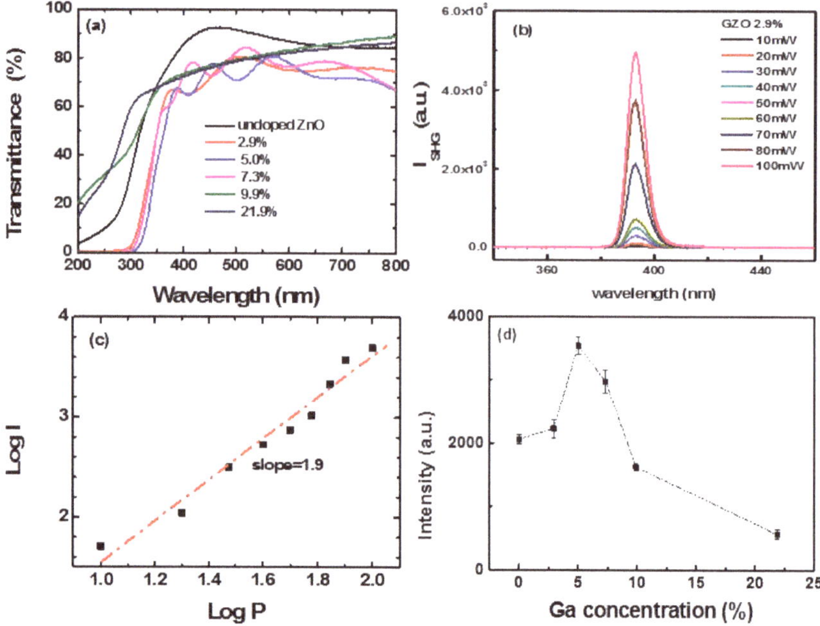

Figure 4. (**a**) Transmission spectra of the GZO films; (**b**) the spectra of SHG signal originated from 2.9% doped GZO film under different incident laser power; (**c**) the output SHG intensity as a function of the incident laser power. (**d**) SHG intensity changes as the Ga doping concentration increased. Dotted line is guide to the eye.

The influence of Ga doping on SHG intensity was investigated for GZO films. Figure 4d shows the SHG intensity (@ 390 nm) of the GZO films, with different doping levels, under the same excitation laser power. We can see that the intensity of transmitted SHG signal varied with gallium doping concentrations. It clearly indicates that the SHG intensity is sensitive to the doping concentration. The SHG intensity from the ZnO films improved by 5.0% and 7.3% in Ga doping. While, the SHG intensity showed a non-monotonic increase with the increasing Ga doping. The maximum values in SHG intensity were obtained in the 5.0% doped GZO films, which was about 1.7 times of the SHG intensity in the undoped ZnO films. We attributed the SHG enhancement to the influence of the centrosymmetry. Doping in the ZnO films significantly influenced the centrosymmetry of the ZnO, which could lead to the enhancement of the effective second-order susceptibility (d_{eff}) in ZnO films and hence SHG effects. From the above crystal structure analysis, a notable modification of lattice constant in the GZO nanofilms was observed. For 7.3%, 9.9%, and 21.9%, the doped GZO nanofilms showed a decrease in SHG, compared with 5.0% doped GZO films. Similar results in the doping enhancement effects of SHG have been found in other materials, such as Eu doped ZnO nanowires, urea doped tristhioureazinc (II) sulfate, and oxalic acid doped ADP crystals [50]. Firstly, the SHG was enhanced with the increase in doping. However, with further increasing doping concentrations, the SHG enhancement effect decreased. We attributed this to the poor crystal quality induced by heavy doping and the influence of O vacancies. Furthermore, for heavy doped ZnO, the doped ions were segregated at the grain boundaries and not distributed in the crystalline matrix to enhance the non-linearity.

The SHG intensity as a function of the excitation laser polarization and its dependence on the doping in GZO films were then studied. Figure 5a–f show the polar pattern evolution of the nanofilms with the increase in doping concentrations. From the figures, we can see that without doping, the SHG intensity I shows a circular shape. Once gallium was doped, the SHG pattern distorted significantly. The polar plots present a symmetrical shape of two-lobe. The apparent tilting of the lob in Figure 5c–e is attributed to the inaccuracy in the rotation of the polarizer in experiments. The dots indicate the experimental data and the red solid curves show the theoretical fitting. The fitting formula was as follows [51,52]:

$$I = A_0(\cos^2\beta + B_0\sin^2\beta + C_0(\sin\beta\cos\beta))^2 \tag{2}$$

where I is intensity of SHG, A_0, B_0 and C_0 are fitting parameters, β is polarization angle. The c axis of ZnO nanofilm is along the z-direction (shown in Figure 1b). Then for the c-axis oriented ZnO films, the second harmonic generalization electrical field, P_x, P_y, P_z, can be described by the following matrix [24],

$$\begin{bmatrix} P_x \\ P_y \\ P_z \end{bmatrix} = \begin{bmatrix} 0 & 0 & 0 & 0 & d_{15} & 0 \\ 0 & 0 & 0 & d_{15} & 0 & 0 \\ d_{31} & d_{31} & d_{33} & 0 & 0 & 0 \end{bmatrix} \begin{bmatrix} E_x^2 \\ E_y^2 \\ E_z^2 \\ 2E_yE_z \\ 2E_xE_z \\ 2E_xE_y \end{bmatrix} \tag{3}$$

where E_x, E_y and E_z are the incident electric fields in the direction of x, y, and z, respectively. And d_{ij} is the contracted notation of the second-order susceptibility tensor, which is a 3×6 matrix. The angle between the y and y_c direction is defined as the polarization angle, θ, see Figure 1b. The obtained patterns are relative to the characteristic of the wurtzite crystal structure.

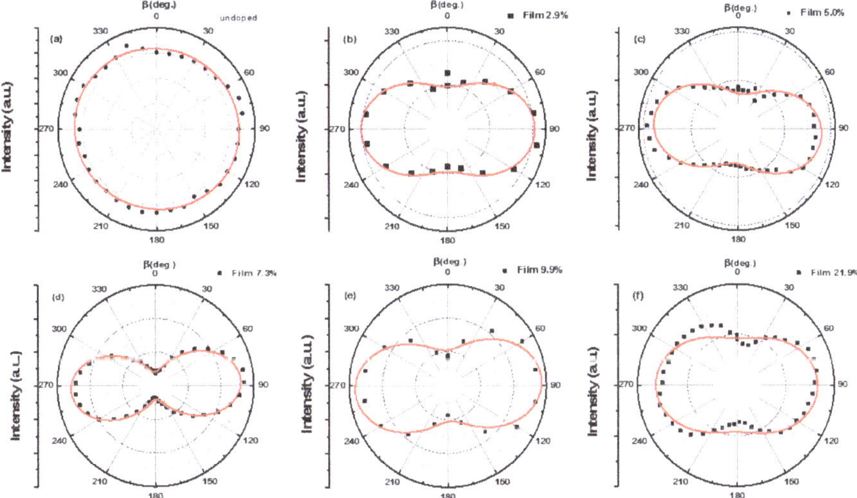

Figure 5. The normalized SHG intensity as the polarization angle, θ, changed for GZO films with different doping concentrations (**a**) 0; (**b**)2.9%; (**c**) 5.0%;(**d**) 7.3%; (**e**) 9.9%; (**f**) 21.9%. The experimental data are shown as dots and the solid curves show the theoretical fits.

In our experiments, the objective with NA 0.55 were used to collect the SHG signal. The corresponding collective angle was about 30 degree. So the SHG signal detected was caused by P_x, P_y and P_z, the whole intensity can be indicated as following,

$$I \propto P_x^2 + P_y^2 + P_z^2 \tag{4}$$

Due to the centrosymmetry, the polarimetric pattern of undoped ZnO shows a circular shape. When the doping concentration in ZnO nanofilms increased, the crystal structure changed, as confirmed by the XRD results. As a consequence of this structural change, the d_{ij} components change accordingly by doping. According to the reference [53], the s-polarized and p-polarized SHG signal can be fitted as following,

$$E_s^{2\omega} \propto A d_{15} \cos\beta \sin\beta \tag{5}$$

$$E_p^{2\omega} \propto (B d_{31} + C d_{33} + D d_{15})\sin^2\beta + E d_{31}\cos^2\beta \tag{6}$$

where $E_s^{2\omega}$ and $E_p^{2\omega}$ are the s-polarized and p-polarized SHG electrical field, separately; A–E are fitting parameters, which are related to the Fresnel transmission coefficients for fundamental beams and the corresponding refraction angles in the film for the fundamental and second harmonic wavelengths. Because d_{33} is much larger than d_{31} and d_{15} in ZnO, we conclude that the intensity ratio (between β is 90° and 0°) is most related to d_{33}/d_{31} [49,54,55].

In order to analyze the relationship between the dependence on the incident laser polarization and the crystal distortion induced by Ga doping, the intensity ratio of SHG between $\beta = 90°$ and 0° in GZO nanofilms was plotted, see Figure 6a. Obviously, the intensity ratio showed a dramatic increase as the doping in ZnO nanofilm increased to 7.3%. To further increase the doping concentration, the ratio decreased. At the same time, the variation of c/a ratio in the doped GZO nanofilms, estimated by XRD, was shown in Figure 6b, which shows a similar trend with the variation of the SHG intensity ratio. We also noticed that the shift of the diffraction peak (Figure 3b) in the nanofilms was in the range from 0.1° to 0.4°. The corresponding variation of the lattice spacing was estimated to be in the range of 0.01 nm to 0.06 nm. The results indicate that crystal distortion can significantly affect the second order susceptibility tensor. The intensity ratio of SHG between $\beta = 90°$ and 0° in GZO nanofilms is related to the absolute value of d_{33}/d_{31}. It increased as the doping concentration increased up to 7.3% at doping, with the corresponding increase in the c/a ratio.

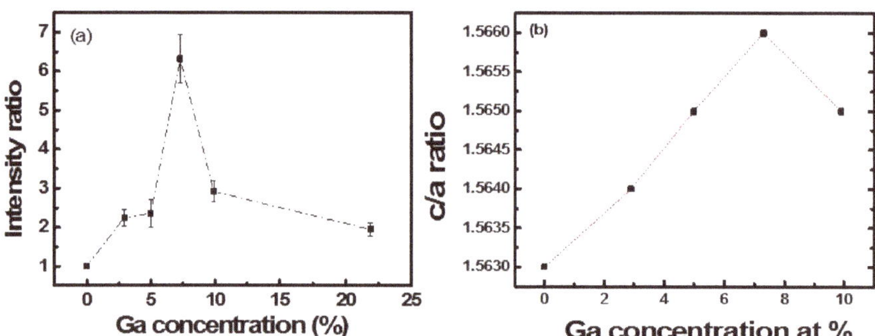

Figure 6. (**a**) Plot of the SHG intensity ratio between $\beta = 90°$ and 0° (black dots) for GZO films as a function of Ga doping; (**b**) Plot of c/a ratio for GZO films as a function of Ga doping. Dotted lines are guide for the eye.

Therefore, the polarization-dependent SHG method can be used as a sensitive and non-destructive detection method for the crystal distortion induced by doping. Compared with traditional XRD methods, it is compatible with various experiment conditions, such as in atmosphere or in liquids. The results are also helpful for sensitively detection of the doping level using the optical method.

4. Conclusions

The crystal structure distortion and SHG from GZO nanofilms, induced by Ga doping, were studied. The results show that the intensity ratio of SHG between $\beta = 90°$ and 0° shows a similar trend with the change of c/a ratios induced by Ga doping. For 5.0% doped GZO films the SHG intensity increased about 70%. The results indicate that the polarization-dependent SHG method has potential application as a non-destructive and sensitive detection of the crystal structure distortion induced by doping. Compared with XRD, it can be compatible with different experimental conditions. These results are also helpful for the sensitive detection of the doping level using the optical method.

Author Contributions: Conceptualization and methodology, H.L., A.A.H., and K.W.; validation, H.L. and A.A.H.; investigation and resources, H.L., A.A.H., and D.K.; writing—original draft preparation, H.L. and A.A.H.; writing—review and editing, H.L., A.A.H., D.K., K.W., B.W., and P.L.; visualization, H.L.; supervision and project administration, B.W.; funding acquisition, P.L.

Funding: This work was supported by the 973 Programs (2014CB921301), National Natural Science Foundation of China (11204097, 11674117 and 11804109), and the Doctoral Fund of Ministry of Education of China (20130142110078).

Conflicts of Interest: The authors declare no conflict of interest.

References

1. Chen, M.; Hu, L.F.; Xu, J.X.; Liao, M.Y.; Wu, L.M.; Fang, X.S. ZnO hollow-sphere nanofilm-based high-performance and low-cost photodetec. *Small* **2011**, *7*, 2449–2453. [PubMed]
2. Zhang, C.H.; Wang, G.F.; Liu, M.; Feng, Y.H.; Zhang, Z.D.; Fang, B. A hydroxyllamine electrochemical sensor based on electro deposition of porous ZnO nanofilms onto carbon nanotubes films modified electrode. *Electroch. Acta* **2010**, *55*, 2835–2840. [CrossRef]
3. Wang, J.X.; Zhou, H.J.; Guo, G.Y.; Tan, J.Q.; Wang, Q.J.; Tang, J.; Liu, W.; Shen, H.; Li, J.H.; Zhang, X.L. Enhanced anti-infective efficacy of ZnO nanoreservoirs through a combination of intrinsic anti-biofilm activity and reinforced innate defense. *ACS Appl. Mater. Inter.* **2017**, *9*, 33609–33623. [CrossRef] [PubMed]
4. Park, H.K.; Kang, J.W.; Na, S.I.; Kim, D.Y.; Kim, H.K. Characteristics of indium-free GZO/Ag/GZO and AZO/Ag/AZO multilayer electrode grown by dual target DC sputtering at room temperature for low-cost organic photovoltaics. *Sol. Energy Mater. Sol. Cells* **2009**, *93*, 1994–2002. [CrossRef]
5. Correia, F.C.; Bundaleski, N.; Teodoro, O.M.N.D.; Correia, M.R.; Rebouta, L.; Mendes, A.; Tavares, C.J. XPS analysis of ZnO:Ga films deposited by magnetron sputtering: Substrate bias effect. *Appl. Surf. Sci.* **2018**, *458*, 1043–1049. [CrossRef]
6. Liu, W.S.; Hsieh, W.T.; Chen, S.Y.; Huang, C.S. Improvement of CIGS solar cells with high performance transparent conducting Ti-doped GaZnO thin films. *Sol. Energy* **2018**, *174*, 83–96. [CrossRef]
7. Long, H.; Bao, L.J.; Habeeb, A.A.; Lu, P.X. Effects of doping concentration on the surface plasmonic resonances and optical nonlinearities in AGZO nanotriangle arrays. *Opt. Quant. Electron.* **2017**, *49*, 345. [CrossRef]
8. Stijn, V.C.; Zachary, J.S.; Olivier, D.; Carmen, B.; Sebastian, W.H.; Thierry, V.; Monique, A.V. Morphology and structure of ZIF-8 during crystallisation measured by dynamic angle resolved second harmonic scattering. *Nat. Commun.* **2018**, *9*, 3418.
9. Stefan, K.; Hannu, H.; Thierry, V.; Koen, C. Role of donor and acceptor substituents on the nonlinear optical properties of gold nanoclusters. *J. Phys. Chem. C* **2018**, *122*, 4019–4028.
10. Morales-Saavedra, O.G.; Castaneda, L. Second harmonic generation of fluorine-doped zinc oxide thin films grown on soda-lime glass substrates by a chemical spray technique. *Opt. Commun.* **2007**, *269*, 370–377. [CrossRef]
11. Li, X.H.; Liu, W.W.; Song, Y.L.; Zhang, C.; Long, H.; Wang, K.; Wang, B.; Lu, P.X. Enhancement of the second harmonic generation from WS2 monolayers by cooperating with dielectric microspheres. *Adv. Opt. Mater.* **2019**, *7*, 1801270. [CrossRef]
12. Li, L.; Lan, P.F.; Zhu, X.S.; Huang, T.F.; Zhang, Q.B.; Manfred, L.; Lu, P.X. Reciprocal-space-trajectory perspective on high-harmonic generation in solids. *Phys. Rev. Lett.* **2019**, *122*, 193901. [CrossRef] [PubMed]
13. Lo, K.Y.; Huang, Y.J.; Huang, J.Y.; Feng, Z.C.; Fenwick, W.E.; Pan, M.; Ferguson, I.T. Reflective second harmonic generation from ZnO thin films: A study on the Zn–O Bonding. *Appl. Phys. Lett.* **2007**, *90*, 161904. [CrossRef]
14. Zhao, D.; Ke, S.L.; Hu, Y.H.; Wang, B.; Lu, P.X. Optical bistability of graphene embedded in parity-time-symmetric photonic lattices. *J. Opt. Soc. Am. B* **2019**, *36*, 1731–1737. [CrossRef]
15. Wickberg, A.; Kieninger, C.; Sürgers, C.; Schlabach, S.; Mu, X.K.; Koos, C.; Wegener, M. Second-harmonic generation from ZnO/Al2O3 nanolaminate optical metamaterials grown by atomic-layer deposition. *Adv. Opt. Mater.* **2016**, *4*, 1203–1208. [CrossRef]

16. Ren, M.L.; Agarwal, R.; Nukala, P.; Liu, W.J.; Agarwal, R. Nanotwin detection and domain polarity determination via optical second harmonic generation polarimetry. *Nano Lett.* **2016**, *16*, 4404–4409. [CrossRef]

17. Wang, Y.F.; Liao, L.M.; Hu, T.; Luo, S.; Wu, L.; Wang, J.; Zhang, Z.; Xie, W.; Sun, L.X.; Kavokin, A.V.; et al. Exciton-Polariton Fano Resonance Driven by Second Harmonic Generation. *Phys. Rev. Lett.* **2017**, *118*, 063602. [CrossRef]

18. Weber, N.; Protte, M.; Walter, F.; Georgi, P.; Zentgraf, T.; Meier, C. Double resonant plasmonic nanoantennas for efficient second harmonic generation in zinc oxide. *Phys. Rev. B* **2017**, *95*, 205307. [CrossRef]

19. Kim, C.J.; Brown, L.; Graham, M.W.; Hovden, R.; Havener, R.W.; McEuen, P.L.; Muller, D.A.; Park, J. Stacking order dependent second harmonic generation and topological defects in h-BN bilayers. *Nano Lett.* **2013**, *13*, 5660–5665. [CrossRef]

20. Li, Y.L.; Rao, Y.; Mak, K.F.; You, Y.; Wang, S.Y.; Dean, C.R.; Heinz, T.F. Probing symmetry properties of few-layer MoS$_2$ and h-BN by optical second-harmonic generation. *Nano Lett.* **2013**, *13*, 3329–3333. [CrossRef]

21. Wei, Y.M.; Yu, Y.; Wang, J.; Liu, L.; Ni, H.Q.; Niu, Z.C.; Li, J.T.; Wang, X.H.; Yu, S.Y. Structural discontinuity induced surface second harmonic generation in single, thin zinc-blende GaAs nanowires. *Nanoscale* **2017**, *9*, 16066–16072. [CrossRef] [PubMed]

22. Hu, H.B.; Wang, K.; Long, H.; Liu, W.W.; Wang, B.; Lu, P.X. Precise determination of the crystallographic orientations in single ZnS nanowires by second-harmonic generation microscopy. *Nano Lett.* **2015**, *15*, 3351–3357. [CrossRef] [PubMed]

23. Tom, H.W.K.; Heinz, T.F.; Shen, Y.R. Second-harmonic reflection from silicon surfaces and its relation to structural symmetry. *Phys. Rev. Lett.* **1983**, *51*, 1983–1986. [CrossRef]

24. Brixius, K.; Beyer, A.; Mette, G.; Güdde, J.; Dürr, M.; Stolz, W.; Volz, K.; Höfer, U. Second-harmonic generation as a probe for structural and electronic properties of buried GaP/Si(0 0 1) interfaces. *J. Phys-Condens. Matter* **2018**, *30*, 484001. [CrossRef] [PubMed]

25. Neeman, L.; Ben-Zvi, R.; Rechav, K.; Popovitz-Biro, R.; Oron, D.; Joselevich, E. Crystallographic mapping of guided nanowires by second harmonic generation polarimetry. *Nano Lett.* **2017**, *17*, 842–850. [CrossRef] [PubMed]

26. Han, X.B.; Wang, K.; Long, H.; Hu, H.B.; Chen, J.W.; Wang, B.; Lu, P.X. Highly sensitive detection of the lattice distortion in single bent ZnO nanowires by second-harmonic generation microscopy. *ACS Photonics* **2016**, *3*, 1308–1314. [CrossRef]

27. Long, H.; Bao, L.J.; Wang, K.; Liu, S.H.; Wang, B. Local-field enhancement of optical nonlinearities in the AGZO nanotriangle array. *Opt. Mater.* **2016**, *60*, 571–576. [CrossRef]

28. Dickson, K.M.; Long, H.; Xing, X.Y.; Njoroge, S.; Wang, K.; Wang, B.; Lu, P.X. Gigahertz acoustic vibrations of Ga-doped ZnO nanoparticle array. *Nanotechnology* **2019**, *30*, 305201.

29. Jassim, N.M.; Wang, K.; Han, X.B.; Long, H.; Wang, B.; Lu, P.X. Plasmon assisted enhanced second-harmonic generation in singlehybrid Au/ZnS nanowires. *Opt. Mater.* **2017**, *64*, 257–261. [CrossRef]

30. Chen, J.W.; Wang, K.; Long, H.; Han, X.B.; Hu, H.B.; Liu, W.W.; Wang, B.; Lu, P.X. Tungsten disulfide–gold nanohole hybrid metasurfaces for nonlinear metalenses in the visible region. *Nano Lett.* **2018**, *18*, 1344–1350. [CrossRef]

31. Perillat-Merceroz, G.; Thierry, R.; Jouneau, P.H.; Ferret, P.; Feuillet, G. Compared growth mechanisms of Zn-polar ZnO nanowires on O-polar ZnO and on sapphire. *Nanotechnology* **2012**, *23*, 125702. [CrossRef] [PubMed]

32. Zhan, J.X.; Dong, H.X.; Sun, S.L.; Ren, X.D.; Liu, J.J.; Chen, Z.H.; Lienau, C.; Zhang, L. Surface-energy-driven growth of ZnO hexagonal microtube optical resonators. *Adv. Opt. Mater.* **2016**, *4*, 126–134. [CrossRef]

33. Huang, M.L.; Yan, Y.; Feng, W.H.; Weng, S.X.; Zheng, Z.Y.; Fu, X.Z.; Liu, P. Controllable tuning various ratios of ZnO Polar facets by crystal seed-assisted growth and their photocatalytic activity. *Cryst. Growth Des.* **2014**, *14*, 2179–2186. [CrossRef]

34. Kishimoto, Y.; Nakagawara, O.; Seto, H.; Koshido, Y.; Yoshinoet, Y. Improvement in moisture durability of ZnO transparent conductive films with Ga heavy doping process. *Vacuum* **2009**, *83*, 544–547. [CrossRef]

35. Shinde, S.D.; Deshmukh, A.V.; Date, S.K.; Sathe, V.G.; Adhi, K.P. Effect of Ga doping on micro/structural, electrical and optical properties of pulsed laser deposited ZnO thin films. *Thin Solid Films* **2011**, *520*, 1212–1217. [CrossRef]

36. Fortunato, E.; Raniero, L.; Silva, L.; Goncalves, A.; Pimentel, A.; Barquinha, P.; A'guas, H.; Pereira, L.; Goncalves, G.; Ferreira, I.; et al. Highly stable transparent and conducting gallium-doped zinc oxide thin films for photovoltaic applications. *Sol. Energy Mater. Sol. Cells* **2008**, *92*, 1605–1610. [CrossRef]

37. Vispute, R.D.; Talyansky, V.; Choopun, S.; Sharma, R.P.; Venkatesan, T.; He, M.; Tang, X.; Halpern, J.B.; Spencer, M.G.; Li, Y.X.; et al. Heteroepitaxy of ZnO on GaN and its implications for fabrication of hybrid optoelectronic devices. *Appl. Phys. Lett.* **1988**, *73*, 348. [CrossRef]

38. Helene, S.; Alain, D.; Manuel, G. Investigation of Ga substitution in ZnO powder and opto-electronic properties. *Inorg. Chem.* **2010**, *49*, 6853–6858.

39. Charles, M.; Cosmas, M.M.; Albert, J. Highly conductive and transparent Ga-doped ZnO thin films deposited by chemical spray pyrolysis. *Optik* **2016**, *127*, 8317–8325.

40. Seyda, H.; Fadil, I.; Ramazan, T.S.; Cem, C.; Mohamed, S.; Abdullah, Y.; Tülay, S. Monitoring the characteristic properties of Ga-doped ZnO by Raman spectroscopy and atomic scale calculations. *J. Mol. Struct.* **2019**, *1180*, 505–511.

41. Yao, I.C.; Lee, D.Y.; Tseng, T.Y.; Lin, P. Fabrication and resistive switching characteristics of high compact Ga-doped ZnO nanorod thin film devices. *Nanotechnology* **2012**, *23*, 145201. [CrossRef] [PubMed]

42. Dhara, S.; Imakita, K.; Mizuhata, M.; Fujii, M. Europium doping induced symmetry deviation and its impact on the second harmonic generation of doped ZnO nanowires. *Nanotechnology* **2014**, *25*, 225202. [CrossRef] [PubMed]

43. Vanpoucke, D.E.P. Comment on 'Europium doping induced symmetry deviation and its impact on the second harmonic generation of doped ZnO nanowires'. *Nanotechnology* **2014**, *25*, 458001. [CrossRef] [PubMed]

44. Wiff, J.P.; Kinemuchi, Y.; Kaga, H.; Ito, C.; Watari, K. Correlations between thermoelectric properties and effective mass caused by lattice distortion in Al-doped ZnO ceramics. *J. Eur. Ceram. Soc.* **2009**, *29*, 1413–1418. [CrossRef]

45. Kumar, N.; Najmaei, S.; Cui, Q.; Ceballos, F.; Ajayan, P.M.; Lou, J.; Zhao, H. Second harmonic microscopy of monolayer MoS_2. *Phys. Rev. B* **2013**, *87*, 161403. [CrossRef]

46. Johnson, J.C.; Yan, H.Q.; Schaller, R.D.; Petersen, P.B.; Yang, P.D.; Saykally, R.J. Near-field imaging of nonlinear optical mixing in single zinc oxide nanowires. *Nano Lett.* **2002**, *2*, 279–283. [CrossRef]

47. Michele, M. Nonlinear optical response of a two-dimensional atomic crystal. *Opt. Lett.* **2016**, *41*, 187–190.

48. Mohammad, M.; Adam, C.; Feruz, G. Nonlinear optical susceptibility of atomically thin WX_2 crystals. *Opt. Mater.* **2019**, *88*, 30–38.

49. Maria, C.L.; Marco, C. Second harmonic generation from ZnO films and nanostructures. *Appl. Phys. Rev.* **2015**, *2*, 031302.

50. Bhagavannarayana, G.; Parthiban, S.; Subbiah, M. An interesting correlation between crystalline perfection and second harmonic generation efficiency on KCl- and oxalic acid-doped ADP crystals. *Cryst. Growth Des.* **2008**, *8*, 446–451. [CrossRef]

51. Butet, J.; Bachelier, G.; Russier-Antoine, I.; Jonin, C.; Benichou, E.; Brevet, P.F. Interference between selected dipoles and octupoles in the optical second harmonic generation from spherical gold nanoparticles. *Phys. Rev. Lett.* **2010**, *105*, 077401. [CrossRef] [PubMed]

52. Bachelier, G.; Russier-Antoine, I.; Benichou, E.; Jonin, C.; Brevet, P.F. Multipolar second-harmonic generation in noble metal nanoparticles. *J. Opt. Soc. Am. B* **2008**, *25*, 955–959. [CrossRef]

53. Liu, S.W.; Weerasinghe, J.L.; Liu, J.; Weaver, J.; Chen, C.L.; Donner, W.; Xiao, M. Reflective second harmonic generation near resonance in the epitaxial Al-doped ZnO thin film. *Opt. Express* **2007**, *15*, 10666–10671. [CrossRef] [PubMed]

54. Zhang, X.Q.; Tang, Z.K.; Kawasaki, M.; Ohtomo, A.; Koinuma, H. Second harmonic generation in self-assembled ZnO microcrystallite thin films. *Thin Solid Films* **2004**, *450*, 320–323. [CrossRef]

55. Chan, S.W.; Barille, R.; Nunzi, J.M.; Tam, K.H.; Leung, Y.H.; Chan, W.K.; Djurišić, A.B. Second harmonic generation in zinc oxide nanorods. *Appl. Phys. B* **2006**, *84*, 351–355. [CrossRef]

 *nanomaterials*

Article

Nonlinear Optical Studies of Gold Nanoparticle Films

Anuradha Rout [1], Ganjaboy S. Boltaev [1], Rashid A. Ganeev [1,*], Yue Fu [1],
Sandeep Kumar Maurya [1], Vyacheslav V. Kim [1], Konda Srinivasa Rao [1] and Chunlei Guo [1,2,*]

[1] The Guo China-US Photonics Laboratory, State Key Laboratory of Applied Optics, Changchun Institute of Optics, Fine Mechanics and Physics, Chinese Academy of Sciences, Changchun 130033, China; anuradharout@ciomp.ac.cn (A.R.); ganjaboy_boltaev@mail.ru (G.S.B.); 15541103381@163.com (Y.F.); sandeep@ciomp.ac.cn (S.K.M.); mik750594@rambler.ru (V.V.K.); ksrao@ciomp.ac.cn (K.S.R.)

[2] The Institute of Optics, University of Rochester, Rochester, NY 14627, USA

* Correspondence: rashid_ganeev@mail.ru (R.A.G.); guo@optics.rochester.edu (C.G.)

Received: 31 December 2018; Accepted: 7 February 2019; Published: 19 February 2019

Abstract: Gold films are widely used for different applications. We present the results of third- and high-order nonlinear optical studies of the thin films fabricated from Au nanoparticle solutions by spin-coating methods. These nanoparticles were synthesized by laser ablation of bulk gold in pure water using 200 ps, 800 nm pulses. The highest values of the nonlinear absorption coefficient (9×10^{-6} cm W^{-1}), nonlinear refractive index (3×10^{-11} cm^2 W^{-1}), and saturation intensity (1.3×10^{10} W cm^{-2}) were achieved using 35 fs, 400 nm pulses. We also determined the relaxation time constants for transient absorption (220 fs and 1.6 ps) at 400 nm. The high-order harmonic generation was studied during propagation of 35 fs, 800 nm pulses through the plasma during the ablation of gold nanoparticle film and bulk gold. The highest harmonic cutoff (29th order) was observed in the plasma containing gold nanoparticles.

Keywords: gold thin film; nonlinear absorption; nonlinear refraction; transient absorption; nanoparticles; high-order harmonics

---

1. Introduction

The quantum confinement effect allows the distinguishing of the parameters of nanoparticles with regard to the bulk materials. Among the metals suited for nanoparticle preparation for optoelectronics and nonlinear optics, one can distinguish silver [1,2], copper [3,4], and gold [5]. The further search for prospective materials in nanoparticle formation, their preparation, and application are of considerable importance. Meantime, the unique properties of low-dimensional materials have ignited numerous studies of their characteristics [6–8].

An interest in nanoparticle-containing films is developing due to their enhanced nonlinear optical response [9]. Such films have attracted interest due to their potential applications in optoelectronics as optical switches and optical limiters. The development of new thin film compounds containing both semiconductor and metal nanoparticles (NPs) allows for further enhancement of the nonlinear optical characteristics of such structures. During the last decade, there has been an interest in the nonlinear optical features of chalcogenide thin films [10–13]. The investigations of these films have shown their prospects as optical limiters.

Most of previous nonlinear optical studies of different materials in a strong electromagnetic field were performed using solutions and thick (of the order of a few micrometers) films. At the same time, it is interesting to investigate thin (of order of a hundred nanometers) films. Particularly, gold nanocomposites have tremendous applications in various fields due to the influence of their surface plasmon resonance (SPR) on the optical properties [14]. In the past decade, researchers have demonstrated the potential applications of gold nanoparticles (NPs) using different lasers [15,16].

In those studies, the NPs were produced by laser ablation of the gold bulk targets. Laser ablation is an efficient method for the synthesis of NPs using different pulse durations, pulse energies, as well as various liquids [17]. In most cases, the deionized water was used during the ablation to produce the gold NPs. Several studies have revealed variable nonlinear optical properties of gold NPs, nanorods, and thin films using different laser pulses [18–22]. Particularly, the optical nonlinearities of Au NP arrays, which were determined by performing Z-scan measurements using a femtosecond laser (800 nm, 50 fs), were reported in [21]. The third-order nonlinear optical properties of gold NPs embedded in Al_2O_3, ZnO, and SiO_2 at the wavelength of 532 nm using the nanosecond Nd:YAG laser were analyzed in [22]. Previously, the high-order harmonics from gold NPs were studied in [23]. One can assume that the plasma produced on the thin films can enhance the efficiency of high-order harmonics due to the influence of NPs.

In this paper, we analyze the third-order nonlinear optical properties and transient absorption of gold NP thin film as well as demonstrate the high-order harmonic generation (HHG) from the ablated gold NP thin film.

2. Experimental Arrangements

Figure 1a shows the scheme for the synthesis of Au NPs in solution by irradiation of the bulk gold target immersed in deionized water using 800 nm, 200 ps, 1 kHz pulses. The laser beam was focused by a 100 mm focal length lens on the bulk gold. Typically, the energy density of laser radiation on the metal surface was in the order of 10 J/cm^2. The irradiation of metal surface resulted in the fast removal of the material confined to the laser spot. The sample was displaced with regard to the laser beam, using a translating stage to avoid the formation of deep holes. The ablation was done for 10 min at continuous stirring. The Au film was prepared by evaporation of the aqueous suspension containing gold nanoparticles and then used as the target to perform the experiments. The thickness of thin film was examined by scanning electron microscope (SEM) (HITACHI S-4800, Tokyo, Japan) to be approximately 100 nm.

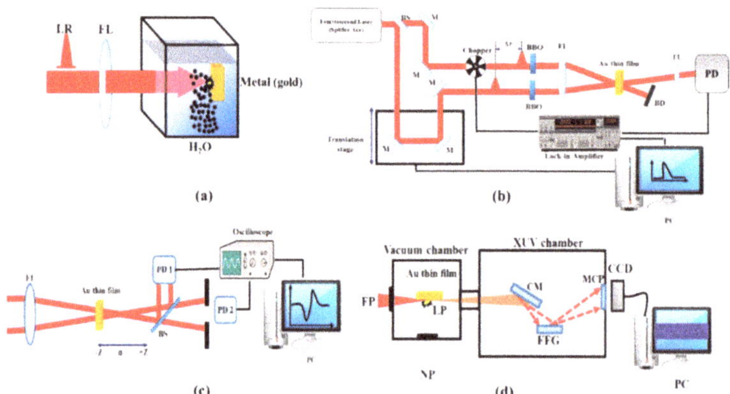

Figure 1. (**a**) Schematic of laser ablation. LR, laser radiation; FL, focal lens. (**b**) Schematic of the transient absorption measurements. BS, beam splitter; M, mirrors; FL, focal lenses; PD, photo diode; BD, beam dumper; BBO, barium borate crystal; PC, personal computer; (**c**) Schematic of Z-scan. FL, focusing lens; PD1, photodiode 1; PD2, photodiode 2. (**d**) High-order harmonic generation (HHG) setup. FP, converting femtosecond pulses; NP, heating nanosecond pulses; LP, laser plasma; CM, cylindrical gold-coated mirror; FFG, flat field grating; MCP, microchannel plate; CCD, charge-coupled device camera.

A noncollinear degenerate pump–probe technique was employed to measure the transient absorption (TA) in Au NP films (Figure 1b). A TA study was performed using the 400 nm radiation obtained by second harmonic generation of 800 nm, 35 fs pulses using a 2 mm thick barium borate (BBO) crystal. Prior to conversion of 800 nm to 400 nm, the laser pulse was split into two pulses using a 30:70 beamsplitter. Pump and probe pulses were focused to the sample using a 300 mm focal length lens. The transmitted probe radiation was detected by an ultrafast Si photodiode (DET025AL, Thorlabs, China) connected to a lock-in-amplifier to measure the transmittance of the probe pulse with respect to the position of the motorized stage. The lock-in-amplifier was externally triggered by the optical chopper running at 300 Hz.

Figure 1c shows the Z-scan scheme, which was used for the third-order nonlinear optical studies in Au NP thin films. The laser radiation (800 nm, 30 fs, 1 kHz) was focused on the sample by a 40 cm focal length lens. The thin film was placed near to the focus of the beam. The sample was scanned along the beam direction by a computer-controlled translation stage. After passing through the sample the transmitted beam was then split by a 50:50 beamsplitter. The closed aperture (CA) and open aperture (OA) Z-scans at 30 nJ probe energy were used to characterize the nonlinear absorption and refraction of the Au NP thin film.

The HHG in the plasmas produced during the ablation of Au NP thin film was performed using the setup shown in Figure 1d. The driving femtosecond pulses (800 nm, 30 fs, 1 kHz) propagated through the plasma formed by the nanosecond heating pulses (1064 nm, 5 ns, 10 Hz) at different delays between the heating and driving pulses. The harmonic yield was maximized by adjusting the position of the target. The generated high-order harmonics were analyzed by an extreme ultraviolet (XUV) spectrometer and detected by a microchannel plate with phosphor screen. The harmonic spectrum from the phosphor screen was imaged by a charge-coupled device (CCD) camera.

3. Results and Discussion

3.1. Low-order Nonlinearities of Au NP Film

SEM images and histograms of the Au NPs prepared by ablation of bulk gold using picosecond pulses are presented in Figure 2a. The inset in Figure 2a shows the size distribution of Au NPs, which covers the 10–90 nm range with mean size 30 nm. The sample shown in Figure 2a was prepared by drying the drop of Au NP solution on the Si wafer or glass. Then this sample was analyzed by SEM. A few existing empty places in the SEM image had the sizes (<200 nm) significantly smaller than the area used for absorption measurements (5×5 mm^2) of the thin (~100 nm) gold film deposited on the silica glass plate. The presence of those tiny holes causes the insignificant variation of a whole spectral pattern. The absorption spectrum remained the same in different parts of deposited film due to the averaging of absorbance measured along the large area.

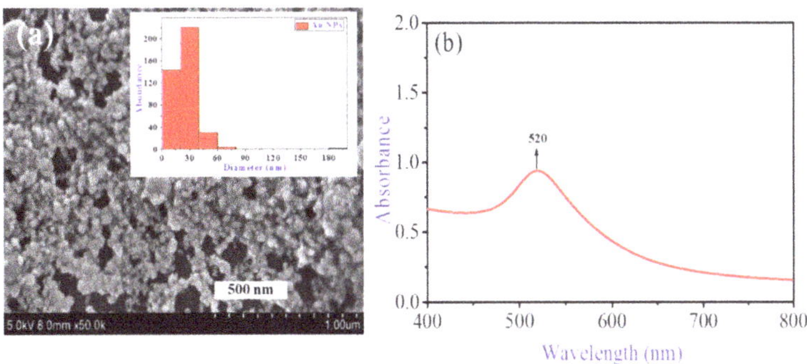

Figure 2. (**a**) SEM image and size distribution of Au NPs. (**b**) Absorption spectrum of Au NP thin film.

The absorption spectrum of thin film was analyzed in the range of 400 to 800 nm. The absorbance measurements were made using a 100 nm thin film. Figure 2b shows the absorption spectrum of Au NP thin film. The observed surface plasmon resonance of Au NPs was at 520 nm.

The Z-scan has proven to be a most versatile technique for the measurements of the lowest-order optical nonlinearities of different materials. Two schemes (OA and CA) are frequently applied to measure the nonlinear absorptive and refractive properties of matter. The use of these two configurations is a prerequisite for the accurate measurements of nonlinear refractive index (γ) and nonlinear absorption coefficient (β). Though the latter parameter can also be retrieved using the CA scheme, the accuracy of those measurements is lower compared with the OA scheme. The beam width of the probe radiation in the focal plane was 76 μm (full width at half maximum), i.e., a thousand times larger than the sizes of nanoparticles (30 nm). Correspondingly, the nonlinear optical response was accumulated from the large amount of particles presented inside the focal spot area. What we see in the SEM hardly can be classified as the aggregates. Actually, they are the separated nanoparticles placed close to each other. The thickness of the film (100 nm) assumes that only a few (actually 3 to 4) NPs comprise the whole active path of this film. Thus, the response occurs from a large amount of separate Au NPs acting as the nonlinear refractive/absorptive material.

The Z-scan of the normalized transmittance caused by the influence of saturable absorption (SA) and reverse saturable absorption (RSA) is described as [24,25]:

$$T_{SA,RSA}(z) = \left(1 - \frac{q}{2\sqrt{2}}\right) \times \left(1 + \frac{I_0}{I_{sat}\,(1+x^2)}\right). \tag{1}$$

Here $q = \beta I_0 L_{eff}/(1 + z^2/z_0{}^2)$, $x = z/z_0$, $z_0 = k(w_0)^2/2$ is the Rayleigh length, $k = 2p/\lambda$ is the wave number, I_0 is the peak intensity in the focal plane, $L_{eff} = (1-\exp(-\alpha_0 L))/\alpha_0$ is the effective length of the medium, w_0 is the beam waist radius at the $1/e^2$ level of intensity distribution, α_0 is the linear absorption coefficient, L is the thickness of sample, and I_{sat} is the saturated intensity of the medium.

In the case of the CA Z-scan, the normalized transmittance of nonlinear refraction and absorption (NRA) is given as [26]:

$$T_{NRA}(z) = 1 + \frac{2\left(-\rho x^2 + 2x - 3\rho\right)}{(x^2+1)(x^2+9)}\Delta\varnothing_o. \tag{2}$$

Here $\rho = \beta/2k\gamma$ and $\Delta\Phi_0 = k\gamma I_0 L_{eff}$ is the phase change due to nonlinear refraction. The investigation of the nonlinear optical characteristics of Au NP thin film was carried out using the CA and OA Z-scans under excitation by 400 nm, 30 fs probe pulses at the same pulse energy (30 nJ). Figure 3 shows the OA and CA Z-scan curves of thin film. In the case of OA, the Au NP thin film showed SA, while close to focal area it demonstrated the RSA.

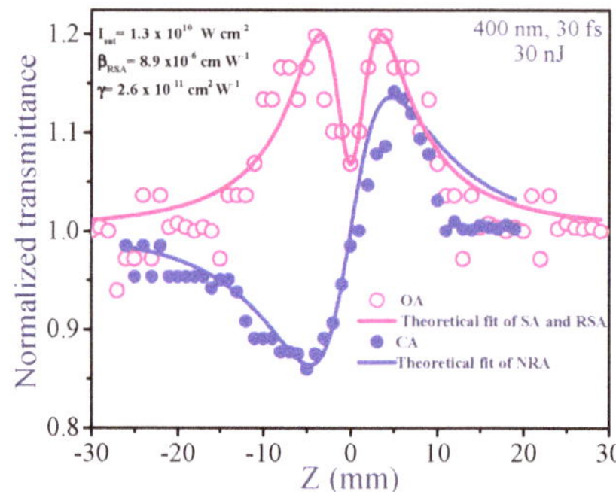

Figure 3. Open aperture (OA) and closed aperture (CA) Z-scan curves of thin film measured using the 400 nm, 30 fs pulses. SA is saturable absorption; RSA is reverse saturable absorption; NRA is nonlinear refraction and absorption.

The fitting of the curve comprising the influence of saturable absorption and reverse saturable absorption was accomplished using Equation (1), similarly to the technique described in [24,25]. This equation contains the saturation intensity (I_{sat}), which was determined during the fitting procedure, alongside the nonlinear absorption coefficient of reverse saturable absorption. The measured saturated intensity was 1.3×10^{10} W cm^{-2}. β_{RSA} was calculated from the fitting curve to be 9×10^{-6} cm W^{-1}. This value is one of largest reported in the case of thin films measurements using different materials.

In the case of CA measurements, the self-focusing was observed. The nonlinear refractive index was calculated from the fitting of the Equation (2) with the experimental CA curve. The fitting allowed for the determination of two parameters ($\Delta\phi_0 = k\gamma I_0 L_{eff}$ and $\rho = \beta/2k\gamma$) required for definition of the nonlinear refractive index and nonlinear absorption coefficient. Correspondingly, by knowing $\Delta\Phi_0$, k, I_0, and L_{eff} one can easily determine γ. The nonlinear refractive index was calculated to be 2.6×10^{-11} cm^2 W^{-1}.

Time evolution of the absorption of probe pulses in the presence of pump pulses in the Au NP film at 400 nm is shown in Figure 4. Prior to the TA measurements of the Au NP film deposited on a glass slide, the TA measurements of the pure glass slide were performed to separate its contribution from the former TA data. The pump–probe profile for thin gold film indicated the process of photobleaching due to excitation of the NPs under irradiation of 52 nJ, 400 nm, 35 fs pump pulses. In general, mechanisms of photoexcitation by femtosecond pulses in a metal include the excitation of electrons to a higher energy state through optical absorption, which leads the nonequilibrium electronic subsystem to relax via redistribution of the energy of excited electrons, typically known as electron–electron scattering interaction. The energy redistribution of the excited electrons occurs through electron–electron scattering within about 100–300 fs, which have been attributed to the electron thermalization process [27,28]. Another process of relaxation for excited electrons occurs via energy transfer of the electron to the lattice in the picosecond time scale due to electron–phonon interaction. In the present study, the employed pulse width (35 fs) of the femtosecond pulses was smaller than the decay time scale of the electron–electron relaxation dynamics in Au NPs. This made it feasible to probe the electron–electron dynamics. The following phenomenological response function was used to determine the time constants associated with electron–electron and electron–phonon interactions [28]:

$$f(t) = H(t)(1 - exp\left(-t/\tau_{th}\right)) \left(exp\left(-t/\tau_{e-ph}\right)\right) \tag{3}$$

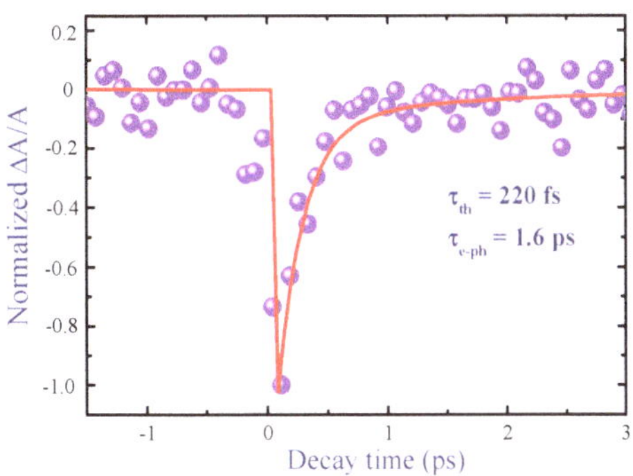

Figure 4. Pump–probe dynamics of Au NP thin film at 400 nm.

Here τ_{th} and τ_{e-ph} are electron thermalization time constant and electron–phonon relaxation time constant, respectively, and $H(t)$ is the Heaviside step function, which is equal to 0 if $t < 0$ and 1 if $t > 0$. The fitting of the TA profile (Figure 4, solid line) allowed determination of these time constants (τ_{th} = 220 fs and = 1.6 ps) at λ = 400 nm.

The relaxation time constant τ_{th} was associated with the electron thermalization process in Au NPs [29]. There are several reports on the study of electron thermalization kinetics, which validate the above-mentioned approach [30–33].

3.2. High-order Harmonic Generation in Au NP Plasmas

The nanostructured materials can significantly enhance the harmonic yield in XUV range, since they have already demonstrated the ability to increase the second- and third-order nonlinear optical processes [23]. Below we report the studies of high-order nonlinear optical processes in the plasma produced on the Au NP thin film deposited on glass substrates. We compared the HHG from the plasmas created on the bulk gold target and Au thin film. The harmonics generated from the plasmas produced on the glass substrates, without thin film, were negligible compared with those from the plasmas produced on the bulk and metal thin films. In two cases, the plasma plumes were ablated using the 1064 nm, 5 ns heating pulses. The moderate intensity of heating pulses allowed the evaporation of the neutral atoms and singly-charged ions from the targets during laser ablation. The advantages of the applications of neutral atoms and singly-charged ions for HHG have been demonstrated in a series of previous HHG studies in the laser-produced plasmas [34–36]. The optimal plasma plume was created by moving the target with regard to femtosecond beam propagation.

Figure 5a shows the harmonic spectra from the plasmas produced on the bulk Au and Au NP-contained thin films at 200 ns delay time between the heating and driving pulses. It was demonstrated that the harmonic intensity was significantly (approximately five times) enhanced in the case of Au NP thin film with regard to bulk Au. The experiments were carried out at different delays between the heating nanosecond and driving femtosecond pulses. Figure 5b shows the intensity variation of the eleventh harmonic with respect to different delay times. The maximum intensity of harmonics was achieved in the range of 200–500 ns delay times. During the ablation of thin film at optimal delays, the generation of the strong harmonics up to the 29th order was achieved. However, after crossing the delay of 800 ns the yield of harmonics was notably decreased.

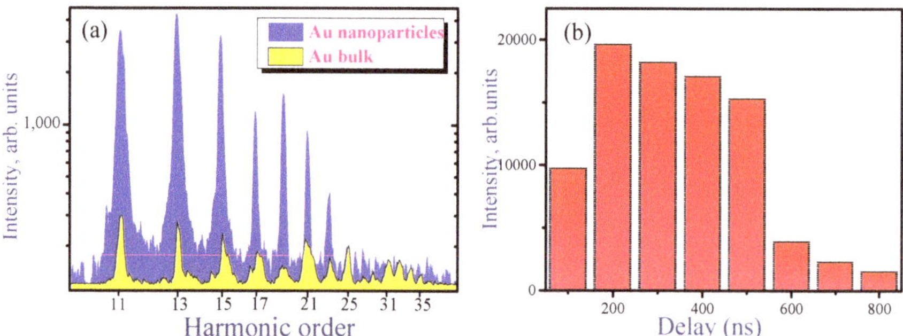

Figure 5. (**a**) High-order harmonic spectra generated in the plasmas produced on the Au thin film and bulk Au. (**b**) Dependence of 11th harmonic yield on the delay between heating and driving pulses.

The HHG from gas clusters has been reported in a few earlier studies [37–41]. Moreover, the harmonic generation in the plasmas containing plasmonic particles of different materials has also been analyzed to demonstrate the advantages of the proposed approach of HHG amendment [42–44]. In the plasma HHG studies, the nanoparticles were synthesized during laser ablation in a vacuum prior to formation of the NP-containing plasma plume or their appearance in the plasma was accomplished by ablation of commercially available nanoparticles.

The present research has demonstrated the preparation of gold nanoparticles during ablation of solid material in the liquid environment, deposition of synthesized NPs on the glass and silicon wafer, and ablation of thin deposited film in a vacuum to produce the plasma containing nanoparticles of narrow size distribution for harmonic generation at different delays between heating and driving pulses. Among them the most important novelty is the analysis of the role of the delay between the heating picosecond pulses creating nanoplasma and the driving femtosecond pulses generating harmonics on the HHG conversion efficiency. All previous HHG studies with nanoparticle-containing plasmas and gases were carried out at either fixed delay between heating and driving pulses (in the former case) or fixed distance from the nozzle of the gas jet. In previous plasma HHG studies, short delays of up to 100 ns between the heating and driving pulses were employed. It was not clear how heavy nanoparticle species could influence the processes of frequency conversion, because there was no proof of their presence in the interaction region with the driving laser (i.e., at the distance of a few hundred micrometers from the target surface), since one can expect their arrival in the region of the femtosecond laser beam propagation a few tens of microsecond from the beginning of ablation. One explanation was based on the disintegration of larger species into small clusters and monoatomic species, which probably could reach the interaction area at the short delays employed. However, no sufficient confirmation of this assumption has been provided.

To match the propagation of the driving pulse and the highest concentration of the studied group of multi-atomic species with a much larger delay, which cannot be achieved by optical methods, one therefore has to use the electronic methods of the adjustment of the delay between the heating and the driving pulses. The application of two electronically separated pulses from different lasers synchronized by a digital delay generator allows analysis of the involvement of various multi-particle species in the HHG process. Application of this approach for HHG in multi-particle plasmas, alongside with other methods of harmonic enhancement, requires the analysis of the dynamics of ablated species spreading out from the target to temporally match them with the propagation of driving femtosecond pulses through the plasma. One can expect the arrival of 30 nm particles in the region of the femtosecond laser beam propagation a few tens of microseconds from the beginning of ablation. Meanwhile, our studies demonstrated the optimization of delays leading to the growth of harmonic yield in the case of Au NP plasma at around 200 ns. Below, we address this difference in the expected and actual optimal delay between heating and driving pulses in the case of Au NP-contained plasma.

In the case of a thermalized ablation plume, the average arrival times can be assigned to different cluster sizes. The delay between heating and driving pulses at which the harmonic yield reaches its maximum should scale as a square root of the atomic or molecular weight of the constituents. The ejection of lighter clusters from NPs allows them to reach the region of the driving beam earlier than heavier species. Therefore, NPs comprised of n atoms should appear in the interaction zone $n^{0.5}$ times later compared to single atoms, molecules, or ions of the sulfides. Our additional studies revealed that, for bulk gold ablation, the maximum harmonic yield from single gold atoms and ions occurred at a delay of about 180–300 ns. Meantime, the Au NPs allowed efficient generation at about 200–400 ns delay (Figure 5b), which is approximately equal to the delay in the case of Au atoms and ions. Furthermore, attempts to observe HHG at the delays of up to 50 μs (i.e., at the expected delay for thermalized larger nanoparticles) did not show any harmonic emission. Thus, our studies demonstrated that NPs arrive at the area of interaction with the femtosecond laser beam notably earlier than one would expect for a thermalized ablation plume. In other words, all gold NPs acquire, from the very beginning, a similar kinetic energy and spread out from the surface with velocities approximately similar to those of the single gold atoms and ions ablating from bulk material. This conclusion reconciles the similarity in the optimal delays for HHG from bulk and NPs targets of the same material (Au).

It was suggested in [45] during their studies of third harmonic generation in plasma plumes that, at laser ablation characterized by the blast mechanism of laser–matter interaction, a similar average kinetic energy $E = mv^2/2$ could characterize all plasma components of the same elemental composition. Thus, the average arrival time assigned to the particles containing different amounts of identical atoms will be approximately the same, contrary to the case of slowly produced thermalized plasma. Our studies of the high-order nonlinear optical processes occurring in the plasmas confirm this assumption. The difference in "optimal" delays between heating and driving pulses is related to the difference in the velocities of particles, which depends on the atomic masses of the components of NPs. Similar conclusions have been reached during recent studies of HHG from the metal sulfide quantum dots [46].

4. Conclusions

We have prepared gold nanoparticle-contained thin film by evaporation of the Au NP suspension, which was synthesized by laser ablation. The Au NP thin film was characterized using the UV-visible absorption spectra and SEM analysis. We have demonstrated the strong nonlinear absorption (9×10^{-6} cm W^{-1}) in this film at 400 nm. The thin film exhibited a switch from SA to RSA at stronger excitation. The relaxation time of SA was measured to be 1.6 ps. The high-order harmonic generation was analyzed in the plasma containing gold nanoparticles. The HHG efficiency in that case was ten times higher compared with the case of bulk gold ablation. For the first time, the effective application of the ablated 100 nm thin film of gold nanoparticles for harmonic generation in the 27–115 nm range was achieved.

Author Contributions: Investigation—A.R., Y.F., S.K.M., V.V.K., and K.S.R.; Writing—original draft, G.S.B.; Writing—review and editing, R.A.G. and C.G.

Funding: The research was supported by the National Key Research and Development Program of China (2017YFB1104700, 2018YFB1107202), National Natural Science Foundation of China (NSFC, 91750205, 61774155), Bill & Melinda Gates Foundation (OPP1157723) of the US, Chinese Academy of Sciences President's International Fellowship Initiative (Grant No. 2018VSA0001), Scientific Research Project of the Chinese Academy of Sciences (QYZDB-SSW-SYS038), and the Key Program of the International Partnership Program of CAS (181722KYSB20160015).

Conflicts of Interest: The authors declare no conflict of interest.

References

1. Ganeev, R.A.; Baba, M.; Ryasnyansky, A.I.; Suzuki, M.; Kuroda, H. Characterization of optical and nonlinear optical properties of silver nanoparticles prepared by laser ablation in various liquids. *Opt. Commun.* **2004**, *240*, 437–448. [CrossRef]

2. Ganeev, R.A.; Ryasnyansky, A.I.; Stepanov, A.L.; Usmanov, T. Saturated absorption and nonlinear refraction of silicate glasses doped with silver nanoparticles at 532 nm. *Opt. Quantum Electron.* **2004**, *36*, 949–960. [CrossRef]

3. Falconieri, M.; Salvetti, G.; Cattaruzza, E.; Gonella, F.; Mattei, G.; Mazzoldi, P.; Piovesan, M.; Battaglin, G.; Polloni, R. Large third-order optical nonlinearity of nanocluster-doped glass formed by ion implantation of copper and nickel in silica. *Appl. Phys. Lett.* **1998**, *73*, 288–290. [CrossRef]

4. Ryasnyansky, A.; Palpant, B.; Debrus, S.; Ganeev, R.; Stepanov, A.; Can, N.; Buchal, C.; Uysal, S. Nonlinear optical absorption of ZnO doped with copper nanoparticles in the picosecond and nanosecond pulse laser field. *Appl. Opt.* **2005**, *44*, 2839–2845. [CrossRef] [PubMed]

5. Debrus, S.; Lafait, J.; May, M.; Pinçon, N.; Prot, D.; Sella, C.; Venturini, J. Z-scan determination of the third-order optical nonlinearity of gold:silica nanocomposites. *J. Appl. Phys.* **2000**, *88*, 4469–4475. [CrossRef]

6. Alvarez-Fregoso, O.; Mendoza-Alvarez, J.G.; Zelaya-Angel, O. Quantum confinement in nanostructured CdNiTe composite thin films. *J. Appl. Phys.* **1997**, *82*, 708–711. [CrossRef]

7. Baskoutas, S.; Poulopoulos, P.; Karoutsos, V.; Angelakeris, M.; Flevaris, N.K. Strong quantum confinement effects in thin zinc selenide films. *Chem. Phys. Lett.* **2006**, *417*, 461–464. [CrossRef]

8. Muller, E.A.; Johns, J.E.; Caplins, B.W.; Harris, C.B. Quantum confinement and anisotropy in thin-film molecular semiconductors. *Phys. Rev. B* **2011**, *83*, 165422. [CrossRef]

9. Rao, C.N.R.; Kulkarni, G.U.; Thomas, P.J.; Edwards, P.P. Metal nanoparticles and their assemblies. *Chem. Soc. Rev.* **2000**, *29*, 27–35. [CrossRef]

10. Fazio, E.; Hulin, D.; Chumash, V.; Michelotti, F.; Andriesh, A.M.; Bertolotti, M. On-off resonance femtosecond non-linear absorption of chalcogenide glassy films. *J. Non. Cryst. Solids* **1994**, *168*, 213–222. [CrossRef]

11. Anshu, K.; Sharma, A. Study of Se based quaternary SePb(Bi,Te) chalcogenide thin films for their linear and nonlinear optical properties. *Optik (Stuttg.)* **2016**, *127*, 48–54. [CrossRef]

12. Michelotti, F.; Bertolotti, M.; Chumash, V.; Andriesh, A. Chalcogenide glass thin films: Z-Scan measurements of refractive index changes. *Proc SPIE* **1992**, *1773*, 423–432.

13. Dawar, A.L.; Shishodia, P.K.; Chauhan, G.; Joshi, J.C.; Jagadish, C.; Mathur, P.C. Effect of UV exposure on optical properties of amorphous as(2)s(3) thin films. *Appl. Opt.* **1990**, *29*, 1971–1973. [CrossRef] [PubMed]

14. Olesiak-banska, J.; Gordel, M.; Kolkowski, R.; Matczyszyn, K.; Samoc, M. Third-order nonlinear optical properties of colloidal gold nanorods. *J. Phys. Chem. C* **2012**, *116*, 13731. [CrossRef]

15. Lv, J.; Jiang, L.; Li, C.; Liu, X.; Yuan, M.; Xu, J.; Zhou, W.; Song, Y.; Liu, H.; Li, Y.; et al. Large third-order optical nonlinear effects of gold nanoparticles with unusual fluorescence enhancement. *Langmuir* **2008**, *24*, 8297–8302. [CrossRef] [PubMed]

16. Storhoff, J.J.; Lucas, A.D.; Garimella, V.; Bao, Y.P.; Müller, U.R. Homogeneous detection of unamplified genomic DNA sequences based on colorimetric scatter of gold nanoparticle probes. *Nat. Biotechnol.* **2004**, *22*, 883–887. [PubMed]

17. Philip, R.; Kumar, G.R. Picosecond optical nonlinearity in monolayer-protected gold, silver, and gold-silver alloy nanoclusters. *Phys. Rev. B* **2000**, *62*, 160–166. [CrossRef]

18. Tajdidzadeh, M.; Zakaria, A.B.; Abidin Talib, Z.; Gene, A.S.; Shirzadi, S. Optical nonlinear properties of gold nanoparticles synthesized by laser ablation in polymer solution. *J. Nanomater.* **2017**, *2017*, 4803843. [CrossRef]

19. Smith, D.D.; Yoon, Y.; Boyd, R.W.; Campbell, J.K.; Baker, L.A.; Crooks, R.M. z-scan measurement of the nonlinear absorption of a thin gold film. *J. Appl. Phys.* **1999**, *86*, 11. [CrossRef]

20. West, R.; Wang, Y.; Goodson, T., III. Nonlinear absorption properties in novel gold nanostructured topologies. *J. Phys. Chem. B* **2003**, *107*, 15. [CrossRef]

21. Wang, K.; Long, H.; Fu, M.; Yang, G.; Lu, P. Intensity-dependent reversal of nonlinearity sign in a gold nanoparticle array. *Opt. Lett.* **2010**, *35*, 1560–1562. [CrossRef] [PubMed]

22. Ryasnyanskiy, A.I.; Palpant, B.; Debrus, S.; Pal, U.; Stepanov, A. Third-order nonlinear-optical parameters of gold nanoparticles in different matrices. *J. Lumin.* **2007**, *127*, 181–185. [CrossRef]

23. Ganeev, R.A.; Suzuki, M.; Baba, M.; Ichihara, M.; Kuroda, H. Low- and high-order nonlinear optical properties of Au, Pt, Pd, and Ru nanoparticles. *J. Appl. Phys.* **2008**, *103*. [CrossRef]

24. Sheik-bahae, M.; Said, A.A.; Van Stryland, E.W. High-sensitivity, single-beam n2 measurements. *Opt. Lett.* **1989**, *14*, 955. [CrossRef] [PubMed]

25. Chapple, P.B.; Staromlynska, J.; Hermann, J.A.; Mckay, T.J.; Mcduff, R.G. Single-beam Z-Scan: Measurement techniques and analysis. *J. Nonlinear Opt. Phys. Mater.* **1997**, *06*, 251–293. [CrossRef]

26. Liu, X.; Guo, S.; Wang, H.; Hou, L. Theoretical study on the closed-aperture Z-scan curves in the materials with nonlinear refraction and strong nonlinear absorption. *Opt. Commun.* **2001**, *197*, 431–437. [CrossRef]

27. Link, S.; Burda, C.; Wang, Z.L.; El-Sayed, M.A. Electron dynamics in gold and gold-silver alloy nanoparticles: The influence of a nonequilibrium electron distribution and the size dependence of the electron-phonon relaxation. *J. Chem. Phys.* **1999**, *111*, 1255–1264. [CrossRef]

28. Voisin, C.; Christofilos, D.; Loukakos, P.A.; Del Fatti, N.; Vallee, F.; Lerme, J.; Gaudry, M.; Cottancin, E.; Pellarin, M.; Broyer, M. Ultrafast electron-electron scattering and energy exchanges in noble-metal nanoparticles. *Phys. Rev. B* **2004**, *69*, 195416.

29. Varnavski, O.P.; Goodson, T.; Mohamed, M.B.; El-Sayed, M.A. Femtosecond excitation dynamics in gold nanospheres and nanorods. *Phys. Rev. B Condens. Matter Mater. Phys.* **2005**, *72*, 1–9. [CrossRef]

30. Guo, L.; Xu, X. Ultrafast spectroscopy of electron-phonon coupling in gold. *J. Heat Transf.* **2014**, *136*, 122401.

31. Kolomenskii, A.A.; Mueller, R.; Wood, J.; Strohaber, J.; Schuessler, H.A. Femtosecond electron-lattice thermalization dynamics in a gold film probed by pulsed surface plasmon resonance. *Appl. Opt.* **2013**, *52*, 7352–7359.

32. Bauer, C.; Abid, J.-P.; Girault, H.H. Role of adsorbates on dynamics of hot-electron (type I and II) thermalization within gold nanoparticles. *C. R. Chim.* **2006**, *9*, 261–267. [CrossRef]

33. Masia, F.; Langbein, W.; Borri, P. Measurement of the dynamics of plasmons inside individual gold nanoparticlesusing a femtosecond phase-resolved microscope. *Phys. Rev. B* **2012**, *85*, 235403. [CrossRef]

34. Ganeev, R.A.; Suzuki, M.; Baba, M.; Ichihara, M.; Kuroda, H. Ablation of boron carbide for high-order harmonic generation of ultrafast pulses in laser-produced plasma. *Opt. Commun.* **2016**, *370*, 6–12. [CrossRef]

35. Ganeev, R.A.; Bom, L.B.E.; Wong, M.C.H.; Brichta, J.-P.; Bhardwaj, V.R.; Redkin, P.V.; Ozaki, T. High-order harmonic generation from C60-rich plasma. *Phys. Rev. A* **2009**, *80*, 043808. [CrossRef]

36. Ganeev, R.A.; Naik, P.A.; Singhal, H.; Chakera, J.A.; Kumar, M.; Joshi, M.P.; Srivastava, A.K.; Gupta, P.D. High order harmonic generation in carbon nanotube-containing plasma plumes. *Phys. Rev. A* **2011**, *83*, 013820. [CrossRef]

37. Donnelly, T.D.; Ditmire, T.; Neuman, K.; Perry, M.D.; Falcone, R.W. High-order harmonic generation in atom clusters. *Phys. Rev. Lett.* **1996**, *76*, 2472–2475. [CrossRef]

38. Tisch, J.W.G.; Ditmire, T.; Fraser, D.J.; Hay, N.; Mason, M.B.; Springate, E.; Marangos, J.P.; Hutchinson, M.H.R. Investigation of high-harmonic generation from xenon atom clusters. *J. Phys. B* **1997**, *30*, L709–L714. [CrossRef]

39. Vozzi, C.; Nisoli, M.; Caumes, J.-P.; Sansone, G.; Stagira, S.; De Silvestri, S.; Vecchiocattivi, M.; Bassi, D.; Pascolini, M.; Poletto, L.; et al. Cluster effects in high-order harmonics generated by ultrashort light pulses. *Appl. Phys. Lett.* **2005**, *86*, 111121. [CrossRef]

40. Hu, S.X.; Xu, Z.Z. Enhanced harmonic emission from ionized clusters in intense laser pulses. *Appl. Phys. Lett.* **1997**, *71*, 2605–2607. [CrossRef]

41. Ruf, H.; Handschin, C.; Cireasa, R.; Thiré, N.; Ferré, A.; Petit, S.; Descamps, D.; Mével, E.; Constant, E.; Blanchet, V.; Fabre, B.; Mairesse, Y. Inhomogeneous high harmonic generation in krypton clusters. *Phys. Rev. Lett.* **2013**, *110*, 083902. [CrossRef] [PubMed]

42. Ganeev, R.A.; Suzuki, M.; Baba, M.; Ichihara, M.; Kuroda, H. High-order harmonic generation in Ag nanoparticle-containing plasma. *J. Phys. B* **2008**, *41*, 045603. [CrossRef]

43. Ganeev, R.A.; Suzuki, M.; Baba, M.; Ichihara, M.; Kuroda, H. Low- and high-order nonlinear optical properties of $BaTiO_3$ and $SrTiO_3$ nanoparticles. *J. Opt. Soc. Am. B* **2008**, *25*, 325–333. [CrossRef]

44. Singhal, H.; Ganeev, R.A.; Naik, P.A.; Chakera, J.A.; Chakravarty, U.; Vora, H.S.; Srivastava, A.K.; Mukherjee, C.; Navathe, C.P.; Deb, S.K.; Gupta, P.D. In-situ laser induced silver nanoparticle formation and high order harmonic generation. *Phys. Rev. A* **2010**, *82*, 043821. [CrossRef]

45. De Nalda, R.; López-Arias, M.; Sanz, M.; Oujja, M.; Castillejo, M. Harmonic generation in ablation plasmas of wide bandgap semiconductors. *Phys. Chem. Chem. Phys.* **2011**, *13*, 10755–10761. [CrossRef] [PubMed]

46. Ganeev, R.A.; Boltaev, G.S.; Kim, V.V.; Zhang, K.; Zvyagin, A.I.; Smirnov, M.S.; Ovchinnikov, O.V.; Redkin, P.V.; Wöstmann, M.; Zacharias, H.; et al. Effective high-order harmonic generation from metal sulfide quantum dots. *Opt. Express* **2018**, *26*, 35013–35025. [CrossRef]

 *nanomaterials*

Article

An Eight-Channel C-Band Demux Based on Multicore Photonic Crystal Fiber

Dror Malka * and Gilad Katz

Faculty of Engineering, Holon Institute of Technology (HIT), Holon 5810201, Israel; giladka@hit.ac.il
* Correspondence: drorm@hit.ac.il

Received: 29 August 2018; Accepted: 15 October 2018; Published: 17 October 2018

Abstract: A novel eight-channel demux device based on multicore photonic crystal fiber (PCF) structures that operate in the C-band range (1530–1565 nm) has been demonstrated. The PCF demux design is based on replacing some air-hole areas with lithium niobate and silicon nitride materials over the PCF axis alongside with the appropriate optimizations of the PCF structure. The beam propagation method (BPM) combined with Matlab codes was used to model the demux device and optimize the geometrical parameters of the PCF structure. The simulation results showed that the eight-channel demux can be demultiplexing after light propagation of 5 cm with a large bandwidth (4.03–4.69 nm) and cross-talk (−16.88 to −15.93 dB). Thus, the proposed device has great potential to be integrated into dense wavelength division multiplexing (DWDM) technology for increasing performances in networking systems.

Keywords: photonic crystal fiber; demultiplexer; dense wavelength division multiplexing

1. Introduction

Dense wavelength division multiplexing (DWDM) is a system [1,2] that is used to integrate information from different sources over one fiber, while each source carried on its own divided light wavelength at the same time. DWDM has the ability to divide sources up into 80 ports, and allow more information to be multiplexed into a light-stream that is transferred on one fiber. Demux is an essential device in the DWDM system, and its main functionality is to divide signals from one input port into multiple ports. The demux device has several advantages such as a low bit error rate [3], a high data rate, a large bandwidth, low cross-talk [4], and less propagation delay. Therefore, researchers have shown the potential of designing demux-based waveguide techniques such as silicon photonics [5], Y-branch [6], multimode interference (MMI) [7–9], Mach–Zehnder interferometers [10,11], MMI in slot waveguide structures [12–14], etc.

Photonic crystal fiber (PCF) is a powerful waveguide that is based on a microstructured arrangement of materials of different refractive indexes [15]. The background material is usually pure silica, and the low-index regions are air-holes that are located along the fiber length. Several works have demonstrated the great potential of using PCF structures in comparison to conventional fibers [16–18]. The main benefit of designing a demux device based on a PCF structure is its ability to integrate different materials that have a high difference in their refractive index values. This is because the light guiding mechanism in PCF is based on the bandgap and modified total internal reflection (MTIR). Another advance is the ability to achieve a lower coupling length, especially in the case of closer coupled ports (cores) [19].

Several techniques have demonstrated how it is possible to couple light between closer coupled ports (cores) in a PCF structure through methods such as changing the PCF index profile by replacing some air-hole regions with pure silica along the fiber length [19,20] and using different air-hole sizes in the PCF structure [21–24].

The C-band range is set from 1530 nm to 1565 nm in the wavelength, and it is the most efficient and useful range in the optical communication field [14]. The advance of the C-band is the ability to transmit data with a high bitrate over a long distance since the C-band supports DWDM and optical amplifier fiber technologies [25].

In this work, we demonstrated a 1×8 wavelength demux in a PCF structure that split eight wavelengths in the C-band range. The operating wavelengths are between 1530–1565 nm with a spacing of 5 nm between two wavelengths. The light coupling between the closer coupled cores was obtained by replacing some air-hole areas with lithium niobate ($LiNbO_3$) and silicon nitride (Si_3N_4) materials along the fiber length.

Numerical investigations were carried out on the locations of the $LiNbO_3$ and Si_3N_4s layers and the key geometrical parameters of the multicore PCF structure to obtain high efficiency demultiplexing between the operating wavelengths. The demux PCF structure was analyzed and simulated using a beam propagation method (BPM) and Matlab codes. This device can be useful to increase the data bitrate in the C-band using a DWDM system.

To the best of our knowledge, this paper is the first to study the controlling of light propagation direction by integrating $LiNbO_3$ and Si_3N_4 rods over a multicore PCF structure. This new study was utilized to design a new demultiplexer based on a multicore PCF structure. Thus, the novelty is that the demultiplexer operated within the multicore PCF structure without using additional optical components. This can lead to a new DWDM technique that can be utilized to reduce the costs and size of the optical communication system.

2. Materials and Methods

Figure 1a–c show the full refractive index profile structure of the demux PCF design on the XZ plane (y = 0 cm), XY plane (z = 0 cm), and XY plane (z = 5 cm), respectively. In these figures, the background material is pure silica, and it is marked in a light blue color; the air-hole areas are marked in a purple color, the $LiNbO_3$ areas are marked in a red color, and the Si_3N_4 areas are marked in a yellow color. The geometrical parameters of the PCF structure are d, Λ (pitch), and z. In Figure 1a,d represents the hole diameter of the air-holes, pitch represents the distance between two air-holes, and z represents the light propagation axis.

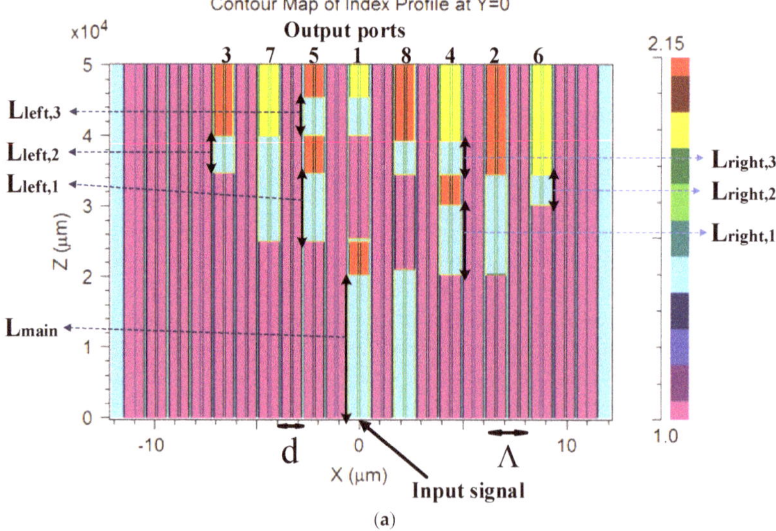

(a)

Figure 1. *Cont.*

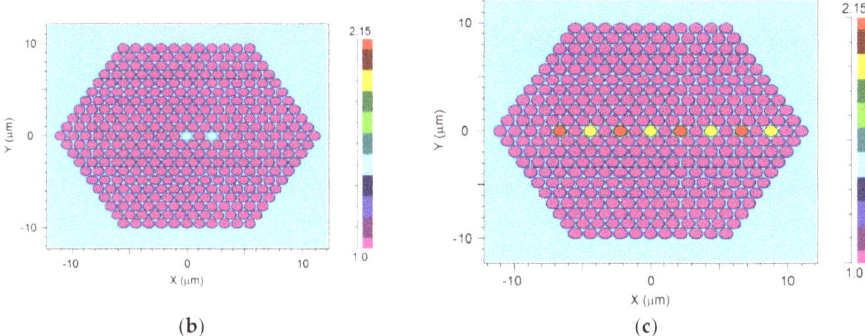

Figure 1. Refractive index profile of the 1 × 8 wavelength demux: (**a**) XZ plane at y = 0 cm. (**b**) XY plane at z = 0 cm. (**c**) XY plane at z = 5 cm.

The principle work of the PCF demux device is based on light confinement inside the channel (core) and controlling the light coupling length size between two closer ports (cores). These effects can be obtained by replacing some air-hole regions with a high-index material along the fiber length. Furthermore, the coupling length value can be shifted by changing the operating wavelength. Thus, the suitable coupling length value between two closer ports can be found by optimized the PCF geometrical parameters (z, Λ, d) and the materials' layer locations over the light propagation axis (z).

In our design, the light-guiding mechanism is based on the bandgap and the MTIR, which means that light can be coupled between two closer ports that have the same refractive index value (light blue color areas in Figure 1a), and a strong light confinement can be obtained by a port that has a high index value (red and yellow areas in Figure 1a).

It can be noticed from Figure 1a that the areas in which light can be coupled between closer silica (light blue color) ports (cores) are located as follows: port 1 and port 8 over the Z-axis range of 0–2.03 cm (L_{main}), port 8 and port 4 over the Z-axis ranges of 2.03–2.1 cm and 3.445–3.93 cm ($L_{right,3}$), port 1 and port 5 over the Z-axis range of 2.5–2.545 cm and 4–4.545 cm ($L_{left,3}$), port 4 and port 2 over Z-axis range of 2.03–3.01 cm ($L_{right,1}$), port 5 and port 7 over the Z-axis range of 2.5–3.46 cm ($L_{left,1}$), port 2 and port 6 over the Z-axis range of 3.01–3.44 cm ($L_{right,2}$), port 3 and port 7 over the Z-axis range of 3.47–3.995 cm ($L_{left,2}$). In addition, it can be noticed that each port at the output has a light confinement area, as shown in Figure 1a. Figure 1b,c show the input port at z = 0 (port 1) and eight output ports at z = 5 cm (red and yellow colors).

Table 1 shows the refractive index material values for the C-band range [26–28], and it is important to mention that the Si_3N_4 and $LiNbO_3$ materials have a very low absorption in this range. Another advance to use these materials is that light cannot be coupled between closer Si_3N_4 and $LiNbO_3$ layers, because they have a small difference in the index value, as shown in Table 1.

Table 1. The multicore photonic crystal fiber (PCF) materials' reflective index values.

λ_m (nm)	1530	1535	1540	1545	1550	1555	1560	1565
$n_{Si_3N_4}$	1.9968	1.9967	1.9966	1.9964	1.9963	1.9961	1.996	1.9959
n_{Silica}	1.4443	1.4442	1.4441	1.4441	1.444	1.444	1.4439	1.4438
n_{LiNbO_3}	2.1481	2.148	2.1478	2.1477	2.1476	2.1474	2.1473	2.1471

The Si_3N_4 and the $LiNbO_3$ layers were used to enable a strong light confinement inside the port that ensures that light cannot be coupled to other closer ports. In the proposed design, there are 11 layers that function as follows: eight output lines (four $LiNbO_3$ layers for ports 2, 3, 5, and 8, and four Si_3N_4 layers for ports 1, 4, 6, and 7) and three selected lines in a similar way to the classical definitions of the digital 1 × 8 demultiplexer.

The three LiNbO$_3$ layers were used to function as the selected waveguide switch that controls the light propagation direction along the fiber length. The first selected switch is located at the Z-axis range between 2.03–2.5 cm, and can demultiplex the light propagation between ports 1, 5, 7, and 3, and ports 8, 4, 2, and 6. The second selected switch is located at the Z-axis range between 3.01–3.445 cm, and can demultiplex the light propagation between ports 8 and 4, and ports 2 and 6. The third selected switch is located at the Z-axis range between 3.46–4 cm, and can demultiplex the light propagation between ports 3 and 7 and ports 5 and 1.

It is important to emphasize that the coupling length of the resonated light between two closer ports is dependent on the key geometrical parameters (d, Λ) and the operating wavelength. Thus, the coupling length can be found using the equation below [29,30]:

$$L^{\lambda}{}_{Coupling} = \frac{\pi}{k_0 \left(n_{symmetric}(d, \Lambda, \lambda) - n_{anti-symmertric}(d, \Lambda, \lambda) \right)} \tag{1}$$

where k_0 is the free-space wave vector, λ is the operating wavelength, and $n_{anti-symmetric}$ and $n_{symmetric}$ are the anti-symmetric and symmetric effective refractive index, respectively.

It is worth mentioning that the coupling length is periodical due to the oscillation between two closer ports.

The conditions for divide eight different wavelengths in the proposed design are given by:

$$
\begin{aligned}
L_{main} &\approx P_1 L^{\lambda_1, \lambda_3, \lambda_5, \lambda_7}_{Coupling} = (P_1 + q_1) L^{\lambda_2, \lambda_4, \lambda_6, \lambda_8}_{Coupling} \\[4pt]
L_{right,1} &\approx P_2 L^{\lambda_8, \lambda_4}_{Coupling} = (P_2 + q_2) L^{\lambda_2, \lambda_6}_{Coupling} \\[4pt]
L_{left,1} &\approx P_3 L^{\lambda_3, \lambda_7}_{Coupling} = (P_3 + q_3) L^{\lambda_5, \lambda_1}_{Coupling} \\[4pt]
L_{right,2} &\approx P_4 L^{\lambda_2}_{Coupling} = (P_4 + q_4) L^{\lambda_6}_{Coupling} \\[4pt]
L_{left,2} &\approx P_5 L^{\lambda_3}_{Coupling} = (P_5 + q_5) L^{\lambda_7}_{Coupling} \\[4pt]
L_{right,3} &\approx P_6 L^{\lambda_8}_{Coupling} = (P_6 + q_6) L^{\lambda_4}_{Coupling} \\[4pt]
L_{left,3} &\approx P_7 L^{\lambda_5}_{Coupling} = (P_7 + q_7) L^{\lambda_1}_{Coupling}
\end{aligned}
\tag{2}
$$

where L_{main} is the fiber length located between core 1 and core 8 (silica areas), $L_{righ,1}$ is the fiber length located between core 2 and core 4 (silica areas), $L_{left,1}$ is the fiber length located between core 7 and core 5 (silica areas), $L_{righ,2}$ is the fiber length located between core 2 and core 6 (silica areas), $L_{left,2}$ is the fiber length located between core 3 and core 7 (silica areas), $L_{righ,3}$ is the fiber length located between core 8 and core 4 (silica areas), $L_{left,3}$ is the fiber length located between core 5 and core 1 (silica areas), $P_{1/2/3/4/5/6/7}$ is a natural number, and $q_{1/2/3/4/5/6/7}$ is an odd number.

For analyzing the performances of our proposed 1 × 8 wavelength demultiplexer multicore PCF, cross-talk (Equation (3)) and loss (Equation (4)) were calculated to observe the ratio between a desirable and undesirable wavelength in a given port and power losses, respectively.

$$C.T_n = \frac{1}{3} \sum_{m=1}^{4} 10 \log\left(\frac{P_m}{P_n} \right) \tag{3}$$

where P_n is the power transmission for the suitable port, and P_m is the interference power transmission from the other ports. The insertion loss is given by:

$$Loss_{dB} = -10 \log\left(\frac{P_{out}}{P_{in}} \right) \tag{4}$$

where P_{out} is the power at the output port, and P_{in} is the power in the input port 1.

3. Results

The 1×8 PCF wavelength demultiplexer structure was simulated using an RSoft Photonics CAD suite software, which is based on the BPM. Figure 2 shows the optimal geometrical parameters (d/Λ) of the 1×8 wavelength demultiplexer multicore PCF structure for the operating wavelengths. It can be noticed from Figure 3 that the optimal value is 0.85 for the operating wavelengths, and the values of d and Λ are 0.97 µm and 1.14 µm.

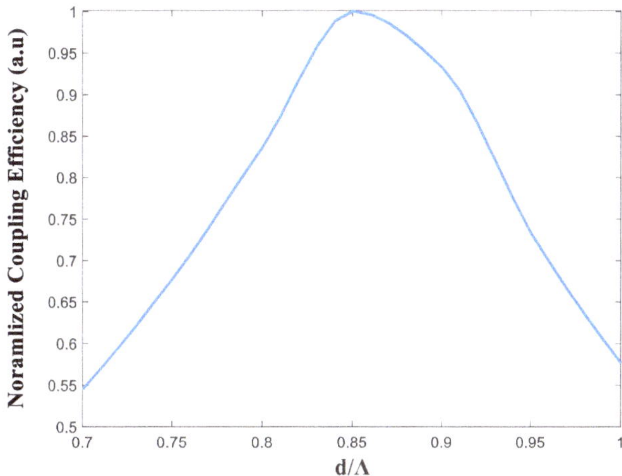

Figure 2. Normalized coupling efficiency as a function of the geometrical parameters of the multicore photonic crystal fiber (PCF) (d/Λ) for the operating wavelengths.

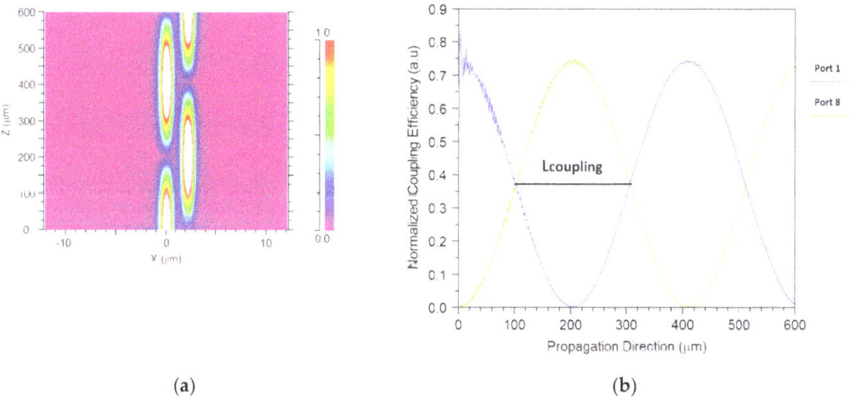

(a) (b)

Figure 3. Energy transfer between core 1 and core 8 for the 1565-nm wavelength. (**a**). Intensity profile. (**b**). Normalized coupling efficiency as a function of the propagation direction.

The coupling length size for each operating wavelength was calculated by simulating the transfer energy between two closer silica cores. Figure 3a,b show the full transfer energy between port 1 and port 8 at the Z-range from 0–600 µm for the 1565–nm operated wavelength. From Figure 3b, it can be noticed that the coupling length value is 203.41 µm. The same simulation was done for each of the operating wavelengths, and the values of the coupling length were found as shown in Table 2.

Table 2. The coupling length values.

λ_m (nm)	1530(λ_1)	1535(λ_2)	1540(λ_3)	1545(λ_4)	1550(λ_5)	1555(λ_6)	1560(λ_7)	1565(λ_8)
$L_{Coupling}$ (μm)	230.89	225.1	222.83	219.12	214.71	211.11	208.34	203.41

By using the results taken from Table 2 and combining them with Equation (2), the locations and the lengths of the silica rods that are suitable for the 1×8 wavelength demultiplexer multicore PCF can be found, and their values are: $L_{main} = 2.03$ cm, $L_{right,1} = 0.98$ cm, $L_{left,1} = 0.96$ cm, $L_{right,2} = 0.43$ cm, $L_{left,2} = 0.525$ cm, $L_{right,3} = 0.485$ cm, and $L_{left,3} = 0.545$ cm. Using these results along with the optimizations over the multicore PCF length, the locations (2.03–2.5cm at the Z-axis in core 1, 3.01–3.445 cm at the Z-axis in core 4, and 3.46–4cm at the Z-axis in core 5) of the three selected switch LiNbO$_3$ layers were found. In addition, the eight output port layers were optimized in order to obtain a compact 1×8 demultiplexer, and their locations are found as shown in Figure 1a.

Figure 4a shows the light propagation of the 1530-nm wavelength in the PCF structure, and its optical path can be described as follows: z = 0–1.9 cm, light coupled between port 1 and port 8; z = 1.9–2.5 cm, light confined in port 1; z = 2.5–2.6 cm, light coupled from port 1 to port 5; z = 2.6–3.5 cm, light coupled between port 5 and port 7; z = 3.5–4 cm, light confined in port 5; z = 4–4.7 cm, light coupled between port 5 and port 1; and z = 4.7–5 cm, light confined in port 1. Figure 4b shows the light propagation of the 1535-nm wavelength in the PCF structure, and its optical path can be described as follows: z= 0–2cm, light coupled between port 1 and port 8; z = 2–2.15 cm, light coupled from port 8 to port 4; z = 2.15–3 cm, light coupled between port 4 and port 2; z = 3–3.5 cm, light coupled between port 2 and port 6; z = 3.5–5cm, light confined in port 2. Figure 4c shows the light propagation of the 1540-nm wavelength in the PCF structure, and its optical path can be described as follows: z = 0–1.9 cm, light coupled between port 1 and port 8; z = 1.9–2.5 cm, light confined in port 1; z = 2.5–2.6 cm, light coupled from port 1 to port 5; z = 2.6–3.5 cm, light coupled between port 5 and port 7; z = 3.6–4 cm, light coupled between port 3 and port 7; z = 4–5 cm, light confined in port 3. Figure 4d shows the light propagation of the 1545-nm wavelength in the PCF structure, and its optical path can be described as follows: z = 0–2 cm, light coupled between port 1 and port 8; z = 2–2.15 cm, light coupled from port 8 to port 4; z = 2.15–3 cm, light coupled between port 4 and port 2; z = 3–3.5 cm, light confined in port 4; z = 3.5–4 cm, light coupled between port 4 and port 8; and z = 4–5 cm, light confined in port 4. Figure 4e shows the light propagation of the 1550-nm wavelength in the PCF structure, and its optical path can be described as follows: z = 0–1.9 cm, light coupled between port 1 and port 8; z = 1.9–2.5 cm, light confined in port 1; z = 2.5–2.6 cm, light coupled from port 1 to port 5; z = 2.6–3.5 cm, light coupled between port 5 and port 7; z = 3.5–4 cm, light confined in port 5; z = 4–4.7 cm, light coupled between port 5 and port 1; and z = 4.7–5 cm, light confined in port 5. Figure 4f shows the light propagation of the 1555-nm wavelength in the PCF structure, and its optical path can be described as follows: z = 0–2 cm, light coupled between port 1 and port 8; z = 2–2.15 cm, light coupled from port 8 to port 4; z = 2.15–3 cm, light coupled between port 4 and port 2; z = 3–3.5 cm, light coupled between port 2 and port 6; and z = 3.5–5 cm, light confined in port 6. Figure 4g shows the light propagation of the 1560-nm wavelength in the PCF structure, and its optical path can be described as follows: z = 0–1.9cm, light coupled between port 1 and port 8; z = 1.9–2.5 cm, light confined in port 1; z = 2.5–2.6 cm, light coupled from port 1 to port 5; z = 2.6–3.5 cm, light coupled between port 5 and port 7; z = 3.6–4 cm, light coupled between port 3 and port 7; and z = 4–5 cm, light confined in port 7. Figure 4h shows the light propagation of the 1565-nm wavelength in the PCF structure and its optical path can be described as follows: z = 0–2 cm, light coupled between port 1 and port 8; z = 2–2.15 cm, light coupled from port 8 to port 4; z = 2.15–3 cm, light coupled between port 4 and port 2; z = 3–3.5 cm, light confined in port 4; z = 3.5–4 cm, light coupled between port 4 and port 8; and z = 4–5 cm, light confined in port 8.

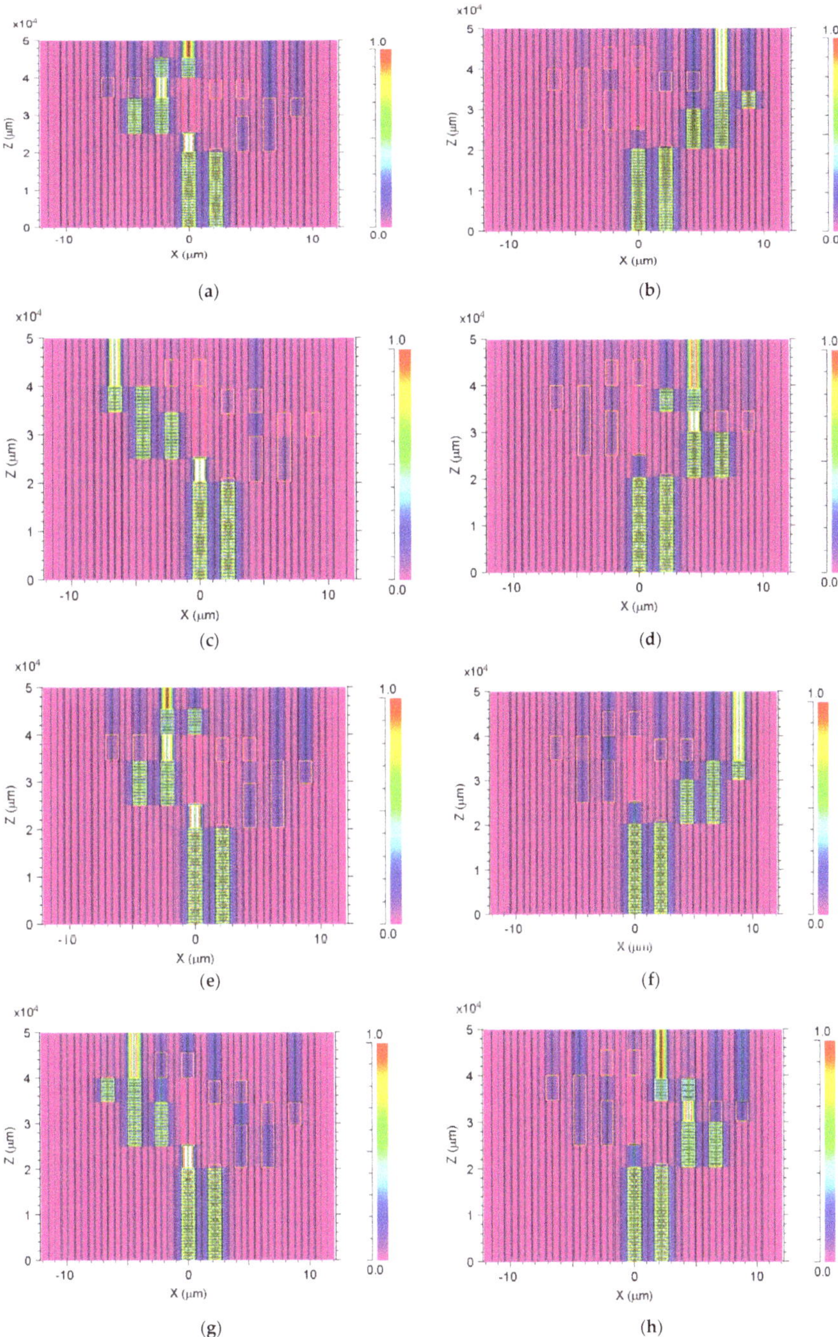

Figure 4. Intensity profile of the 1×8 multimode interference (MMI) wavelength demultiplexer: (**a**) $\lambda_1 = 1530$ nm (port 1). (**b**) $\lambda_2 = 1535$ nm (port 2). (**c**) $\lambda_3 = 1540$ nm (port 3). (**d**) $\lambda_4 = 1545$ nm (port 4). (**e**). $\lambda_5 = 1550$ nm (port 5). (**f**) $\lambda_6 = 1555$ nm (port 6). (**g**) $\lambda_7 = 1560$ nm (port 7). and (**h**) $\lambda_8 = 1565$ nm (port 8).

BPM simulations combined with the Matlab script code were performed to determine the 1×8 wavelength PCF demultiplexer properties. Figure 5 shows the optical bandwidth transmission results for wavelengths around the C-band range (1530–1565 nm).

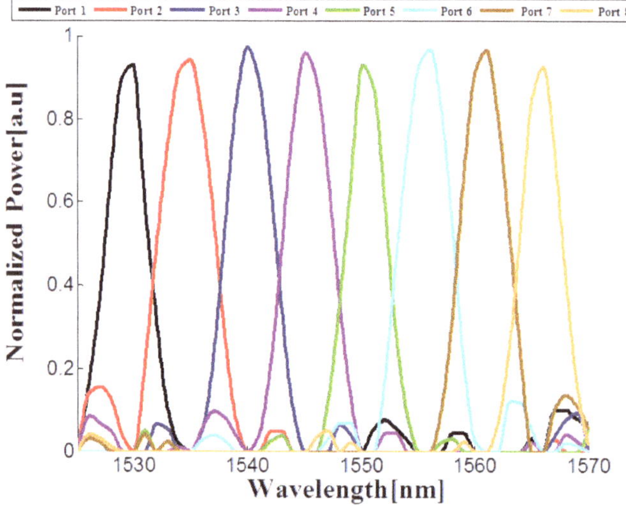

Figure 5. Normalized power as function of the optical signals.

From Figure 5, combined with Equations (3) and (4), the values of the cross-talk, insertion losses and full width maximum (FWHM) can be found. Table 3 shows the values of the cross-talk, bandwidth (FWHM), and loss for each port.

Table 3. Values of the cross-talk, full width maximum (FWHM), and losses for the operating wavelengths.

λ_m (nm)	1530	1535	1540	1545	1550	1555	1560	1565
Port number	1	2	3	4	5	6	7	8
Cross-talk (dB)	−16.65	−16.73	−16.88	−16.81	−16.6	−16.32	−16.08	−15.93
FWHM (nm)	4.23	4.38	4.67	4.69	4.03	4.1	4.62	4.15
Losses (dB)	0.31	0.26	0.2	0.18	0.55	0.31	0.45	0.69

4. Conclusions

This study shows how it is possible to implement the classical 1×8 demultiplexer that is usually based on seven 1×2 demultiplexer units in a cascade tree structure, only with one multicore PCF. In addition, this work shows how to control the light propagation direction inside the multicore PCF using the MTIR and bandage mechanism that enables light to be coupled only in the silica areas, and a strong light confinement in the $LiNbO_3$ and Si_3N_4 areas.

By analysis of the numerical results, it is clear that the main benefit is that the multicore PCF was designed without using additional device elements, which are usually required in order for the device to function as a 1×8 demultiplexer. This benefit can lead to a new design of a compact DWDM system that can be utilized to obtain better performances. However, integrating this type of demultiplexer to the DWDM system will require modifying the mode field diameter (MFD) coming out from the multicore PCF. This is due to the DWDM system usually using silica fibers that have a large MFD compared to the multicore PCF. This issue can be solved by integrated a tapered fiber that can convert the MFD of the 1×8 wavelength demultiplexer to the MFD of the DWDM system.

The 1×8 wavelength demultiplexer multicore PCF can be fabricated using a fiber fabrication facility with hybrid materials. The fabrication process cannot be carried out only by a staking or drawing technique; it also requires a microelectronic technique, such as the lithography technique, that can be used to integrate the $LiNbO_3$ and Si_3N_4 layers over the multicore PCF length.

To conclude, in this work, we have shown that a 1×8 wavelength demultiplexer can be implemented using a multicore PCF structure with integrated $LiNbO_3$ and Si_3N_4 materials.

The simulation results showed that the operated wavelengths of 1530 nm, 1535 nm, 1540 nm, 1545 nm, 1550 nm, 1555 nm, 1560 nm, and 1565 mm, with a short spacing of 5 nm and supporting the whole C-band range, can be separated after a propagation length of 5 cm with a bandwidth range of 4.02–4.69 nm.

The proposed device has low cross-talk ((−16.88)–(−15.93)) dB) with an insertion loss of 0.18–0.69 dB.

Thus, this device has the potential to increase the data bitrate in an optical communications system that works on DWDM technology.

This research can be used in future work to design a powerful and new demultiplexer device based on multicore polymer/silica fiber in a very similar way, and with the right materials' modification.

Although only the 1×8 wavelength demultiplexer configuration is considered in this paper, the demultiplexer can also operate as a multiplexer with a reversed direction of the guided light.

Author Contributions: D.M. envisioned the project. D.M. designed the device and performed simulations with support of G.K. D.M. wrote the manuscript and the reply letter. D.M. made all figures and all authors reviewed the manuscript.

Funding: This research received no external funding.

Conflicts of Interest: The authors declare no conflict of interest.

References

1. Bogdan, H. DWDM fundamentals, components, and applications. *J. Opt. Netw.* **2002**, *1*, 184–185.
2. Gong, J.M.; Zuo, X.; Zhao, Y. The steady SRS analysis theory of DWDM transmission system in single-mode silica fiber. *Opt. Commun.* **2015**, *350*, 257–262. [CrossRef]
3. Charlier, J.; Laurent, B.; Berlioz, P.; Perbos, J.L. Multi/demultiplexer and spectral isolator for optical inter-satellites communications. *Proc. SPIE Opt. Space Commun.* **1989**, *54*, 1131.
4. Malka, D.; Sintov, Y.; Zalevsky, Z. Design of a 1×4 silicon-Alumina wavelength demultiplexer based on multimode interference in slot waveguide structures. *J. Opt.* **2015**, *17*, 125702–125711. [CrossRef]
5. Dalai, P.K.; Sarkar, P. Analysis of silicon waveguide structure for realization of optical MUX/DEMUX circuit: An application of silicon photonics. *Optik* **2016**, *127*, 10569–10574. [CrossRef]
6. Xiang, F.; Lam, Y.G. A TM Y-branch wavelength multi/demultiplexer by K/sup +/and Ag/sup +/ion-exchange for/spl lambda/= 1.31 and 1.55/spl mu/m. *J. Light. Technol.* **1996**, *14*, 1690–1694. [CrossRef]
7. Lin, K.C.; Lee, W.Y. Guided-wave 1.30/1.55 μm wavelength division multiplexer based on multimode interference. *Electron. Lett.* **1996**, *32*, 1259–1261. [CrossRef]
8. Li, B.; Li, G.; Liu, E.; Jiang, Z.; Qin, J.; Wang, X. Low-loss 1×2 multimode interference wavelength demultiplexer in silicon-germanium alloy. *IEEE Photon. Technol. Lett.* **1999**, *11*, 575–577.
9. Tsao, S.L.; Guo, H.C.; Tsai, C.W. A novel 1×2 single-mode 1300/1550nm wavelength division multiplexer with output facet-tilted MMI waveguide. *Opt. Commun.* **2004**, *232*, 371–379. [CrossRef]
10. Goto, N.; Yip, G.L. Y-branch wavelength multi-demultiplexer for λ = 1.30 μm and 1.55 μm. *Electron. Lett.* **1990**, *26*, 102–103. [CrossRef]
11. Tervonen, A.; Poyhonen, P.; Honkanen, S.; Tahkokorpi, M. A guided-wave Mach-Zehnderinterferometer structure for wavelength multiplexing. *IEEE Photonics Technol. Lett.* **1991**, *3*, 516–518. [CrossRef]
12. Xiao, J.; Liu, X.; Sun, X. Design of an ultracompact MMI wavelength demultiplexer in slot waveguide structures. *Opt. Express.* **2007**, *15*, 8300–8308. [CrossRef] [PubMed]
13. Malka, D.; Danan, Y.; Ramon, Y.; Zalevsky, Z. A Photonic 1×4 power splitter based on multimode interference in silicon–gallium-nitride slot waveguide structures. *Materials* **2016**, *9*, 516. [CrossRef] [PubMed]

14. Zaken, B.B.B.; Zanzury, T.; Malka, D. An 8-Channel Wavelength MMI Demultiplexer in Slot Waveguide Structures. *Materials* **2016**, *9*, 881. [CrossRef] [PubMed]

15. Russell, P.S.J. Photonic-crystal fibers. *J. Lightw. Technol.* **2006**, *24*, 4729–4749. [CrossRef]

16. Birks, T.A.; Knight, J.C.; Russel, P.S.J. Endlessly single-mode photonic crystal fiber. *Opt. Lett.* **1997**, *22*, 961–963. [CrossRef] [PubMed]

17. Knight, J.C.; Broeng, J.; Birks, T.A.; Russel, P.S.J. Photonic band gap guidance in optical fiber. *Science* **1998**, *282*, 1476–1478. [CrossRef] [PubMed]

18. Broeng, J.; Mogilevstev, D.; Barkou, S.E.; Bjarklev, A. Photonic crystal fibers: A new class of optical waveguides. *Opt. Fiber Technol.* **1999**, *5*, 305–330. [CrossRef]

19. Elbaz, D.; Malka, D.; Zalevsky, Z. Photonic crystal fiber based 1xN intensity and wavelength splitters/couplers. *Electromagnetics* **2013**, *32*, 209–220. [CrossRef]

20. Malka, D.; Sintov, Y.; Zalevsky, Z. Fiber-laser monolithic coherent beam combiner based on multicore photonic crystal fiber. *Opt. Eng.* **2014**, *54*, 011007. [CrossRef]

21. Koshiba, M. Coupling characteristics of multicore photonic crystal fiber-based 1×4 power splitters. *J. Lightw. Technol.* **2009**, *27*, 2062–2068.

22. Malka, D.; Zalevsky, Z. Multicore photonic crystal fiber based 1×8 two-dimensional intensity splitters/couplers. *Electromagnetics* **2013**, *33*, 413–420. [CrossRef]

23. Malka, D.; Peled, A. Power splitting of 1×16 in multicore photonic crystal fibers. *Appl. Surf. Sci.* **2017**, *417*, 34–39. [CrossRef]

24. Malka, D.; Cohen, E.; Zalevsky, Z. Design of 4×1 power beam combined based on multicore photonic crystal fiber. *Appli. Sci.* **2017**, *7*, 695–699. [CrossRef]

25. Dike, J.N.; Ogbe, D.A. Optimizing the efficiency of fiber-optics technology in telecommunications system. In *Proceedings of the IEEE international conference on Emerging & Sustainable Technologies for power& ICT in Developing Society*; IEEE: Owerri, Nigeria, 2013; pp. 135–142.

26. Tan, C.Z. Determination of refractive index of silica glass for infrared wavelengths by IR spectroscopy. *J. Non-Cryst. Solids* **1998**, *223*, 158–163. [CrossRef]

27. Luke, K.; Okawachi, Y.; Lamont, M.R.E.; Gaeta, A.L.; Lipson, M. Broadband mid-infrared frequency comb generation in a Si_3N_4 microresonator. *Opt. Lett.* **2015**, *40*, 4823–4826. [CrossRef] [PubMed]

28. Zelmon, D.E.; Small, D.L.; Jundt, D. Infrared corrected Sellmeier coefficients for congruently grown lithium niobate and 5 mol.% magnesium oxide-doped lithium niobite. *J. Opt. Soc. Am. B* **1997**, *14*, 3319–3322. [CrossRef]

29. Shoresh, T.; Katanov, N.; Malka, D. 1×4 MMI visible light wavelength demultiplexer based on GaN slot waveguide structures. *Photonics Nanostruct. Fundam. Appl.* **2018**, *30*, 45–49. [CrossRef]

30. Saitoh, K.; Sato, Y.; Koshiba, M. Coupling characteristics of dual-core photonic crystal fiber couplers. *Opt. Express* **2003**, *11*, 3188–3194. [CrossRef] [PubMed]

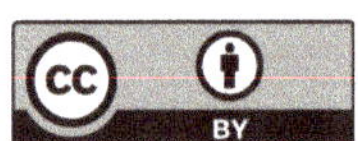

 *nanomaterials*

Article

Effect of Thickness of Molybdenum Nano-Interlayer on Cohesion between Molybdenum/Titanium Multilayer Film and Silicon Substrate

Huahai Shen [1,*,†], Bing Yao [1,†], Jianwei Zhang [2], Xinqiao Zhu [1], Xia Xiang [2], Xiaosong Zhou [1,*] and Xiaotao Zu [2]

1   Institute of Nuclear Physics and Chemistry, China Academy of Engineering Physics, Mianyang 621900, China; yaob1974@sina.com (B.Y.); zhuxinqiao@zju.edu.cn (X.Z.)
2   School of Physics, University of Electronic Science and Technology of China, Chengdu 610054, China; jianweizhang19@163.com (J.Z.); xiaxiang@uestc.edu.cn (X.X.); xtzu@uestc.edu.cn (X.Z.)
*   Correspondence: huahaishen@caep.cn (H.S.); zlxs77@126.com (X.Z.); Tel.: +86-816-2483364 (H.S.); +86-816-2497405 (X.Z.)
†   These authors contributed equally to this work.

Received: 26 March 2019; Accepted: 11 April 2019; Published: 16 April 2019

Abstract: Titanium (Ti) film has been used as a hydrogen storage material. The effect of the thickness of a molybdenum (Mo) nano-interlayer on the cohesive strength between a Mo/Ti multilayer film and a single crystal silicon (Si) substrate was investigated by X-ray diffraction (XRD), scanning electron microscopy (SEM), and nano-indenter. Four groups of Si/Mo/Ti multilayer films with different thicknesses of Mo and Ti films were fabricated. The XRD results showed that the introduction of the Mo layer suppressed the chemical reaction between the Ti film and Si substrate. The nano-indenter scratch results demonstrated that the cohesion between the Mo/Ti film and Si substrate decreased significantly with increasing Mo interlayer thickness. The XRD stress analysis indicated that the residual stress in the Si/Mo/Ti film was in-plane tensile stress which might be due to the lattice expansion at a high film growth temperature of 700 °C and the discrepancy of the thermal expansion coefficient between the Ti film and Si substrate. The tensile stress in the Si/Mo/Ti film decreased with increasing Mo interlayer thickness. During the cooling of the Si substrate, a greater decrease in tensile stress occurred for the thicker Mo interlayer sample, which became the driving force for reducing the cohesion between the Mo/Ti film and Si substrate. The results confirmed that the design of the Mo interlayer played an important role in the quality of the Ti film grown on Si substrate.

Keywords: titanium film; interlayer; cohesion; residual stress; nano-indenter

1. Introduction

Thin film materials maintain the fantastic properties of bulk materials [1–6] and have the significant advantages of being more economic [7], having small device size [4,5] and having the new physical performance of two-dimensional material [8–10]. Thin film materials have been used as functional devices in the applications of semi-conductors [2,8–10], metal hydrides [1,4,5,11–13], and solar cells [14,15], etc. It is more conducive to the combination of multiple functional materials to form multilayer film materials [16,17], which can not only realize the unique characteristics of each material, but also give play to the characteristics of new interface materials [18–20] and improve their comprehensive performance [19–23]. Therefore, the design and fabrication of new multilayer functional materials have attracted attention as hot issues in current research [23]. Postolnyi et al. [23] has proposed that a lower individual layer thickness with smaller grain size and more interlayer interfaces can significantly improve the mechanical properties of CrN/MoN multilayers. Li et al. [18]

has found that smaller layer thickness is beneficial in reducing the irradiation hardening of Fe/W multilayers. Pshyk et al. [24] synthesized multilayer TiAlSiYN/MoN coatings that had improved mechanical properties in comparison to a single-layer TiAlSiYN nanocomposite film. The cohesive strength between multilayer and substrate is another crucial problem in multilayer film design and is affected by substrate roughness [25], substrate temperature [25,26], different thermal expansion coefficient in the film-substrate couple [27,28], and film thickness [29–31], etc. A novel nanoindentation technique has been proposed and was employed to test the nanomechanical properties of multilayered Al_2O_3/TiO_2 nanolaminates [32]. A theoretical method based on density functional calculation (DFT) has been developed to study film structure stability and the cohesion between the thin film and substrate [33]. Goyenola et al. [33] has addressed stable compounds of fullerene-like sulpho-carbide and obtained their geometry optimizations and cohesive energies. Ren et al. [34] investigated the strain and cohesive energy of a TiN film on an Al(001) substrate using the DFT method and found that the TiN film could be steadily deposited on the AlN(001) interface.

Titanium film has the advantages of good hydrogen absorption performance, high thermal stability, and low room temperature equilibrium pressure. Ti film has been used as a metal hydride in the solid phase and has played an important role in the fields of hydrogen storage [35], solar thermal energy storage [36], and nuclear energy storage [4,37–39]. It has been reported that the surface morphology and grain size of Ti film on a Mo substrate was severely affected by the heterogeneity of polycrystalline Mo [4,38,40]. The hydrogenation performances of Ti films have been strongly correlated to their grain size and density of grain boundary. Thus, the design of new substrates for high quality Ti film growth is in demand. Single crystal Si is a candidate substrate material due to its properties of high thermal and electrical conductivity [5,11,41]. However, the chemical reaction between Ti and Si to form a Ti-Si complex at high substrate temperature restricts its application [42,43]. Shen et al. [13] have reported the formation of an Er-Si complex in Er film grown on Si. It has been demonstrated by Parish et al. [5,11] that the reaction between the Er film and Si substrate was suppressed by the introduction of a Mo interlayer between the Er film and Si substrate.

Screening and designing membrane material systems and film thicknesses are of great importance for the fabrication of high quality multilayer materials [18,23]. However, extensive study on the relationship between the cohesion of the film-substrate couple and film thickness has rarely been reported to date [29–31]. Based on the important application prospects of multilayer films in many fields [16,18,22,23], this work focuses on the influence of a Mo interlayer on the cohesion of an Si-based Ti film. The cohesion between the Ti film and the Si substrate has been discussed in relation to the residual stress [6,44,45] of the Ti film and the different thermal expansion coefficients of the Ti and Si materials [27,28].

2. Experimental Details

2.1. Film Growth and Sample Preparation

Ti films and Mo interlayers were electron beam evaporated onto polished single crystal Si (110) substrates by using source materials of Ti bulk (PrMat, Shanghai, China, 99.99% purity, 25 mm in diameter) and Mo bulk (PrMat, Shanghai, China, 99.99% purity, 25 mm in diameter) [4,13]. Before deposition, the Si substrates were ultrasonically cleaned in ethanol and acetone and outgassed at 700 °C for 2 h in a chamber with a base vacuum pressure better than 2×10^{-4} Pa. The sample holder was rotated during deposition to obtain uniform films and the distance between source materials and sample holder was ~20 cm. The Mo interlayer was first grown on the Si substrate at a temperature of 500 °C, followed by Ti film evaporation on the Mo layer to form Si/Mo/Ti multilayer films. The Ti films were grown at a temperature of 700 °C to avoid the severe oxidation of the Ti-based getter material [4,46]. The evaporation rates detected by the IC5 thin film deposition controller were 15 nm/s and 2 nm/s for the Ti and Mo films, respectively. Four kinds of Si/Mo/Ti films were fabricated with different thicknesses of Mo interlayers by controlling the deposition time from 1 min to 8 min. The deposition parameters and sample designation of the as-grown multilayer films are summarized in Table 1. The substrate temperature was naturally cooled to room temperature immediately after the Mo and Ti film evaporations.

Table 1. Sample designations of Si/Mo/Ti multilayer films with different Mo deposition times and thicknesses of Mo and Ti films measured by scanning electron microscopy (SEM).

Sample Designation	Mo-1	Mo-2	Mo-4	Mo-8
Deposition time of Mo films (min)	1	2	4	8
Thickness of Mo films (nm)	54.3	103.7	139.8	331.5
Thickness of Ti films (nm)	692.0	657.0	654.3	693.2

2.2. Sample Characterizations

The surface morphology of the multilayer films was acquired by SEM using a Zeiss Auriga workstation (Carl Zeiss Microscopy GmbH, Jena, Germany) equipped with X-ray energy dispersive spectroscopy (EDS) and a gallium focused ion beam (FIB). The elemental distributions of Ti, Mo, and Si elements on the film surface and along the cross-sectional structure were collected by EDS using an Oxford X-Max[N] 150 mm^2 (Oxford Instruments, Abingdon, UK). Cross-sectional SEM samples of the Si/Mo/Ti films were prepared to measure the thicknesses of the Mo and Ti films by using an advanced FIB method working in the SEM facility [47]. A large stair-step trench was milled out at the interested area from the film surface to expose the cross-sectional structures of the Si/Mo/Ti films. The crystal structures of the Si/Mo/Ti films were analyzed by XRD using an X'Pert PRO MPD (PANalytical B.V., Almelo, The Netherlands) with Cu Kα irradiation working at 45 kV and 40 mA. The residual stress of the Si/Mo/Ti films was analyzed by a classical $\sin^2\psi$ method using XRD equipment [12,23]. Ten θ-2θ scans were performed at a fixed ψ position by tilting the ψ angle from 0° (the sample normal direction) to 36°. The Ti (112) diffraction peak positioned at ~76.3° is preferred for residual stress analysis considering the peak intensity and accuracy of stress results. Therefore, each θ–2θ scan was collected from 74.5° to 79.5° using the typical parameters of 0.013° in step size and 80 s in count time.

The cohesion between the Mo/Ti film and Si substrate was tested using an Agilent G200 Nano Indenter (Agilent, Santa Clara, United States) working with the scratch mode [48–50]. The scratch tests involved dragging a nano-indenter across the surface of the film with a continuously increasing normal load until surface buckling occurred at the critical load. The scratch experiment was explored by moving the indenter along the film surface for the first time to check the surface smoothness, followed by scratching the film with a linearly increasing normal force to a maximal load of 150 mN over a total scratch length of 500 μm [49]. The distance between the scratches was set to 100–600 μm and the scratch velocity was set to 50 μm/s. The SEM and EDS analyses were carried out on five scratch tracks for each sample to evaluate the cohesion between the film and substrate.

3. Results and Discussions

Based on the Mo interlayer and deposition times of 1 min, 2 min, 4 min, and 8 min, the as-grown Si/Mo/Ti multilayer films on single crystal Si (110) substrates were designated Mo-1, Mo-2, Mo-4, and Mo-8, respectively. Figure 1 shows the SEM images of the Mo-1, Mo-2, Mo-4, and Mo-8 samples, where the insets are the corresponding optical images. As shown in Figure 1a,b, it is clear that the Mo-1 and Mo-2 samples were of high quality, with very smooth surfaces and nanometer grain sizes. When the deposition time of the Mo interlayer was increased to 4 min and 8 min, the surface deformation started to appear for the Mo-4 sample and the obvious surface delamination occurred in the center area of the Mo-8 sample (labeled by arrow, shown in Figure 1c,d).

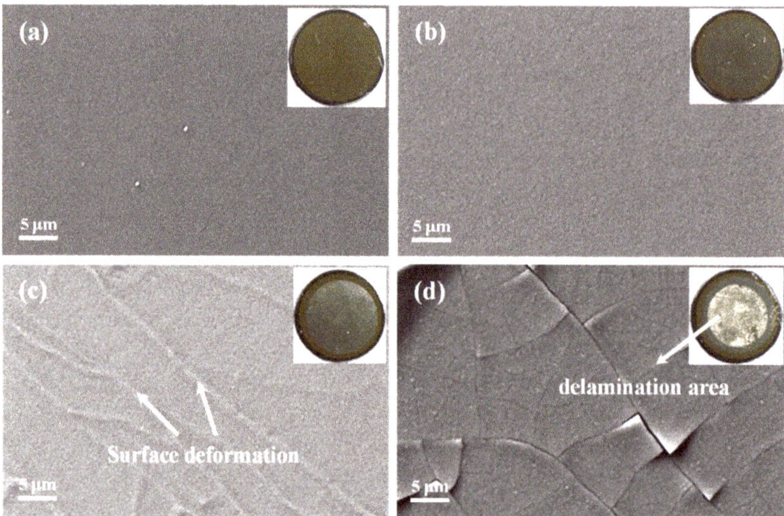

Figure 1. SEM and optical microscopic images of as-grown Si/Mo/Ti multilayer films with different Mo transition layer thicknesses for (**a**) Mo-1, (**b**) Mo-2, (**c**) Mo-4, and (**d**) Mo-8. The Mo-1 and Mo-2 samples exhibited smooth surfaces while the Mo-4 sample started to display surface deformation and the Mo-8 sample showed obvious surface delamination.

Figure 2 shows the XRD patterns of the Mo-1, Mo-2, Mo-4, and Mo-8 samples. The 2θ range from 42° to 55° is omitted to clearly present the diffraction peaks of the Ti and Mo crystals, since the Si (110) peak at 47.28° is much stronger than the other peaks. All the diffraction peaks could be well indexed as Ti and Mo have hexagonal-close packed (HCP) and body-centered cubic (BCC) crystal structures [4]. As shown in Figure 2, all four samples presented a strong preferred crystallographic orientation of Ti (101). Two tiny peaks positioned at 57.76° and 72.87° were indexed as Mo (200) and Mo (211) [4]. The relative intensities of the Mo peaks increased significantly with the Mo deposition time, revealing that the Mo layer thickness increased with deposition time. It is notable that no obvious peak of the Ti-Si compound [42,43] could be found in the XRD patterns of all four samples, demonstrating that the introduction of the Mo interlayer suppressed the chemical reaction between the Ti film and Si substrate, although the substrate temperature was higher than the onset temperature of 620 °C for the Ti-Si chemical reaction reported by Bensch et al. [11,43]. The Er-Si compounds started to form at a temperature above 350 °C [13] and could be well prevented by the introduction of an Mo interlayer with a thickness of about 100 nm [5]. The above results indicate that the pure Ti films were successfully grown on Si substrates, and the thickness of the Mo interlayer had a strong impact on the surface morphology and cohesion between the Mo/Ti film and Si substrate.

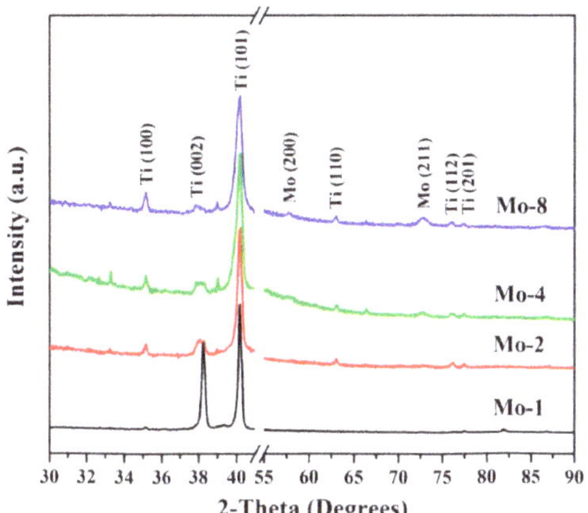

Figure 2. (Color online) X-ray diffraction (XRD) patterns of Si/Mo/Ti multilayer films with different Mo transition layer thicknesses. All four samples show strong Ti (101) preferred crystal orientation.

Figure 3a shows a representative cross-sectional SEM image of the Mo-2 sample, where the Si substrate, Mo interlayer, and Ti film could be identified and labeled. It is clear that the grain size of the Ti film is in nanometers and ranges from ~200 nm to ~400 nm. Figure 3b–d displays the EDS mapping results of the Ti, Mo, and Si element distributions seen in Figure 3a. As shown in Figure 3c, the signal of the Mo X-ray in the Mo interlayer is remarkably stronger than that in the Ti film, confirming the fabrication of the Si/Mo/Ti multilayer film. The thicknesses of the Mo interlayer and Ti film could be accurately measured from the cross-sectional SEM image. Table 1 tabulates the measured results of the Mo interlayer and Ti film thicknesses of the Mo-1, Mo-2, Mo-4, and Mo-8 samples. The thickness of the Mo interlayer increased from 54.3 nm to 331.5 nm with increasing Mo deposition time from 1 min to 8 min, which is in accord with the XRD results. The thicknesses of the Ti films for all four samples stayed between 650 nm to 700 nm with nearly the same deposition time. The EDS mapping results also demonstrated that a pure Ti film had been grown without the precipitation of other compounds like the Ti-Si compound [42,43].

To verify whether or not the surface delamination occurred during the Ti film deposition, another Mo-8a sample was fabricated with the same parameters as the Mo-8 sample, except that the thickness of the Ti film increased to around 1500 nm. Figure 4a,b show an SEM image of the Mo-8a sample which was taken from the center delamination area and the corresponding Si element EDS mapping result. It is clear that the dark areas (labeled by arrows) in Figure 4a indicate a strong Si signal, revealing that the Mo/Ti multilayer film completely peeled off from the Si substrate in those areas on the Mo-8a sample surface. Figure 4c,d exhibit a cross-sectional SEM image of the Mo-8a sample and an EDS line scan of the Ti, Mo, and Si elements along the line labeled in Figure 4c, respectively. The EDS line scan result illustrates that the Mo interlayer is absent from the multilayer film. It was demonstrated that the Ti film bonded well to the Mo film, which induced the full delamination of the Mo/Ti multilayers from the Si substrate. However, the Mo/Ti film was not firmly bonded on the Si substrate, which might have been due to the small roughness of the polished single crystal Si substrate. These results support the conclusion that the Ti film was directly deposited on the Si substrate after the Mo/Ti film started to delaminate from the Si substrate during the Ti film deposition.

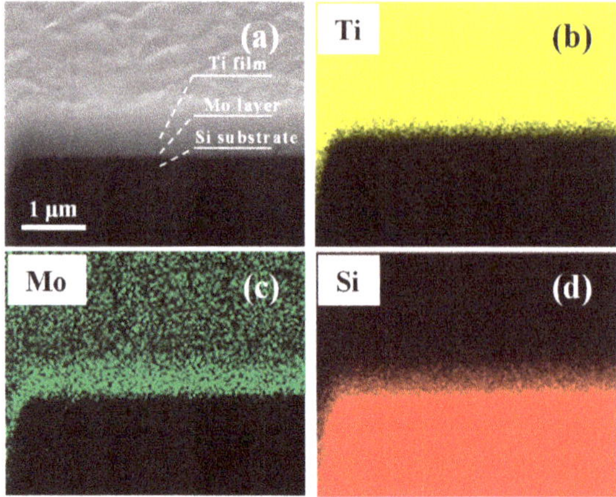

Figure 3. (Color online) (**a**) Cross-sectional SEM image of Mo-2 sample. (**b–d**) The Ti, Mo, and Si element distribution of (a) shows a clear Mo transition layer between the Ti film and Si substrate.

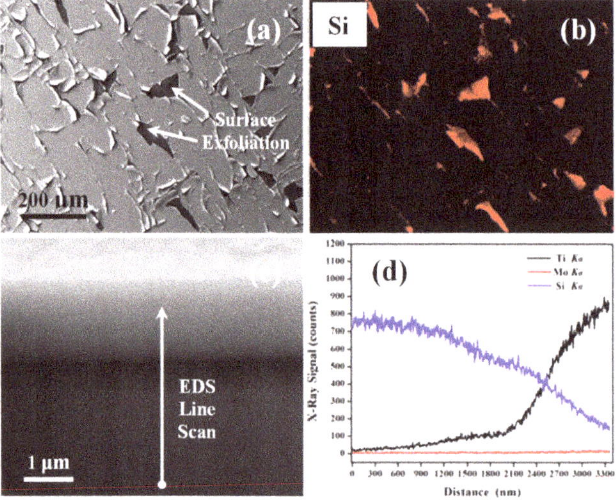

Figure 4. (Color online) (**a**) SEM image of Mo-8a sample taken from the delamination area and (**b**) the Si element distribution in (a). (**c**) Cross-sectional SEM image obtained from Mo layer absent area and (**d**) energy dispersive spectroscopy (EDS) line scan of Ti, Mo, and Si elements along Line 1 in (c).

The SEM images in Figure 5 show the five scratch tracks induced by the nano-indenter on each of the Mo-1, Mo-2, Mo-4, and Mo-8 samples. For the Mo-8 sample, the scratch experiments were explored on the area that was free of surface delamination. The scratch tests were performed from left to right with a linear increasing normal force to a maximum distance of 500 μm. For each scratch track, the width of the track steadily increased with the increasing displacement of the nano-indenter into the film surface when the film was tightly bonded to the Si substrate. However, the width and the shape of the track were sharply enlarged or the film was completely peeled off from the substrate once the film started to delaminate [49]. As shown in Figure 5a, it was clear that a slight exfoliation was induced by the nano-indenter at the end of the scratch tracks for the Mo-1 sample. With the Mo deposition time increased to 2 min, the surface exfoliation became stronger for the Mo-2 sample while a few Mo/Ti film particles could be found around the scratch tracks. It is interesting to note that the surface buckling occurred at the center of the scratch distance for the Mo-4 sample. Obviously, a severe surface exfoliation with a great number of film particles could be observed at the center of the scratch distance for the Mo-8 sample. Figure 6a displays the magnified three scratch tracks on the surface of the Mo-8 sample, and Figure 6b–d shows the EDS mapping of the Ti, Mo, and Si element distributions seen in Figure 6a. The signals of the Ti and Mo elements disappeared while the signal of the Si element changed to be become stronger in the surface exfoliation area. The EDS mapping results demonstrate that the multilayer of the Ti film and the Mo interlayer was tightly bonded and was entirely exfoliated from the Si substrate during the scratch test.

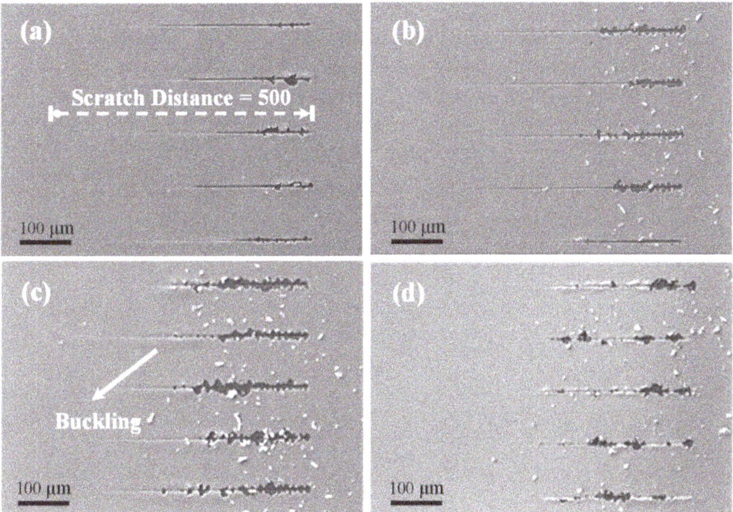

Figure 5. SEM images of the five scratch tracks performed on each of the (**a**) Mo-1, (**b**) Mo-2, (**c**) Mo-4, and (**d**) Mo-8 samples. The scratch tracks were explored from left to right in these SEM images to a maximal scratch distance of 500 μm.

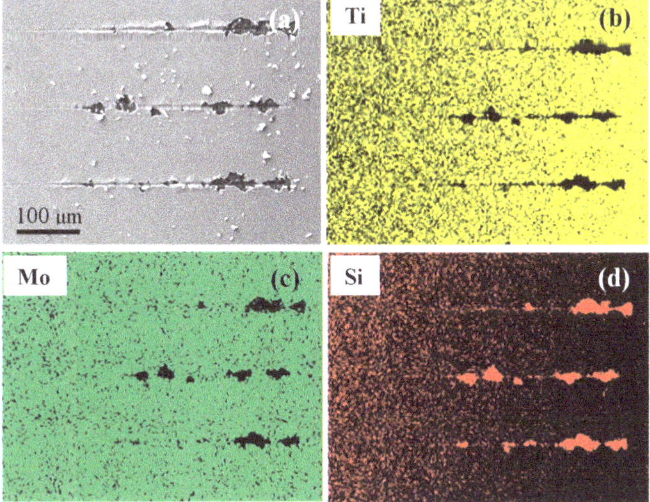

Figure 6. (Color online) (**a**) A magnified SEM image of three scratch tracks on the Mo-8 sample and the (**b**) Ti, (**c**) Mo, and (**d**) Si element distributions of (**a**).

Figure 7 depicts the relationship curves between the displacement into the surface and scratch distance recorded during the scratch experiments performed on the Mo-1, Mo-2, Mo-4, and Mo-8 samples. All four samples had a smooth surface with slight fluctuations in displacement in the first round of the scratch scan, which were beneficial for acquiring accurate scratch results. As for the Mo-1 and Mo-2 samples, the displacement steadily increased with increasing normal force load and scratch distance where the Mo/Ti films remained attached to the Si substrate. Once the film exfoliation had taken place, the displacement sharply increased with the scratch distance. For the Mo-4 and Mo-8 samples, the displacement started to fluctuate remarkably with the scratch distance when the film started to exfoliate. The nano-indenter-induced film delamination from the scratch track was an ideal way to evaluate the film adhesion on the substrate [31,49,51]; indentation methods generally rely on the onset scratch distance to cause the film to exfoliate. The average scratch distances and displacements, where the Mo/Ti film started to delaminate from the Si substrate, were recorded for all scratch tracks and have been tabulated in Table 2. It is notable that the film exfoliation occurred at the displacement where the scratch nano-indenter was away from the interface between the Mo interlayer and Si substrate. The weak cohesion between the Mo/Ti multilayer and Si substrate might have been due to the small roughness of the Si substrate. The scratch distance decreased from 301.1 ± 29.6 nm to 224.5 ± 22.7 nm with the increasing Mo interlayer thickness, suggesting that the thicker Mo interlayer reduced the cohesion between the Mo/Ti film and the Si substrate.

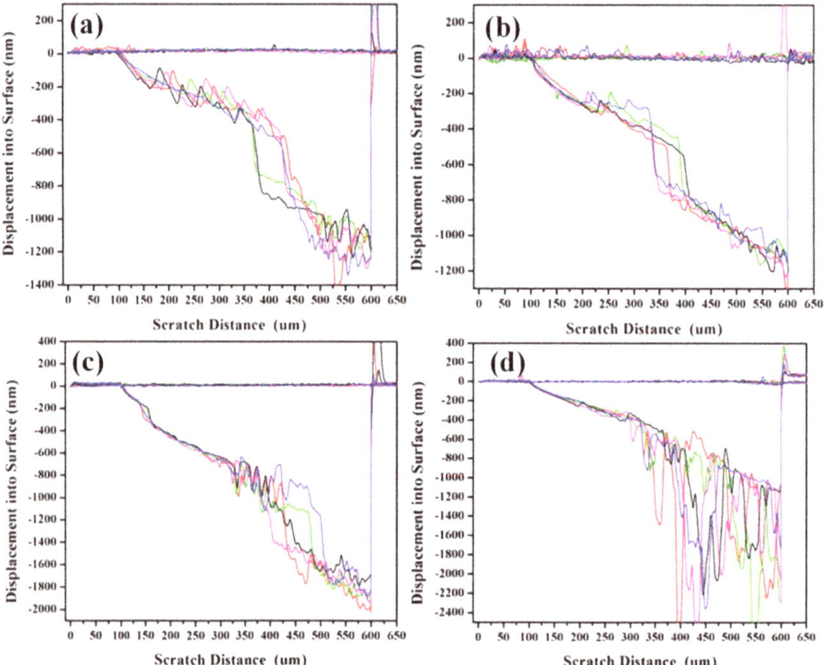

Figure 7. (Color online) The relationship between displacement into surface and scratch distance recorded during the scratch experiments performed for the (**a**) Mo-1, (**b**) Mo-2, (**c**) Mo-4, and (**d**) Mo-8 samples.

Table 2. The average scratch distance and displacement into the surface of five scratch tracks on the Si/Mo/Ti multilayer film surface where the Mo/Ti film started to delaminate from the Si substrate. The standard deviation was statistically calculated from the five scratch tracks for each sample.

Sample Number	Mo-1	Mo-2	Mo-4	Mo-8
Scratch distance (nm)	301.1 ± 29.6	262.1 ± 26.7	227.4 ± 8.7	224.5 ± 22.7
Displacement into surface (nm)	469.4 ± 51.0	433.8 ± 92.1	689.3 ± 27.6	419.6 ± 63.0

Figure 8a shows the XRD patterns as a function of ψ angle for the representative Mo-1 sample. Two diffraction peaks of Ti (112) and Ti (201) can be found in the XRD patterns with a 2θ range between 74.5° and 79.5° [4]. The Ti (112) peak shifted towards a lower 2θ angle with increasing ψ angle, suggesting that the Mo-1 sample had an in-plane tensile stress. The intensity of the Ti (112) peak decreased with increasing ψ angle, which was ascribed to the smaller number of grains detected at higher ψ angles. The relationships between the 2θ and ψ angles for the Mo-1, Mo-2, Mo-4, and Mo-8 samples were acquired. To clearly distinguish the relationship between residual stress and the Mo interlayer thickness, curves of $\Delta d/d$ versus $\sin^2\psi$ with different slopes were plotted for all four samples and are shown in Figure 8b. All date points for each of samples were linearly fitted, with the slope positively relating to the residual stress. It is apparent that the slopes of the fitted lines decreased with increasing Mo layer thickness. The residual stress values were calculated by the 2θ-$\sin^2\psi$ method, which has been reported elsewhere [12,23]. The Young's modulus and Poisson's ratio parameters used were 120.2 MPa and 0.361 [52], respectively. Table 3 summarizes the calculated residual stresses of all four Si/Mo/Ti multilayer films. It is interesting to note that the residual stress decreases from 686.4 ± 40.6 MPa to 257.6 ± 14.0 MPa with an increase in Mo interlayer thickness from 54.3 nm to 331.5 nm. It is well known that the residual stress of thin film strongly correlates to the film growth mechanism [44,53], substrate temperature [26], and the discrepancy in thermal expansion coefficient between film and substrate [53]. However, the thickness of the Mo interlayer would be the critical factor affecting the residual stresses in the Si/Mo/Ti multilayer films.

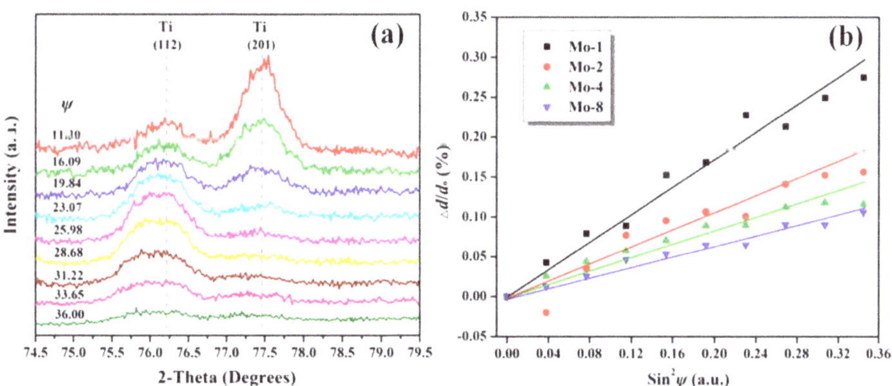

Figure 8. (Color online) (**a**) XRD patterns of the Mo-1 sample showing the peak shift of Ti (112) as a function of ψ angle. (**b**) In-plane residual stress derived from the XRD patterns in (a) using the $\sin^2\psi$ method for the Mo-1, Mo-2, Mo-4, and Mo-8 samples.

Table 3. The residual stresses of the Si/Mo/Ti multilayer films determined by XRD using the $\sin^2\psi$ method. The standard deviation was obtained by fitting the data points of $\Delta d/d$ verse $\sin^2\psi$ using the linear least square method.

Sample Number	Mo-1	Mo-2	Mo-4	Mo-8
Residual stress (MPa)	686.4 ± 40.6	395.1 ± 34.6	294.0 ± 22.2	257.6 ± 14.0

According to the XRD patterns and EDS maps shown in Figures 2 and 3, the Mo films with different thicknesses were normally grown on the Si substrates despite the smooth surface of the substrates. The Ti film could also be successfully grown on the Mo film if the Mo interlayer thickness was below 139.8 nm. However, the Mo/Ti multilayers started to exfoliate from the Si substrates during the subsequent Ti film deposition, which was due to the weak cohesion between the film and substrate once the Mo interlayer thickness was thicker than 139.8 nm, as demonstrated by the EDS result of the Mo-8a sample as shown in Figure 4. It was supposed that the Mo interlayer and Ti film were both in thermal expansion states, since the substrate temperature was high, at 700 °C. The thermal expansion coefficients of Si, Mo, and Ti are 4.0×10^{-6}/K [27], 4.8×10^{-6}/K, and 8.6×10^{-6}/K [54], respectively. The Ti film delaminated from the substrate during the Ti film deposition, which was most probably due to the higher thermal expansion coefficient of Ti compared to those of Mo and Si. The thermal expansion of the Ti film was constrained by the Mo interlayer caused the Ti film delamination [28,53]. The following Ti film was grown directly on the Si substrate to form the Ti-Si complex [42,43], which alternately reduced the quality of the Ti film. The growth of high-quality Mo film on Si was attributed to their similar thermal expansion coefficients.

Dauskardt et al. [55] has suggested that the debonding of multilayer thin film structures is driven by residual stresses, including the intrinsic growth stresses produced during deposition and the thermal expansion mismatch stresses. On the other hand, the thickness of the Mo interlayer was also one important factor affecting the cohesion between the Mo/Ti film and Si substrate, with the thicker Mo layer reducing their cohesion. Chason et al. [44] have reported that compressive stress keeps increasing with increasing film thickness, while Xi et al. [56] have found that residual stress transforms from compressive to tensile and that tensile stress increases with increasing film thickness. Leterrier et al. [29] have verified that the cohesive strength and crack onset strains of SiO_x film on polymer substrate decrease with increasing coating thickness. Roshangias et al. [31] have disclosed that the adhesion of TiW coatings on Si substrates decrease significantly with the increasing thickness of coatings. It can be concluded from these results that the compressive or tensile stress increases with film thickness, originating from the Ti film growth and the discrepancy of the thermal expansion coefficients of the Si, Mo, and Ti materials, which reduces the adhesive strength between the Ti film and Si substrate.

After deposition, the as-grown Si/Mo/Ti multilayer films were naturally cooled in a high vacuum environment of 2×10^{-4} Pa. The Ti and Mo films started to shrink during cooling. Similarly, the shrinkage of the Ti film must have been larger than that of the Mo film due to the higher thermal expansion coefficient of Ti metal. The shrinkage of the Ti film induced the generation of tensile residual stress [6,44], which was proven by XRD stress analysis. It has been demonstrated by Dehm et al. [28] that the difference in thermal expansion coefficients can induce thermal compressive stress in films if the thermal expansion coefficient of film material is higher than that of the substrate. However, the residual stress becomes tensile with the occurrence of isolated cluster coalescence [6,44] and the subsequent increasing of film thickness [28]. Chason et al. [44] and Floro et al. [45] have demonstrated that residual stress in films undergoes typical compressive and tensile steps. Film stress is compressive during the early stage of nucleation and later becomes tensile, which is associated with the formation of grain boundaries. The Mo/Ti multilayer film delaminates from the Si substrate if the film is not tightly bonded to the substrate. The XRD stress results demonstrated that the Si/Mo/Ti samples with thicker Mo interlayers have lower residual tensile stress. This could be explained by the fact that the released residual stress provided the driving force for the film exfoliation and the film exfoliation relaxed the residual tensile stress of the Ti film [55]. The above results suggest that the thermal expansion coefficient of the film and the substrate materials and the design of the film thickness are important factors affecting the film quality in the design of multilayer films.

4. Conclusions

In conclusion, pure Ti films were successfully grown on single crystal Si substrates by the introduction of Mo interlayers. The design of an Si/Mo/Ti multilayer film suppressed the chemical reaction between the Ti film and Si substrate. It was found that the thickness of the Mo interlayer played an important role in the cohesion between the Mo/Ti multilayer film and the Si substrate, which significantly decreased with increasing Mo interlayer thickness. All the Si/Mo/Ti multilayer films presented in-plane tensile residual stresses which might be due to the lattice expansion at a high film growth temperature of 700 °C and the discrepancy in thermal expansion coefficients between the Ti film and Si substrate. The tensile stress, derived from the XRD stress analysis results, decreased from 686.4 ± 40.6 MPa to 257.6 ± 14.0 MPa when the Mo interlayer thickness increased from 54.3 nm to 331.5 nm. The decreased tensile stress during the Si substrate cooling process provided the driving force to reduce the cohesion between the Mo/Ti film and Si substrate for the thicker Mo interlayer samples.

Author Contributions: H.S. directed the experiments and wrote the original manuscript. B.Y., J.Z., and X.Z. prepared the samples and performed the XRD, SEM, and nano-indentation measurements. X.Z., X.X., and X.Z. revised the manuscript.

Funding: This research was funded by the President Foundation of the China Academy of Engineering Physics (Grant No. YZJJLX2018003) and the National Natural Science Foundation of China (Grant No. 21601168).

Conflicts of Interest: The authors declare no conflict of interest.

References

1. Kalisvaart, W.P.; Niessen, R.A.H.; Notten, P.H.L. Electrochemical hydrogen storage in MgSc alloys: A comparative study between thin films and bulk materials. *J. Alloys Compd.* **2006**, *417*, 280–291. [CrossRef]
2. Pugh, R.D.; Sabochick, M.J.; Luke, T.E. A comparison of ferroelectric aging in bulk and thin-film materials. *J. Appl. Phys.* **1992**, *72*, 1049–1055. [CrossRef]
3. Yoder, K.B.; Stone, D.S.; Lin, J.C.; Hoffmann, R.A. Indentation Creep of Molybdenum: Comparison Between Thin Film and Bulk Material. *MRS Proc.* **2011**, *356*, 651. [CrossRef]
4. Zhou, X.S.; Chen, G.J.; Peng, S.M.; Long, X.G.; Liang, J.H.; Fu, Y.Q. Thermal desorption of tritium and helium in aged titanium tritide films. *Int. J. Hydrogen Energy* **2014**, *39*, 11006–11015. [CrossRef]
5. Parish, C.M.; Snow, C.S.; Kammler, D.R.; Brewer, L.N. Processing effects on microstructure in Er and ErD$_2$ thin-films. *J. Nucl. Mater.* **2010**, *403*, 191–197. [CrossRef]
6. Detor, A.J.; Hodge, A.M.; Chason, E.; Wang, Y.; Xu, H.; Conyers, M.; Nikroo, A.; Hamza, A. Stress and microstructure evolution in thick sputtered films. *Acta Mater.* **2009**, *57*, 2055–2065. [CrossRef]
7. Elsheikh, A.H.; Sharshir, S.W.; Ahmed Ali, M.K.; Shaibo, J.; Edreis, E.M.A.; Abdelhamid, T.; Du, C.; Haiou, Z. Thin film technology for solar steam generation: A new dawn. *Sol. Energy* **2019**, *177*, 561–575. [CrossRef]
8. Tian, K.; Baskaran, K.; Tiwari, A. Growth of two-dimensional WS$_2$ thin films by pulsed laser deposition technique. *Thin Solid Films* **2018**, *668*, 69–73. [CrossRef]
9. Venkata Subbaiah, Y.P.; Saji, K.J.; Tiwari, A. Atomically Thin MoS$_2$: A Versatile Nongraphene 2D Material. *Adv. Funct. Mater.* **2016**, *26*, 2046–2069. [CrossRef]
10. Novoselov, K.S.; Mishchenko, A.; Carvalho, A.; Castro Neto, A.H. 2D materials and van der Waals heterostructures. *Science* **2016**, *353*, 9439. [CrossRef]
11. Parish, C.M.; Snow, C.S.; Brewer, L.N. The manifestation of oxygen contamination in ErD$_2$ thin films. *J. Mater. Res.* **2009**, *24*, 1868–1879. [CrossRef]
12. Rodriguez, M.A.; Browning, J.F.; Frazer, C.S.; Snow, C.S.; Tissot, R.G.; Boespflug, E.P. Unit cell expansion in ErT$_2$ films. *Powder Diffr.* **2007**, *22*, 118–121. [CrossRef]
13. Shen, H.H.; Peng, S.M.; Long, X.G.; Zhou, X.S.; Liu, J.H.; Sun, K.; Yang, L.; Sun, Q.Q.; Zu, X.T. Influence of growth parameters on the microstructures of erbium films deposited on Si(111) substrates. *Vacuum* **2012**, *86*, 2075–2081. [CrossRef]
14. Zhu, H.; Dong, Z.; Niu, X.; Li, J.; Shen, K.; Mai, Y.; Wan, M. DC and RF sputtered molybdenum electrodes for Cu(In,Ga)Se$_2$ thin film solar cells. *Appl. Surf. Sci.* **2019**, *465*, 48–55. [CrossRef]

15. Liang, G.-X.; Luo, Y.-D.; Hu, J.-G.; Chen, X.-Y.; Zeng, Y.; Su, Z.-H.; Luo, J.-T.; Fan, P. Influence of annealed ITO on PLD CZTS thin film solar cell. *Surf. Coat. Technol.* **2019**, *358*, 762–764. [CrossRef]

16. Beltiukov, A.N.; Stashkova, E.V.; Boytsova, O.V. Anodic oxidation of Al/Ge/Al multilayer films. *Appl. Surf. Sci.* **2018**, *459*, 583–587. [CrossRef]

17. Çölmekçi, S.; Karpuz, A.; Köçkar, H. Total film thickness controlled structural and related magnetic properties of sputtered Ni/Cu multilayer thin films. *J. Magn. Magn. Mater.* **2019**, *478*, 48–54. [CrossRef]

18. Li, N.; Fu, E.G.; Wang, H.; Carter, J.J.; Shao, L.; Maloy, S.A.; Misra, A.; Zhang, X. He ion irradiation damage in Fe/W nanolayer films. *J. Nucl. Mater.* **2009**, *389*, 233–238. [CrossRef]

19. Singh, J.P.; Lim, W.C.; Gautam, S.; Asokan, K.; Chae, K.H. Swift heavy ion irradiation induced effects in Fe/MgO/Fe/Co multilayer. *Mater. Des.* **2016**, *101*, 72–79. [CrossRef]

20. Callisti, M.; Karlik, M.; Polcar, T. Competing mechanisms on the strength of ion-irradiated Zr/Nb nanoscale multilayers: Interface strength versus radiation hardening. *Scr. Mater.* **2018**, *152*, 31–35. [CrossRef]

21. Du, S.; Zhang, K.; Wen, M.; Ren, P.; Meng, Q.; Hu, C.; Zheng, W. Tribochemistry dependent tribological behavior of superhard TaC/SiC multilayer films. *Surf. Coat. Technol.* **2018**, *337*, 492–500. [CrossRef]

22. Zhou, Q.; Ren, Y.; Du, Y.; Hua, D.P.; Han, W.C.; Xia, Q.S. Adhesion energy and related plastic deformation mechanism of Cu/Ru nanostructured multilayer film. *J. Alloys Compd.* **2019**, *772*, 823–827. [CrossRef]

23. Postolnyi, B.O.; Beresnev, V.M.; Abadias, G.; Bondar, O.V.; Rebouta, L.; Araujo, J.P.; Pogrebnjak, A.D. Multilayer design of CrN/MoN protective coatings for enhanced hardness and toughness. *J. Alloys Compd.* **2017**, *725*, 1188–1198. [CrossRef]

24. Pshyk, A.; Coy, E.; Kempiński, M.; Iatsunskyi, I.; Załęski, K.; Pogrebnjak, A.; Jurga, S. Microstructure, phase composition and mechanical properties of novel nanocomposite (TiAlSiY)N and nano-scale (TiAlSiY)N/MoN multifunctional heterostructures. *Surf. Coat. Technol.* **2018**, *350*, 376–390. [CrossRef]

25. Mellali, M.; Fauchais, P.; Grimaud, A. Influence of substrate roughness and temperature on the adhesion/cohesion of alumina coatings. *Surf. Coat. Technol.* **1996**, *81*, 275–286. [CrossRef]

26. Pershin, V.; Lufitha, M.; Chandra, S.; Mostaghimi, J. Effect of substrate temperature on adhesion strength of plasma-sprayed nickel coatings. *J. Therm. Spray Technol.* **2003**, *12*, 370–376. [CrossRef]

27. Takimoto, K.; Fukuta, A.; Yamamoto, Y.; Yoshida, N.; Itoh, T.; Nonomura, S. Linear thermal expansion coefficients of amorphous and microcrystalline silicon films. *J. Non-Cryst. Solids* **2002**, *299-302*, 314–317. [CrossRef]

28. Dehm, G.; Weiss, D.; Arzt, E. In situ transmission electron microscopy study of thermal-stress-induced dislocations in a thin Cu film constrained by a Si substrate. *Mater. Sci. Eng. A* **2001**, *309-310*, 468–472. [CrossRef]

29. Leterrier, Y.; Andersons, J.; Pitton, Y.; Månson, J.-A.E. Adhesion of silicon oxide layers on poly(ethylene terephthalate). II: Effect of coating thickness on adhesive and cohesive strengths. *J. Polym. Sci. Pol. Phys.* **1997**, *35*, 1463–1472. [CrossRef]

30. Kumar, D.D.; Kumar, N.; Kalaiselvam, S.; Thangappan, R.; Jayavel, R. Film thickness effect and substrate dependent tribo-mechanical characteristics of titanium nitride films. *Surf. Interfaces* **2018**, *12*, 78–85. [CrossRef]

31. Roshangias, A.; Pelzer, R.; Khatibi, G.; Steinbrenner, J. Thickness dependency of adhesion properties of TiW thin films. In Proceedings of the 2014 IEEE 16th Electronics Packaging Technology Conference (EPTC), Singapore, 3–5 December 2014; pp. 192–195.

32. Coy, E.; Yate, L.; Kabacińska, Z.; Jancelewicz, M.; Jurga, S.; Iatsunskyi, I. Topographic reconstruction and mechanical analysis of atomic layer deposited Al_2O_3/TiO_2 nanolaminates by nanoindentation. *Mater. Des.* **2016**, *111*, 584–591. [CrossRef]

33. Goyenola, C.; Gueorguiev, G.K.; Stafström, S.; Hultman, L. Fullerene-like CS_x: A first-principles study of synthetic growth. *Chem. Phys. Lett.* **2011**, *506*, 86–91. [CrossRef]

34. Ren, Y.; Liu, X. Strain and Cohesive Energy of TiN Deposit on Al(001) Surface: Density Functional Calculation. *Int. J. Nanosci.* **2016**, *15*, 1650017. [CrossRef]

35. Jiménez, C.; Garcia-Moreno, F.; Pfretzschner, B.; Klaus, M.; Wollgarten, M.; Zizak, I.; Schumacher, G.; Tovar, M.; Banhart, J. Decomposition of TiH_2 studied in situ by synchrotron X-ray and neutron diffraction. *Acta Mater.* **2011**, *59*, 6318–6330. [CrossRef]

36. Corgnale, C.; Hardy, B.; Motyka, T.; Zidan, R.; Teprovich, J.; Peters, B. Screening analysis of metal hydride based thermal energy storage systems for concentrating solar power plants. *Renew. Sustain. Energy Rev.* **2014**, *38*, 821–833. [CrossRef]

37. Xiaosong, Z.; Xinggui, L.; Lin, Z.; Shuming, P.; Shunzhong, L. X-ray diffraction analysis of titanium tritide film during 1600 days. *J. Nucl. Mater.* **2010**, *396*, 223–227. [CrossRef]

38. Wang, H.; Peng, S.; Zhou, X.; Long, X.; Shen, H. Evolution of 3He bubble microstructure in TiT_2 films during aging. *J. Nucl. Mater.* **2018**, *509*, 700–706. [CrossRef]

39. Shen, H.H.; Peng, S.M.; Zhou, X.S.; Sun, K.; Wang, L.M.; Zu, X.T. Microstructure evolution of zircaloy-4 during Ne ion irradiation and annealing: An in-situ TEM investigation. *Chin. Phys. B* **2014**, *23*, 036102. [CrossRef]

40. Zhou, X.S.; Liu, Q.; Zhang, L.; Peng, S.M.; Long, X.G.; Ding, W.; Cheng, G.J.; Wang, W.D.; Liang, J.H.; Fu, Y.Q. Effects of tritium content on lattice parameter, ^{3}He retention, and structural evolution during aging of titanium tritide. *Int. J. Hydrogen Energy* **2014**, *39*, 20062–20071. [CrossRef]

41. Snow, C.S.; Browning, J.F.; Bond, G.M.; Rodriguez, M.A.; Knapp, J.A. ^{3}He bubble evolution in ErT_2: A survey of experimental results. *J. Nucl. Mater.* **2014**, *453*, 296–306. [CrossRef]

42. Juang, M.H.; Cheng, H.C. Novel annealing scheme for fabricating high-quality Ti-silicided shallow n+p junction by P^+ implantation into thin Ti films on Si substrate. *Appl. Phys. Lett.* **1992**, *60*, 1579–1581. [CrossRef]

43. Bensch, W.; Pamler, W. The formation of titanium silicides by rapid thermal processing: An XRD and AES study. *React. Solids* **1989**, *7*, 249–262. [CrossRef]

44. Chason, E.; Sheldon, B.W.; Freund, L.B.; Floro, J.A.; Hearne, S.J. Origin of compressive residual stress in polycrystalline thin films. *Phys. Rev. Lett.* **2002**, *88*, 156103. [CrossRef] [PubMed]

45. Floro, J.A.; Chason, E.; Cammarata, R.C.; Srolovitz, D.J. Physical Origins of Intrinsic Stresses in Volmer–Weber Thin Films. *MRS Bull.* **2011**, *27*, 19–25. [CrossRef]

46. Shen, H.H.; Peng, S.M.; Long, X.G.; Zhou, X.S.; Yang, L.; Zu, X.T. The effect of substrate temperature on the oxidation behavior of erbium thick films. *Vacuum* **2012**, *86*, 1097–1101. [CrossRef]

47. Shen, H.H.; Peng, S.M.; Xiang, X.; Naab, F.N.; Sun, K.; Zu, X.T. Proton irradiation effects on the precipitate in a Zr–1.6Sn–0.6Nb–0.2Fe–0.1Cr alloy. *J. Nucl. Mater.* **2014**, *452*, 335–342. [CrossRef]

48. Yang, X.; Qiu, Z.; Li, X. Investigation of scratching sequence influence on material removal mechanism of glass-ceramics by the multiple scratch tests. *Ceram. Int.* **2019**, *45*, 861–873. [CrossRef]

49. Kleinbichler, A.; Pfeifenberger, M.J.; Zechner, J.; Wöhlert, S.; Cordill, M.J. Scratch induced thin film buckling for quantitative adhesion measurements. *Mater. Des.* **2018**, *155*, 203–211. [CrossRef]

50. Meneses-Amador, A.; Jiménez-Tinoco, L.F.; Reséndiz-Calderon, C.D.; Mouftiez, A.; Rodríguez-Castro, G.A.; Campos-Silva, I. Numerical evaluation of scratch tests on boride layers. *Surf. Coat. Technol.* **2015**, *284*, 182–191. [CrossRef]

51. Lee, A.; Clemens, B.M.; Nix, W.D. Stress induced delamination methods for the study of adhesion of Pt thin films to Si. *Acta Mater.* **2004**, *52*, 2081–2093. [CrossRef]

52. Gale, W.F.; Totemeier, T.C. Elastic properties, damping capacity and shape memory alloys. In *Smithells Metals Reference Book*, Eighth Edition; Gale, W.F., Totemeier, T.C., Eds.; Butterworth-Heinemann: Oxford, UK, 2004; pp. 1–45.

53. Hsueh, C.H. Thermal stresses in elastic multilayer systems. *Thin Solid Films* **2002**, *418*, 182–188. [CrossRef]

54. Speight, J.G. Inorganic Chemistry. In *Lange's Handbook of Chemistry*, 16th ed.; McGraw-Hill: New York, NY, USA, 2005; pp. 1.128–1.131.

55. Dauskardt, R.H.; Lane, M.; Ma, Q.; Krishna, N. Adhesion and debonding of multi-layer thin film structures. *Eng. Fract. Mech.* **1998**, *61*, 141–162. [CrossRef]

56. Xi, Y.; Gao, K.; Pang, X.; Yang, H.; Xiong, X.; Li, H.; Volinsky, A.A. Film thickness effect on texture and residual stress sign transition in sputtered TiN thin films. *Ceram. Int.* **2017**, *43*, 11992–11997. [CrossRef]

Article

Fabrication, Characterization, and Properties of Poly (Ethylene-Co-Vinyl Acetate) Composite Thin Films Doped with Piezoelectric Nanofillers

Giulia Mariotti and Lorenzo Vannozzi *

The BioRobotics Institute, Scuola Superiore Sant'Anna, 56025 Pontedera (PI), Italy
* Correspondence: lorenzo.vannozzi@santannapisa.it; Tel.: +39-050-883091

Received: 11 July 2019; Accepted: 14 August 2019; Published: 20 August 2019

Abstract: Ethylene vinyl acetate (EVA) is a copolymer comprehending the semi-crystalline polyethylene and amorphous vinyl acetate phases, which potentially allow the fabrication of tunable materials. This paper aims at describing the fabrication and characterization of nanocomposite thin films made of polyethylene vinyl acetate, at different polymer concentration and vinyl acetate content, doped with piezoelectric nanomaterials, namely zinc oxide and barium titanate. These membranes are prepared by solvent casting, achieving a thickness in the order of 100–200 µm. The nanocomposites are characterized in terms of morphological, mechanical, and chemical properties. Analysis of the nanocomposites shows the nanofillers to be homogeneously dispersed in EVA matrix at different vinyl acetate content. Their influence is also noted in the mechanical behavior of thin films, which elastic modulus ranged from about 2 to 25 MPa, while keeping an elongation break from 600% to 1500% and tensile strength from 2 up to 13 MPa. At the same time, doped nanocomposite materials increase their crystallinity degree than the bare ones. The radiopacity provided by the addition of the dopant agents is proven. Finally, the direct piezoelectricity of nanocomposites membranes is demonstrated, showing higher voltage outputs (up to 2.5 V) for stiffer doped matrices. These results show the potentialities provided by the addition of piezoelectric nanomaterials towards mechanical reinforcement of EVA-based matrices while introducing radiopaque properties and responsiveness to mechanical stimuli.

Keywords: nanomaterial; zinc oxide; barium titanate; composite; ethylene vinyl acetate; elastic modulus; toughness; flexural rigidity; radiopacity; piezoelectricity

---

1. Introduction

Recently, nanomaterials are becoming important players in many different technological domains. In fact, nanoelements may enhance the properties of several bulk materials, especially for biomedical applications. The synergy between polymers and inorganic nanofillers can bring to the enhancement of macroscopic properties of neat materials, such as material toughness, thermal resistance, electrical conductivity, optical properties, piezoelectricity, etc. [1]. Specific filler features have been exploited in order to provide bulk materials with responsive and smart properties upon the stimulation offered by an external source for targeted drug delivery purposes [2] as well as for tissue engineering [3–5]. For example, our group has already proved the interaction between ultrasound-mediated stimulation and piezoelectric nanocomposite in the field of drug delivery [6].

The inclusion of nanomaterials within polymeric thin films has been identified as a promising strategy for designing functional nanocomposite materials. The thin film technology enables the fabrication of polymeric membranes which thickness can range from few nanometers up to hundreds of micrometers. The relatively high aspect ratio due to the large surface area (few mm or even cm) with respect to the low thickness makes them compliant matrices able to adjust their shape when adhered to a wide range of topographic surfaces. Such combination of multiple properties (e.g. relatively high flexibility, large surface area, high aspect ratio) makes them unique to be used as drug-loaded platforms, actuators, and sensors [7].

The ethylene vinyl acetate (EVA) co-polymer, which is vinyl acetate (VA) block-copolymerized with ethylene, has increased its appealing for possible application within the biomedical field during recent years [8], as for example in controlled release of drugs [9]. EVA has been tested as main component of drug release devices to be used in harsh human departments as the vagina duct [10] and the gastrointestinal tract [11].

Depending on the chased scope, the addition of nanomaterials within the EVA matrix may allow the modification of original matrix properties. However, these properties depend on different features such as size and shape of nanomaterials, EVA polarity, crystallinity, and degree of dispersion of the nanofiller in the polymeric matrix. For example, EVA at different content of VA has been mixed with montmorillonite, which increases the nanocomposite mechanical properties with respect to the bare polymer [12]. In fact, the authors found that both elastic modulus and mechanical strength are a combined function of the clay concentration, proving an increment of the elastic modulus up to five times. Similarly, the addition of halloysites improves the composite mechanical properties, leading to an increase of the Young's modulus and tensile strength while increasing the nanofiller concentration. Apart from the mechanical behavior, the introduction of clays also enhances both water resistance and permeability to oxygen [13]. Multi-walled carbon nanotubes correspond to another common nanomaterial type widely explored as dopant of polymeric matrices. Such nanofiller can be used to alter both the mechanical behavior and the rheology of EVA in its melt phase. Indeed, EVA can behave as non-Newtonian fluids for higher doping concentrations and as Newtonian fluid for low doping concentrations (up to 0.5% wt.) [14]. Another carbon-based nanomaterial that has been combined with EVA matrices is the graphite [15]. The authors investigated the effect of the nanofiller size and the type of expanded graphite on the thermal behavior and electrical conductivity of EVA nanocomposites. The nanocomposite matrix containing the nanofiller with higher aspect ratio exhibits a stronger strengthening effect due to a higher crystallinity degree. Another type of nanomaterial proposed for doping EVA matrices is the magnetic one. For example, Fe_3O_4 nanoparticles can be dispersed in EVA matrices, leading to a reduction of both elongation at break and impact strength, while enhancing the material hardness with respect to neat EVA [16]. The scientific literature is also filled with plenty of examples in which multiple nanofillers have been added in EVA nanocomposites. For example, multiwalled carbon nanotube and montmorillonite are proposed as dopant agents in EVA nanocomposites [17]. Their synergy is demonstrated by resulting mechanical reinforcement of EVA-based matrices.

Despite the great interest in the study of composite materials doped with different type of clays, carbon-based and metallic nanomaterials, the interaction between EVA polymers and ceramic materials has not been widely explored. Only few research studies report possible relations between EVA and piezoelectric nanomaterials. Among them, surface-modified $BaTiO_3$ nanoparticles have been only exploited to modify the electrical and thermophysical properties of vulcanized EVA [18], as well as ZnO nanoparticles [19]. To the best of our knowledge, the characterization of both mechanical and piezoelectric properties of EVA-based nanocomposite has not been reported, so far.

This paper reports the fabrication and characterization of nanocomposite EVA thin films doped with piezoelectric nanomaterials. These matrices were evaluated in terms of their morphological, mechanical, thermal, radiopaque, and piezoelectric properties. At this scope, a series of EVA copolymers has been investigated (18%, 25%, and 40% of VA), varying their concentration (5% and 10% wt. in toluene) and doping them with two types of piezoelectric nanomaterials, namely barium titanate ($BaTiO_3$) nanoparticles and zinc oxide (ZnO) nanopowder, at different concentrations (10% and 20% wt.).

2. Materials and Methods

2.1. Materials

Polyethylene vinyl acetate (EVA, vinyl acetate content: 18%, 25%, and 40%) and ZnO nanopowder (diameter less than 100 nm) were purchased from Sigma-Aldrich (Merck KGaA, Darmstadt, Germany). $BaTiO_3$ nanoparticles (nominal diameter: 300 nm, purity > 99.9%) were purchased from Nanostructured & Amorphous Materials (Houston, TX, USA). Nanomaterials were imaged through a Dual Beam microscopy workstation (Figure S1).

2.2. Thin Film Fabrication

The EVA copolymers (18%, 25%, and 40% of VA, named EVA18, EVA25, and EVA40) were dissolved in toluene (5% and 10% wt., namely 5EVA and 10EVA) by heating the hot plate up to 60 °C to favor the material dissolution while magnetically stirring the solution. Then, in case of nanocomposite materials, ZnO or $BaTiO_3$ were added to the polymer solution (10% and 20% wt. with respect to the polymer content) and sonicated with an ultrasonic bath for 1 h while keeping the temperature higher than 40 °C. Nanocomposite membranes were prepared by casting. Each solution (3 mL) was casted on a glass Petri dish (diameter: 60 mm) and let to evaporate for 24 h within a chemical hood.

2.3. Thickness and Morphological Evaluation

The thickness was measured by means of a profilometer (KLA Tencor, Milpitas, CA, USA). For each sample type, five independent samples were tested.

The top film surface was imaged by means of a scanning electron microscopy (SEM, EVO MA 15, Carl Zeiss, Oberkochen, Germany). SEM scans were carried out by setting a beam voltage of 10 kV and a current of 30 pA. This allowed investigating the material surface, for identifying the nanomaterial dispersion within the polymer matrix. The acquired images were analyzed in order to calculate the interparticle distance (l) in each composite formulation, according to the following equation [20]:

$$l = d[(\frac{\pi}{6V})^{\frac{1}{3}} - 1],\tag{1}$$

where V and d are the volume fraction and diameter of the nanofiller.

2.4. Mechanical Characterization

Mechanical testing was performed on nanocomposite EVA membranes by means of a traction machine (Instron 2444, load cell ± 1 kN): Material Young's modulus, elongation at break, ultimate strength and toughness were assessed for each sample type. The specimens were stretched at a constant load speed of 10 mm/min following ASTM D638. The elastic modulus was calculated by analyzing the first linear region of the stress-strain curves (deformation up to 10%), according as follows:

$$\sigma = \frac{F}{hb},\tag{2}$$

and

$$\varepsilon = \frac{L - L_0}{L_0},\tag{3}$$

where σ, F, h, b, ε, L, and L_0 are the stress, tensile load, thickness, width, tensile strain, length when stressed and initial length, respectively. The test was repeated on eight different samples for each material formulation.

Flexural rigidity (D) was expressed as:

$$D = \frac{Eh^3}{12(1 - v^2)},\tag{4}$$

where E and v are the Young's modulus and Poisson's ratio (defined as 0.35, [21]), respectively.

2.5. Differential Scanning Calorimetry

The thermal behavior (melting point and crystallinity) of the selected 10EVA formulations was investigated by Differential Scanning Calorimetry (DSC). The thermograms were recorded on a Mettler Toledo DSC1 Star System instrument (Greifensee, Switzerland). For each measurement, ~10 mg of material was placed in a standard aluminum sealed capsule and underwent a specific thermal cycle, described as follows: (1) Heating from −30 to 180 °C at a heating rate of 10 °C/min; (2) cooling to −30 °C at a cooling rate of 10 °C/min; (3) heating from −30 to 180 °C at a heating rate of 10 °C/min. The material melting point was obtained from the second heating run, while crystallinity was calculated with reference to the enthalpy of fusion of the perfect polyethylene crystal (277.1 J/g, [11]). The test was repeated on three different samples for each material formulation.

2.6. Radiopacity Measurement

For the radiopacity test, the exposure parameters were set up at 50–60 kV (tube voltage), 100 mA, and 0.063 s. The wavelength was set to 2.5×10^{-11} m. The object to focus distance was 15 cm. The radiographs were processed, and a digital image of the radiograph was obtained. The signal-to-noise ratio of the tested materials were analyzed using specific imageJ software (ImageJ version 1.51i, National institutes of Health, Bethesda, MD, USA). Numbers between 0 (pure black) and 255 (pure white) were assigned accordingly, and the signal-to-noise ratio was calculated by subdividing the grey value for the background. In this experiment, only 10EVA40 samples with the highest content of ZnO and $BaTiO_3$ were analyzed to verify the radiopacity introduced by the nanofiller type.

2.7. Electromechanical Response Analysis

Two electrodes made of a bilayer of Ti (~10 nm) and Au (~100 nm) were deposited in the opposite planar faces of 10EVA thin films through sputtering. A pressure of 400 kPa has been applied formulation using a cylinder with a diameter of 4 mm to test the electromechanical response of the thin film, connected to a linear rail. To verify the piezoelectric behavior of the nanocomposite membranes, signals were also acquired by inverting the poles of acquisition (Figure S2). The electromechanical response of each nanocomposite EVA matrices was analyzed by using a custom-made circuit, and signals were processed in Matlab (2018a).

2.8. Statistical Analysis

Normal data were reported as average value ± standard deviation. Data were analyzed through a one-way ANOVA with Tukey's post-test (GraphPad Prism v6). Statistically significant differences among sample types were defined through a significance threshold set at 5% (** = $p < 0.01$, * = $p < 0.05$).

3. Results and Discussion

3.1. Thin Film Fabrication: Analysis of Thickness and Nanomaterial Dispersion

Composite EVA thin films were successfully fabricated by film casting in a Petri dish with specific dimensions and controlling the volume of deposition. The results in terms of thickness are summarized in Figure 1.

The achieved thickness ranges from about 100 to 200 μm, without significant differences between the different formulations. Despite testing of EVA solutions at different polymer concentration (5% and 10% wt.), the obtained thicknesses are not statistically different. This demonstrates that the polymer concentration in toluene is not an effective factor for significantly differing the thickness among all the nanocomposite formulations tested. Furthermore, the film casting procedure allowed the fabrication of thin films with a relatively flat surface. The rather high polymer concentrations and the low boiling point of toluene do not allow the formation of macropores during solvent evaporation, which are not present on the surface.

After the fabrication step, SEM images were acquired to qualitatively investigate the dispersion of nanomaterials within the polymeric matrices. A representation of the surface of all 10EVA formulations is reported in Figure 2.

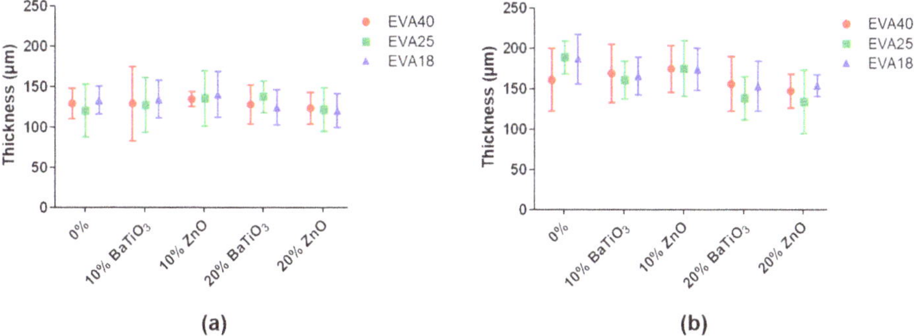

(a)   **(b)**

Figure 1. Thickness of the nanocomposite ethylene vinyl acetate (EVA) thin film for the EVA concentration of 5% wt. (**a**) and 10% wt. (**b**).

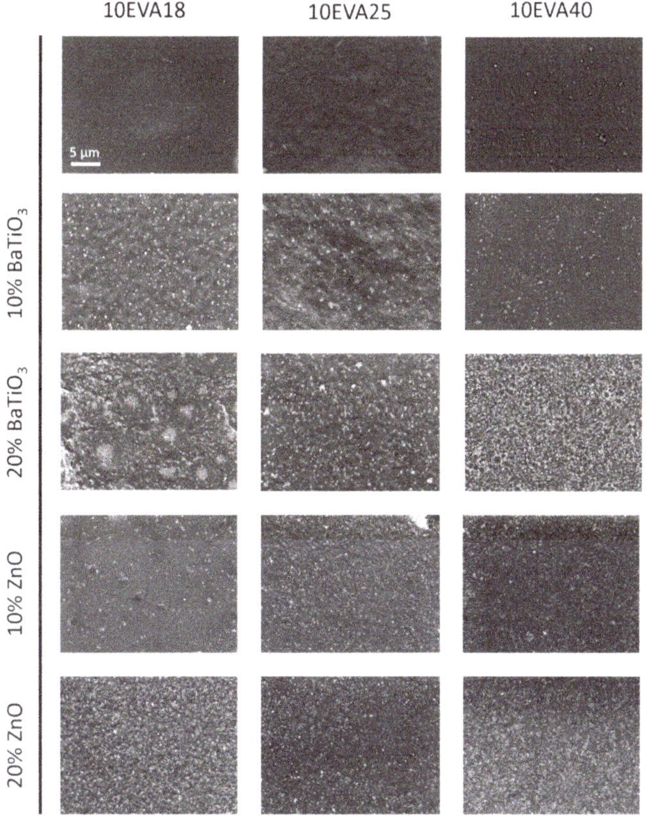

Figure 2. Scanning electron microscopy (SEM) images of each 10EVA formulation.

Both nanomaterials (BaTiO$_3$ and ZnO) result homogeneously dispersed in each EVA formulation without any evident sign of aggregation. In addition, the evaporation of the solvent did not cause any formation of macropores on the material surface, preserving a relatively flat topography. A similar trend was found for all the nanocomposite formulations at 5% of EVA (data not shown). The high shear forces introduced by ultrasonic energy during the nanocomposite polymeric solution were enough to uniformly and stably separate nanomaterials in the solution. Furthermore, a good interfacial interaction between the nanomaterial and the polymer allowed the good dispersion of both dopant agents. From the images, it can be noted the higher number of ZnO elements, approximately three times more the amount of BaTiO$_3$ particles, despite the two nanomaterials were equally weighted during the phase of preparation. The mentioned difference in terms of number of particles is mainly caused by the smaller diameter of ZnO nanopowder (nominally about 1/3 of the diameter of BaTiO$_3$ nanoparticles) because these materials have similar densities (5.61 g/cm^3 for ZnO and 6.02 g/cm^3 for BaTiO$_3$). This was also demonstrated by the estimation of the interparticle distance (Equation (1)), that is 217 nm (10% wt.) and 153 nm (20% wt.) for the ZnO, and 675 nm (10% wt.) and 478 nm (20% wt.) for BaTiO$_3$, respectively.

3.2. Mechanical Properties

Each nanocomposite material formulation underwent traction tests. As already familiar from the scientific literature [22], EVA mechanical properties are influenced by the weight percent of VA. Figure 3a shows representative stress-strain curves of EVA18, EVA25, and EVA40 derived from Equations (2) and (3), with a magnification on the first 20% of deformation (Figure 3b).

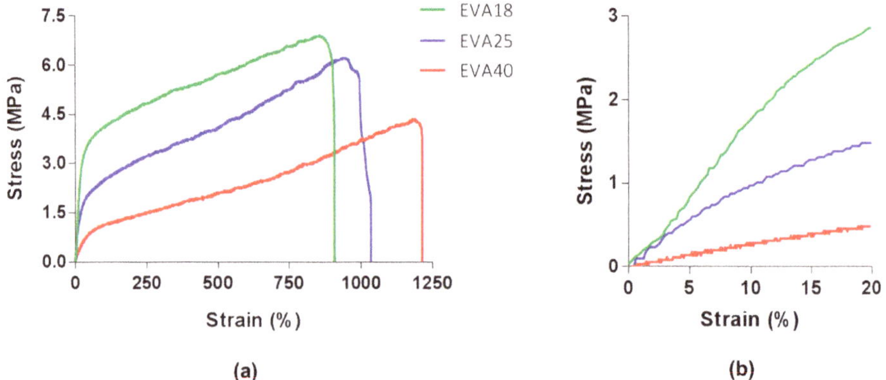

Figure 3. (**a**) Representative stress-strain curves of ethylene vinyl acetate (EVA) formulations at different vinyl acetate (VA) contents, and (**b**) a magnification of the first 20% of deformation.

Tensile tests allowed getting insights on many mechanical parameters of nanocomposite materials. The elastic modulus (Figure 4), the elongation at break (Figure 5), and the tensile strength (Figure 6) were analyzed, and data are summarized in Table 1.

The analysis of the mechanical properties provides interesting understandings on the mechanical behavior of nanocomposite EVA matrices. As shown in Figure 4, the stiffness of the nanocomposite polymer mainly increases by decreasing the content of vinyl acetate within the EVA polymer, thus the polarity of the matrix (from EVA40 to EVA18). Instead, the variation of the polymer concentration (5% and 10% wt. in toluene) does not involve a significant change in the bare material stiffness. A similar result was also reported by Faker et al., who analyzed the mechanical behavior of EVA18 dependently on polymer concentration [23].

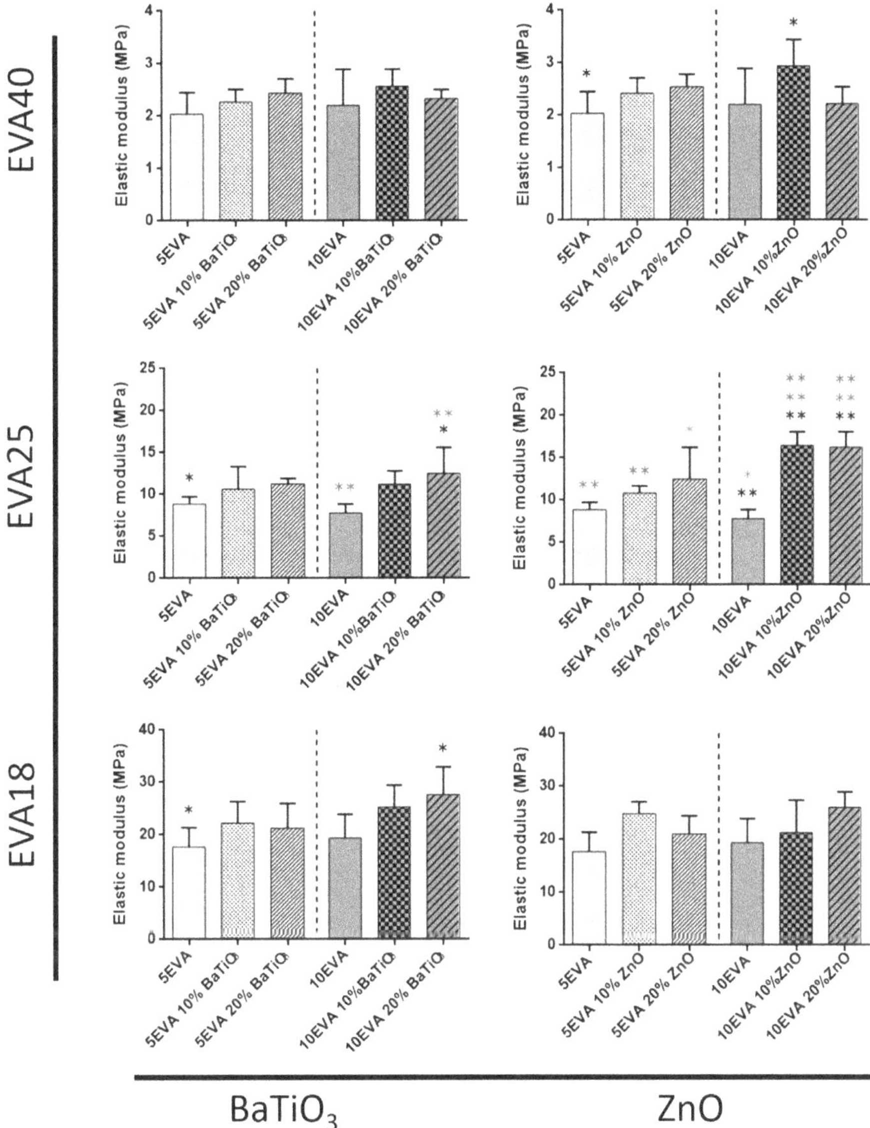

Figure 4. Elastic modulus of nanocomposite EVA-based matrices subdivided for the content of vinyl acetate and the ceramic nanomaterial included. * = $p < 0.05$, ** = $p < 0.01$. Different colors have been used to evidence the statistical significance between the compared cases in each graph.

On the one hand, the introduction of nanomaterials altered the mechanical properties of such nanocomposite matrices, showing in some cases an effective interaction between the polymer and the ceramic nanofiller. Data show a minimal influence due to the addition of nanomaterials while using EVA18 and EVA40 as polymeric matrix. On the other hand, the matrices based on EVA25 are strongly influenced by both nanofillers, leading to significant changes in the elastic modulus up to two times (10EVA25 vs. 10EVA25 10% and 20% ZnO). Generally, material mechanical properties increase in each material formulation.

The same materials were also analyzed in terms of tensile strength, and a summary of the results is reported in Figure 5.

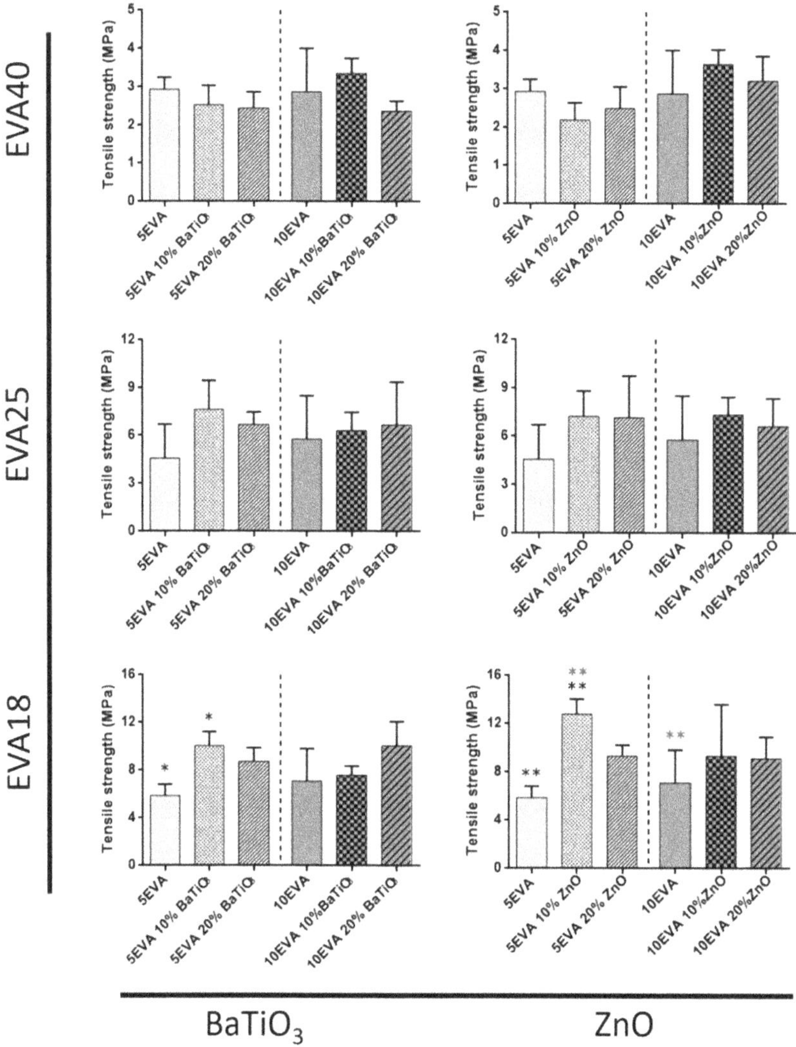

Figure 5. Tensile strength of nanocomposite EVA-based matrices subdivided for the content of vinyl acetate and the ceramic nanomaterial included. * = $p < 0.05$, ** = $p < 0.01$. Different colors have been used to evidence the statistical significance between the compared cases in each graph.

The tensile strength of nanocomposite polymers generally increases with the polarity of the matrix, without any significant effect due to the introduction of nanomaterials. In this case, the only exception corresponds to the case of EVA18, in which the addition of 10% ZnO strongly impacted on the 5EVA18 formulation. On the other hand, the tensile strength slightly decreases in the EVA40 compositions, especially when the 5EVA polymer was tested.

Finally, the elongation at break of all the EVA formulations is reported in Figure 6.

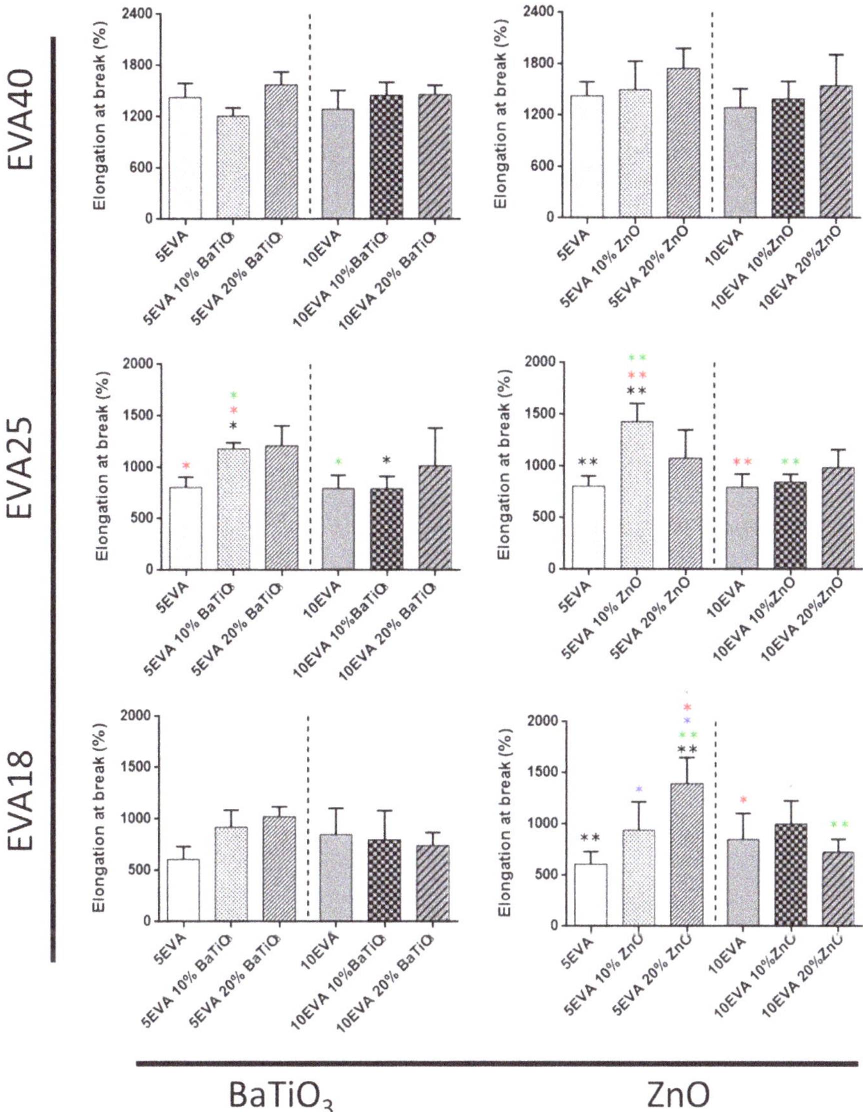

Figure 6. Elongation at break of nanocomposite EVA-based matrices subdivided for the content of vinyl acetate and the ceramic nanomaterial included. * = $p < 0.05$, ** = $p < 0.01$. Different colors have been used to evidence the statistical significance between the compared cases in each graph.

The elongation at break of the nanocomposites generally decreases with the polarity of the matrix. The addition of nanomaterials positively contributed to improve such EVA feature. Indeed, the chain mobility of the polymeric macromolecules is affected by the incorporation of both nanomaterials, leading to a general maintenance of elongation rates. The addition of ZnO in polymeric matrices has been demonstrated to positively influence the elongation at break of the composite materials [24]. Here, the only exception is represented by the softer formulation (EVA40), for which this effect was not significantly visible.

Finally, the stress-strain curves allowed also to get further insights on the toughness of all the material formulations (Table 1). The toughness is generally higher for material formulations made of EVA18 and EVA25, and this feature is further improved by the addition of the nanofiller (up to 100 MPa) probably because of the achievement of a homogenous dispersion and good adhesion between matrix and nanoparticles. The differences are more highlighted while testing the polymer content of 5% (5EVA). For EVA40-based matrices ones, values are not statistically relevant in all cases.

The last analyzed parameter is the flexural rigidity (Equation (4)). This is dependent on the VA content, thus the stiffness, even if the predominant role is assumed by the thickness. For such reason, the flexural rigidity of 10EVA-based thin films result higher than 5EVA ones due to the formation of thicker films. Results are summarized in Table 1.

Table 1. Mechanical analysis on the different material formulations: elastic modulus, tensile strength, elongation at break, toughness, and flexural rigidity.

Polymer Formulation	Elastic Modulus (MPa)	Tensile Strength (MPa)	Elongation at Break (%)	Toughness (MPa)	Flexural Rigidity (MPa)
5EVA18	20.85 ± 5.51	5.89 ± 0.94	611 ± 121	34.44 ± 7.36	4.00×10^{-6}
5EVA18 10% BaTiO$_3$	24.11 ± 1.91	10.05 ± 1.18	922 ± 167	55.54 ± 6.14	5.18×10^{-6}
5EVA18 20% BaTiO$_3$	21.29 ± 4.70	8.75 ± 1.13	1026 ± 95	65.61 ± 16.07	6.59×10^{-6}
5EVA18 10% ZnO	24.86 ± 2.19	12.80 ± 1.24	944 ± 274	79.28 ± 13.48	3.92×10^{-6}
5EVA18 20% ZnO	22.12 ± 3.18	9.33 ± 0.90	1396 ± 256	100.33 ± 15.14	3.48×10^{-6}
10EVA18	20.44 ± 4.84	7.12 ± 2.71	852 ± 254	54.92 ± 37.51	1.20×10^{-5}
10EVA18 10% BaTiO$_3$	25.24 ± 4.18	7.59 ± 0.77	801 ± 281	53.14 ± 27.17	1.09×10^{-5}
10EVA18 20% BaTiO$_3$	26.32 ± 5.69	10.07 ± 2.01	745 ± 125	56.46 ± 14.22	1.07×10^{-5}
10EVA18 10% ZnO	22.94 ± 6.28	9.33 ± 4.27	1002 ± 227	67.72 ± 26.35	9.4×10^{-6}
10EVA18 20% ZnO	24.87 ± 3.81	9.13 ± 1.77	728 ± 124	55.69 ± 17.28	9.04×10^{-6}
5EVA25	8.85 ± 0.86	4.58 ± 2.13	806 ± 103	42.59 ± 21.74	1.46×10^{-6}
5EVA25 10% BaTiO$_3$	10.63 ± 2.74	7.66 ± 1.81	1184 ± 57	83.91 ± 9.80	2.09×10^{-6}
5EVA25 20% BaTiO$_3$	11.25 ± 0.67	6.70 ± 0.78	1215 ± 190	73.05 ± 10.26	2.56×10^{-6}
5EVA25 10%ZnO	10.83 ± 0.82	7.23 ± 1.59	1432 ± 173	90.37 ± 26.86	2.80×10^{-6}
5EVA25 20%ZnO	12.47 ± 3.76	7.16 ± 2.59	1078 ± 272	77.78 ± 37.94	2.14×10^{-6}
10EVA25	7.81 ± 1.06	5.79 ± 2.74	797 ± 128	58.35 ± 25.36	4.98×10^{-6}
10EVA25 10% BaTiO$_3$	11.25 ± 1.58	6.31 ± 1.16	794 ± 122	50.64 ± 17.26	4.44×10^{-6}
10EVA25 20% BaTiO$_3$	12.54 ± 3.10	6.68 ± 2.70	1019 ± 362	66.47 ± 23.16	8.40×10^{-6}
10EVA25 10%ZnO	16.45 ± 1.58	7.35 ± 1.10	847 ± 75	52.35 ± 11.74	3.14×10^{-6}
10EVA25 20%ZnO	16.24 ± 1.81	6.62 ± 1.75	989 ± 172	66.87 ± 13.70	3.70×10^{-6}
5EVA40	2.03 ± 0.41	2.94 ± 0.31	1426 ± 164	39.82 ± 13.84	4.18×10^{-7}
5EVA40 10% BaTiO$_3$	2.27 ± 0.24	2.54 ± 0.50	1212 ± 95	22.19 ± 2.13	4.63×10^{-7}
5EVA40 20% BaTiO$_3$	2.44 ± 0.27	2.45 ± 0.43	1577 ± 147	30.45 ± 4.02	5.60×10^{-7}
5EVA40 10% ZnO	2.42 ± 0.29	2.19 ± 0.45	1498 ± 334	28.74 ± 5.77	4.86×10^{-7}
5EVA40 20% ZnO	2.64 ± 0.05	2.49 ± 0.57	1749 ± 231	38.85 ± 14.84	4.55×10^{-7}
10EVA40	2.20 ± 0.68	2.88 ± 1.13	1292 ± 218	34.73 ± 18.49	8.74×10^{-7}
10EVA40 10% BaTiO$_3$	2.56 ± 0.33	3.35 ± 0.40	1452 ± 155	31.45 ± 5.67	1.17×10^{-6}
10EVA40 20% BaTiO$_3$	2.30 ± 0.17	2.37 ± 0.27	1463 ± 108	24.65 ± 1.19	1.49×10^{-6}
10EVA40 10% ZnO	2.78 ± 0.56	3.64 ± 0.38	1389 ± 204	33.44 ± 2.06	8.78×10^{-7}
10EVA40 20% ZnO	2.22 ± 0.32	3.21 ± 0.64	1546 ± 363	38.91 ± 13.07	6.72×10^{-7}

Generally, an increase of VA concentration (from EVA18 to EVA40) results in decreased stiffness and tensile strength, but an increased elongation at break. The toughness is generally similar between EVA18 and EVA25, while decreasing for the softer formulations (EVA40). Such mechanical properties can be furtherly tuned by adding ceramic nanomaterials. The introduction of nanofillers usually might provide additional and/or peculiar features to the polymeric matrix that are not usual for the polymer, such as mechanical, optical, and piezoelectric ones [25]. In the field of biomaterials, the mechanical reinforcement of polymers due to the use of inorganic nanofillers is of great interest for many applications. The peculiar features of nanomaterials, as a large surface to volume ratio combined with their intrinsic rigidity, enable multiple particle–matrix interactions when dispersed into the polymeric matrix, thus leading to an overall improvement of material properties [26–28]. There are many mechanisms with which nanomaterials can improve the mechanical strength of polymer, as transferring the stress from the matrix to the stiffer filler, thus substituting the softer polymeric components of the polymeric matrix. Indeed, nanomaterials can help in absorbing the energy due to the applied stress, enabling its dispersion in a larger volume of the nanocomposite matrix, thus increasing the material toughness. In previous studies, it has been shown that the Young's modulus, mechanical strength and ductility of barium titanate-doped EVA matrices (40% VA content) increase with increasing $BaTiO_3$ content up to loading levels of 20% vol., while testing loading level of 30% vol., both the mechanical strength and ductility of the nanocomposites decrease in relation with the loading [18]. Another explanation is provided by the concept of interphase. The interphase is a third phase with different properties respect with polymer matrix and nanoparticle phases. The interphase may be formed due to high interfacial areas and strong interfacial interactions between polymer and nanoparticles and may play an important role in their properties. For example, small nanoparticles and large interphase thickness have positive effects on Young's modulus of nanocomposite polymers [29].

In some other cases, mechanical reinforcing of nanocomposite polymers can be provoked by the aggregation of nanofillers, rather than the interfacial adhesion between polymer and nanoparticles [30]. On the other hand, the fabrication of nanocomposite materials may lead to aggregation/agglomeration phenomena of the included nanofiller that can generate defects and stress concentrations, which may sometimes decrease the mechanical properties of composite materials [31]. Those effects are generally increased by increasing the nanofiller content and reducing the filler size. In addition, nanofiller morphology has a very important role in the overall mechanical behavior of nanocomposite materials.

In the proposed nanocomposite thin films, the choice of the components and the preparation procedure allow the fabrication of membranes in which the addition of nanomaterials provoked beneficial effects, in terms of stiffness, elongation at break, and tensile strength. Furthermore, the toughness is significantly improved in the case of EVA18 and EVA25.

A further exploration of the mechanical behavior of nanocomposite EVA thin films was carried out by analyzing the flexural rigidity. The flexural rigidity describes the resistance to bend thin films. Here, the effect was mainly provided by the VA content of each matrix, which led to a significant change in the material stiffness, thus a consequent variation in the flexural rigidity. Indeed, the effect of the thickness, which is relevant for the estimation of the flexural rigidity, was not extremely relevant due to the not significant changes among all the EVA formulations. The analysis of the flexural rigidity has already been correlated with the polymer concentration in thin films, as shown by Hasebe et al. [32]. In such case, the fabrication of thin films with different polymer concentration led to a significant variation of the substrate thickness, which derived the difference in the estimated flexural rigidity.

The adoption of EVA matrices alternatively to the most standard polydimethylsiloxane (PDMS) may involve several advantages for the fabrication of piezoelectric nanocomposite materials. PDMS is mechanically tunable depending on the ratio monomer/curing agent ratio (Young's modulus can range from tens of kPa to MPa); on the other hand, EVA has a decisively higher elongation at break than PDMS (up to 100%). Above all, EVA is also more economic than PDMS, thus representing a cheaper alternative for building nanocomposite materials.

3.3. Thermal Properties

DSC is used to measure the melting point and the crystallinity of nanocomposite EVA polymers. Analyses were performed on 10EVA matrices, focusing on the maximum content of the included nanofiller. An example of the trend of distinct thermograms is reported in Figure 7.

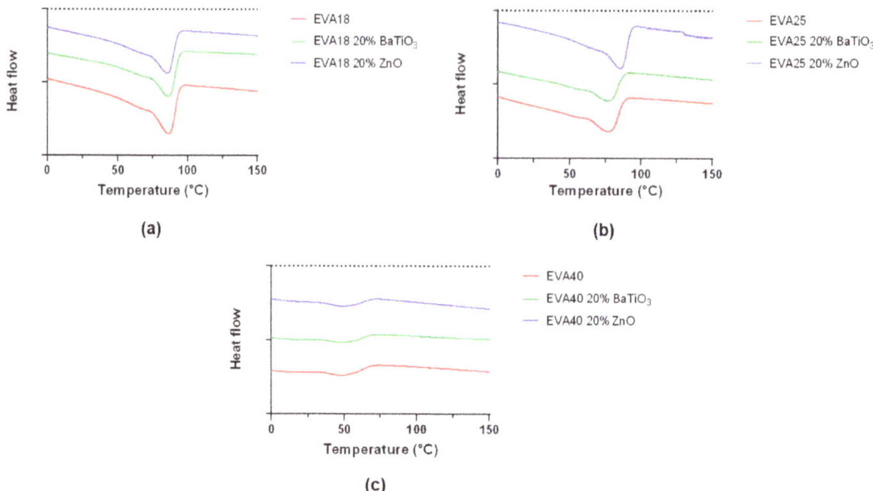

Figure 7. Comparison of the thermograms of ethylene vinyl acetate (EVA) formulations based on: (a) EVA18, (b) EVA25, (c) EVA40.

Thermal investigation shows a clear difference between the EVA polymers at different VA content. In general, EVA exhibits multiple melting endotherms, composed of a slight endotherm started at a lower temperature and a major melting peak at the end and a broad medium peak which overlaps the others. A summary of the results is reported in Table 2.

Table 2. Differential Scanning Calorimetry (DSC) analysis on the different material formulations. Statistical differences are defined by p-value > 0.05 (*) in each EVA subgroup.

Polymer Formulation	Melting Point (°C)	Crystallinity (%)
EVA18	83.1 ± 2.2	23.4 ± 4.1 (*)
EVA18 20% BaTiO$_3$	85.1 ± 0.3	33.8 ± 1.9 (*)
EVA18 20% ZnO	85.5 ± 0.5	29.7 ± 4.5
EVA25	77.3 ± 1.4 (*)	24.8 ± 0.4 (*)
EVA25 20% BaTiO$_3$	78.9 ± 5.3	28.9 ± 5.7
EVA25 20%ZnO	85.2 ± 0.7 (*)	31.7 ± 0.8 (*)
EVA40	46.7 ± 0.3	9.8 ± 0.5
EVA40 20% BaTiO$_3$	46.7 ± 0.8	10.2 ± 0.4
EVA40 20% ZnO	46.1 ± 0.9	8.8 ± 3.2

For the neat polymer, the melting point ranges from 46.7 °C (EVA40) to 83.1 °C (EVA18), while the crystallinity ranges from 9.8% (EVA40) to 23.4% (EVA18). As the content of VA increases, the melting point of the EVA copolymer decreases, because the polyethylene crystallinity is disrupted by the VA component. These values are in accordance with results already showed in the scientific literature [11]. In fact, the incorporation of VA units into the polyethylene backbone chain has the effect to reduce both crystallinity and melting point while increasing the material flexibility, as previously showed by the elastic modulus/elongation at break results (Figures 4 and 6). Interestingly, the introduction of ceramic nanofillers slightly increases both the melting point and the crystallinity of the EVA polymer, with a more relevant effect at smaller content of VA (EVA18 and EVA25). To be more specific, the statistical difference for both melting point and crystallinity between EVA25 and EVA25 with the 20% wt. of ZnO reflects the statistical significance found for the elastic modulus in Figure 2, as well as the statistical difference found for the crystallinity between EVA18 and EVA18 with the 20% wt. of

BaTiO$_3$. It was observed that the crystallization mechanism of nanocomposite polymers may strongly depend on the intrinsic features of the nanofiller and in consequence its dispersion in the polymeric matrix [33]. For example, in well-dispersed nanocomposites the growing lamellae can influence the disposition of the included nanoparticles, thereby broadening interstitials to allow bulk-like lamellae to form [34]. These results may also suggest that nanomaterials can act as nucleation site. In fact, nucleation of crystallization can appear with the inclusion of inorganic nanomaterials, widening the usual confined crystallization offered by the neat polymer [35]. This analysis allows to better clarify the important role of piezoelectric nanofillers for varying bulk properties of composite polymeric matrices. Generally, EVA consists of two phases [36]: An interfacial and more rigid phase, and a very mobile amorphous phase. The introduction of nanofillers may alter the mobility of the amorphous phase between crystalline chains, thus leading to a general increase of the degree of crystallinity of doped formulations. For example, polymer chains close to nanofillers can be stretched and can decrease the conformational entropy of chains. The presence of a rigid interface due to the ceramic origin of nanofiller could drive the segregation of lower molecular weight chains during the thin film formation upon solvent evaporation [27].

3.4. Radiopacity

Polymeric composites can be made radiopaque by the incorporation of piezoelectric nanofillers possessing high atomic numbers such as zinc and barium [37]. The evaluation of radiopacity is shown in Figure 8.

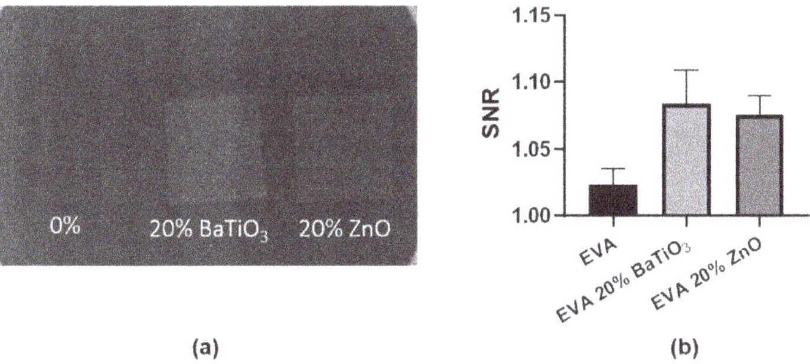

Figure 8. (**a**) Digital image of an X-ray analysis of the EVA-based thin films, and (**b**) analysis of their radiopacity.

According to our results, the addition of both ceramic nanofillers confers radiopacity to the nanocomposite material, increasing the signal-to-noise ratio in comparison with the neat EVA. This result demonstrates that nanocomposite EVA materials can be used to manufacture implantable devices that are radiopaque, making possible their visualization using radiography. Many biomedical devices lack radiopacity, making the visualization and assessment of material within the human body difficult. This may lead to a difficult evaluation of the nanocomposite material fate without using invasive methods [38].

3.5. Electromechanical Response

The electromechanical response of nanocomposite 10EVA matrices was evaluated by applying a pressure on top of the thin film and recording the signal generated by the presence of piezoelectric nanomaterials within the matrix. When the external force is applied to the thin film, each piezoelectric nanomaterial undergoes deformation, generating a net local polarization on it, thus a potential difference. This leads to a piezoelectric voltage that can be detected by electrodes. Results are reported in Figure 9.

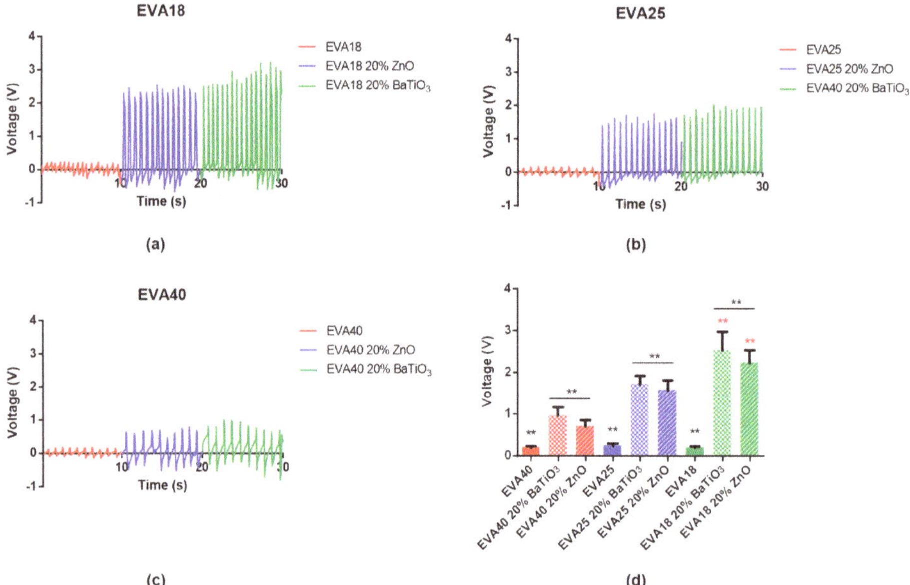

Figure 9. Electromechanical response of nanocomposite ethylene vinyl acetate (EVA) matrices: (**a**) EVA18, (**b**) EVA25, and (**c**) EVA40. (**d**) Summary of the output voltage (** = $p < 0.01$).

In general, the effect of piezoelectric dopant agents on the electromechanical response of nanocomposite EVA matrices is evident. In fact, in absence of nanofiller the output voltages do not overcome 0.2–0.25 V. On the other hand, their presence within the polymeric matrix allows the achievement of increasing output voltages in relation to the material stiffness, ranging from almost 0.71 ± 0.14 V (EVA40 20% ZnO) to 2.55 ± 0.43 V (EVA18 20% BaTiO$_3$). In fact, the highest stiffness of EVA18 allows the transmission of higher stresses to piezoelectric elements that can consequently produce a higher voltage output. Interestingly, the difference in using a different type of dopant is only statistically evident when fabricating stiffer matrices, while in the other cases there are no evident differences.

To the best of our knowledge, there are no reports which aim at demonstrating the electromechanical properties of doped EVA matrices. For example, it is only shown that the increased loading of BaTiO$_3$ nanoparticles (diameter: 100 nm) can improve the conductivity and permittivity of EVA thin films [18]. Our analysis demonstrates how EVA substrates with different content of VA and doping may vary the responsivity to mechanical stresses, with results comparable to those found in the scientific literature for some PDMS-doped matrices [39].

The use of nanocomposite matrices based on EVA may have a wider applicability than Polyvinylidene Fluoride (PVDF) for certain applications. PVDF is a FDA-approved thermoplastic polymer which presents interesting piezoelectric properties, widely investigated for biomedical application [40]. Despite this, PVDF results are decisively stiffer (Young's modulus in the order of GPa) and possess a lower elongation at break (from 25% to 500%). Such features make the PVDF less appropriate in applications in which a certain degree of flexibility, thus a low flexural rigidity value, is required (e.g., sensing in tissues with irregular shapes [41]).

These nanocomposite membranes may find space in many biomedical applications, being EVA FDA-approved. For example, such nanocomposite membranes may find potential applications in self-powered touch sensors [42] or energy harvesters [43]. The biocompatibility and flexibility of EVA makes it a suitable candidate as material to be implanted inside the body for healthy monitoring [44]. Another interesting domain of application is the soft robotics. Soft robots require soft sensors that can be embedded into the robot body without adding rigidity and kinematic limitations [45]. Alternatively, such nanocomposite membranes may be further explored for possible applications in tissue engineering. Indeed, the piezoelectricity induced by the addition of piezoelectric nanofillers could be exploited to generate local electrical charges upon external mechanical stimuli (e.g., ultrasound waves), enabling regenerative phenomena which can help the restoration of functions in piezoelectric tissues, as for example the articular cartilage one [46].

The performance of these nanomembranes could be furtherly improved by acting on the inclusion of material with higher piezoelectric coefficient or increasing the film thicknesses. In fact, the relatively low thickness of EVA matrices (up to 200 μm) could not lead to high output voltages with respect to other nanocomposite matrices with higher thickness (up to 1 mm, [47]). In fact, the voltage output of is a function of its capacitance, as the piezoelectric layer is very thin, there could be high capacitance and low charge. In case of EVA thin films, since the voltage output is linearly correlated with the nanocomposite thickness, an increase of material thickness will lead to increased output for the same applied pressure. Furthermore, EVA matrices cannot be subjected to poling, which can increase the piezoelectric properties of nanomaterials as $BaTiO_3$ of a factor of at least 10 [48]. The use of different nanomaterials, as well as different shapes/sizes may allow the achievement of higher output voltages.

4. Conclusions

Nanocomposite thin films of EVA at different VA content, doped with different loading of $BaTiO_3$ and ZnO, were prepared by solvent casting. These nanocomposite matrices were characterized in terms of morphological, mechanical, thermal, radiopaque, and piezoelectric properties. The tuning of material formulation highlights the possibility to vary thin film mechanical properties, crystallinity, and melting point. The doped EVA composites were also radiopaque, enabling their visualization under x-ray. The electromechanical response induced by the presence of piezoelectric nanomaterials has been verified, demonstrating the achievement of output voltages up to 2.55 V for the doped 18EVA substrates with barium titanate (20% wt.).

The use of piezoelectric nanomaterials as dopant agent in EVA matrices is still rather poorly explored. The combination of EVA with ceramic nanomaterials as piezoelectric nanoparticles may open future scenarios for possible applications in sensing and monitoring as well as drug release systems or engineering of human tissue, being a FDA-approved material.

Supplementary Materials: The following are available online at http://www.mdpi.com/2079-4991/9/8/1182/s1, Figure S1: Dual Beam imaging of zinc oxide nanopowder (left) and barium titanate nanoparticles (right), Figure S2: Voltage output in standard (left) and inverted (right) poles configuration.

Author Contributions: Formal analysis and validation, G.M.; supervision and writing—original draft, L.V.

Funding: This research was funded by the European Commission, through the project ADMAIORA (ADvanced nanocomposite MAterIals fOr in situ treatment and ultRAsound-mediated management of osteoarthritis), funded in the Horizon 2020 framework (Grant number: 814413).

Acknowledgments: The authors acknowledge Leonardo Ricotti for his contribution in the discussion of the results, Francesca Pignatelli for her contribution in performing DSC analysis, and Carlo Filippeschi for his help during clean room procedures.

Conflicts of Interest: The authors declare no conflict of interest.

References

1. Cafarelli, A.; Verbeni, A.; Poliziani, A.; Dario, P.; Menciassi, A.; Ricotti, L. Tuning acoustic and mechanical properties of materials for ultrasound phantoms and smart substrates for cell cultures. *Acta Biomater.* **2017**, *49*, 368–378. [CrossRef]
2. Vannozzi, L.; Iacovacci, V.; Menciassi, A.; Ricotti, L. Nanocomposite thin films for triggerable drug delivery. *Exp. Opin. Drug Deliv.* **2018**, *15*, 509–522. [CrossRef]
3. Bramhill, J.; Ross, S.; Ross, G. Bioactive nanocomposites for tissue repair and regeneration: A review. *Int. J. Environ. Res. Public Health* **2017**, *14*, 66. [CrossRef]

4. Ciofani, G.; Ricotti, L.; Mattoli, V. Preparation, characterization and in vitro testing of poly (lactic-co-glycolic) acid/barium titanate nanoparticle composites for enhanced cellular proliferation. *Biomed. Microdev.* **2011**, *13*, 255–266. [CrossRef]

5. Whulanza, Y.; Battini, E.; Vannozzi, L.; Vomero, M.; Ahluwalia, A.; Vozzi, G. Electrical and mechanical characterisation of single wall carbon nanotubes based composites for tissue engineering applications. *J. Nanosci. Nanotechnol.* **2013**, *13*, 188–197. [CrossRef]

6. Vannozzi, L.; Ricotti, L.; Filippeschi, C.; Sartini, S.; Coviello, V.; Piazza, V.; Pingue, P.; La Motta, C.; Dario, P.; Menciassi, A. Nanostructured ultra-thin patches for ultrasound-modulated delivery of anti-restenotic drug. *Int. J. Nanomed.* **2016**, *11*, 69. [CrossRef]

7. Fujie, T. Development of free-standing polymer nanosheets for advanced medical and health-care applications. *Polym. J.* **2016**, *48*, 773–780. [CrossRef]

8. Osman, A.F.; Alakrach, A.M.; Kalo, H.; Azmi, W.N.W.; Hashim, F. In vitro biostability and biocompatibility of ethyl vinyl acetate (EVA) nanocomposites for biomedical applications. *RSC Adv.* **2015**, *5*, 31485–31495. [CrossRef]

9. Schneider, C.; Langer, R.; Loveday, D.; Hair, D. Applications of ethylene vinyl acetate copolymers (EVA) in drug delivery systems. *J. Control. Release* **2017**, *262*, 284–295. [CrossRef]

10. Shastri, P.V. Toxicology of polymers for implant contraceptives for women. *Contraception* **2002**, *65*, 9–13. [CrossRef]

11. Almeida, A.; Possemiers, S.; Boone, M.; De Beer, T.; Quinten, T.; Van Hoorebeke, L.; Remon, J.P.; Vervaet, C. Ethylene vinyl acetate as matrix for oral sustained release dosage forms produced via hot-melt extrusion. *Eur. J. Pharm. Biopharm.* **2011**, *77*, 297–305. [CrossRef]

12. Chaudhary, D.S.; Prasad, R.; Gupta, R.K.; Bhattacharya, S.N. Morphological influence on mechanical characterization of ethylene-vinyl acetate copolymer–clay nanocomposites. *Polym. Eng. Sci.* **2005**, *45*, 889–897. [CrossRef]

13. Bidsorkhi, H.C.; Adelnia, H.; Pour, R.H.; Soheilmoghaddam, M. Preparation and characterization of ethylene-vinyl acetate/halloysite nanotube nanocomposites. *J. Mater. Sci.* **2015**, *50*, 3237–3245. [CrossRef]

14. Stan, F.; Stanciu, N.V.; Fetecau, C. Melt rheological properties of ethylene-vinyl acetate/multi-walled carbon nanotube composites. *Compos. Part B Eng.* **2017**, *110*, 20–31. [CrossRef]

15. Yousefzade, O.; Hemmati, F.; Garmabi, H.; Mahdavi, M. Thermal behavior and electrical conductivity of ethylene vinyl acetate copolymer/expanded graphite nanocomposites: Effects of nanofiller size and loading. *J. Vinyl Addit. Technol.* **2016**, *22*, 51–60. [CrossRef]

16. Ramesan, M. Fabrication, characterization, and properties of poly (ethylene-co-vinyl acetate)/magnetite nanocomposites. *J. Appl. Polym. Sci.* **2014**, *131*, 7. [CrossRef]

17. Bhuyan, B.; Roy, A.; Srivastava, S.; Mittal, V. Multiwalled carbon nanotube/montmorillonite hybrid filled ethylene-co-vinyl acetate nanocomposites with enhanced mechanical properties, thermal stability, and dielectric response. *Polym. Eng. Sci.* **2018**, *58*, 1155–1165. [CrossRef]

18. Huang, X.; Xie, L.; Jiang, P.; Wang, G.; Liu, F. Electrical, thermophysical and micromechanical properties of ethylene-vinyl acetate elastomer composites with surface modified BaTiO$_3$ nanoparticles. *J. Phys. D Appl. Phys.* **2009**, *42*, 245407. [CrossRef]

19. Sebastian, J.; Thachil, E.T.; Mathen, J.J.; Madhavan, J.; Thomas, P.; Philip, J.; Jayalakshm, M.S.; Mahmud, S.; Ginson, P.J. Enhancement in the electrical and thermal properties of ethylene vinyl acetate (EVA) co-polymer by zinc oxide nanoparticles. *Open J. Compos. Mater.* **2015**, *5*, 79. [CrossRef]

20. Wu, S. Phase structure and adhesion in polymer blends: a criterion for rubber toughening. *Polymer* **1985**, *26*, 1855–1863. [CrossRef]

21. Tomić, N.; Međo, B.; Stojanović, D.; Radojević, V.; Rakin, M.; Jančić-Heinemann, R.; Aleksić, R.R. A rapid test to measure adhesion between optical fibers and ethylene–vinyl acetate copolymer (EVA). *Int. J. Adhesion Adhesives* **2016**, *68*, 341–350. [CrossRef]

22. Henderson, A.M. Ethylene-vinyl acetate (EVA) copolymers: a general review. *IEEE Electr. Insulat. Mag.* **1993**, *9*, 30–38. [CrossRef]

23. Faker, M.; Aghjeh, M.R.; Ghaffari, M.; Seyyedi, S. Rheology, morphology and mechanical properties of polyethylene/ethylene vinyl acetate copolymer (PE/EVA) blends. *Eur. Polym. J.* **2008**, *44*, 1834–1842. [CrossRef]

24. Esthappan, S.K.; Nair, A.B.; Joseph, R. Effect of crystallite size of zinc oxide on the mechanical, thermal and flow properties of polypropylene/zinc oxide nanocomposites. *Compos. Part B Eng.* **2015**, *69*, 145–153. [CrossRef]

25. Jayaraman, A.; Schweizer, K.S. Effective interactions and self-assembly of hybrid polymer grafted nanoparticles in a homopolymer matrix. *Macromolecules* **2009**, *42*, 8423–8434. [CrossRef]

26. Njuguna, J.; Pielichowski, K.; Desai, S. Nanofiller-reinforced polymer nanocomposites. *Polym. Adv. Technol.* **2008**, *19*, 947–959. [CrossRef]

27. Crosby, A.J.; Lee, J.Y. Polymer nanocomposites: The "nano" effect on mechanical properties. *Polym. Rev.* **2007**, *47*, 217–229. [CrossRef]

28. Zare, Y. Modeling the yield strength of polymer nanocomposites based upon nanoparticle agglomeration and polymer–filler interphase. *J. Colloid Interface Sci.* **2016**, *467*, 165–169. [CrossRef]

29. Zare, Y. Development of Halpin-Tsai model for polymer nanocomposites assuming interphase properties and nanofiller size. *Polym Test.* **2016**, *51*, 69–73. [CrossRef]

30. Dorigato, A.; Dzenis, Y.; Pegoretti, A. Filler aggregation as a reinforcement mechanism in polymer nanocomposites. *Mech Mater.* **2013**, *61*, 79–90. [CrossRef]

31. Zare, Y. Study of nanoparticles aggregation/agglomeration in polymer particulate nanocomposites by mechanical properties. *Compos. Part A Appl. Sci. Manuf.* **2016**, *84*, 158–164. [CrossRef]

32. Hasebe, A.; Suematsu, Y.; Takeoka, S.; Mazzocchi, T.; Vannozzi, L.; Ricotti, L.; Fuije, T. Biohybrid Actuators Based on Skeletal Muscle-Powered Microgrooved Ultrathin Films Consisting of Poly(styrene-*block*-butadiene-*block*-styrene). *ACS Biomater. Sci. Eng.* **2019**. [CrossRef]

33. Cyras, V.; D'Amico, D.; Manfredi, L. Crystallization behavior of polymer nanocomposites. In *Crystallization in Multiphase Polymer Systems*; Elsevier: Amsterdam, The Netherlands, 2018; pp. 269–311.

34. Khan, J.; Harton, S.E.; Akcora, P.; Benicewicz, B.C.; Kumar, S.K. Polymer crystallization in nanocomposites: spatial reorganization of nanoparticles. *Macromolecules* **2009**, *42*, 5741–5744. [CrossRef]

35. Müller, K.; Bugnicourt, E.; Latorre, M.; Jorda, M.; Echegoyen Sanz, Y.; Lagaron, J.; Miesbauer, O.; Bianchin, A.; Hankin, S.; Bolz, U.; et al. Review on the processing and properties of polymer nanocomposites and nanocoatings and their applications in the packaging, automotive and solar energy fields. *Nanomaterials* **2017**, *7*, 74. [CrossRef]

36. Bistac, S.; Kunemann, P.; Schultz, J. Crystalline modifications of ethylene-vinyl acetate copolymers induced by a tensile drawing: Effect of the molecular weight. *Polymer* **1998**, *39*, 4875–4881. [CrossRef]

37. Ang, H.Y.; Toong, D.; Chow, W.S.; Seisilya, W.; Wu, W.; Wong, P.; Venkatraman, S.S.; Foin, N.; Huang, Y. Radiopaque fully degradable nanocomposites for coronary stents. *Sci. Rep.* **2018**, *8*, 17409. [CrossRef]

38. Wang, Y.; Van den Akker, N.M.; Molin, D.G.; Gagliardi, M.; Van der Marel, C.; Lutz, M.; Knetsch, L.W.M.; Koole, L.H. A nontoxic additive to introduce X-ray contrast into poly (lactic acid). Implications for transient medical implants such as bioresorbable coronary vascular scaffolds. *Adv. Healthc. Mater.* **2014**, *3*, 290–299. [CrossRef]

39. Park, K.I.; Xu, S.; Liu, Y.; Hwang, G.T.; Kang, S.J.L.; Wang, Z.L.; Lee, K.J. Piezoelectric BaTiO$_3$ thin film nanogenerator on plastic substrates. *Nano Lett.* **2010**, *10*, 4939–4943. [CrossRef]

40. Ribeiro, C.; Sencadas, V.; Correia, D.M.; Lanceros-Méndez, S. Piezoelectric polymers as biomaterials for tissue engineering applications. *Colloids Surfaces B Biointerf.* **2015**, *136*, 46–55. [CrossRef]

41. Khodasevych, I.; Parmar, S.; Troynikov, O. Flexible sensors for pressure therapy: Effect of substrate curvature and stiffness on sensor performance. *Sensors* **2017**, *17*, 2399. [CrossRef]

42. Nanshu, L.; Dae-Hyeong, K. Flexible and stretchable electronics paving the way for soft robotics. *Soft Robot.* **2014**, *1*, 53–62.

43. Hwang, G.T.; Byun, M.; Jeong, C.K.; Lee, K.J. Flexible piezoelectric thin-film energy harvesters and nanosensors for biomedical applications. *Adv. Healthc. Mater.* **2015**, *4*, 646–658. [CrossRef]

44. Yeo, J.C.; Lim, C.T. Emerging flexible and wearable physical sensing platforms for healthcare and biomedical applications. *Microsyst. Nanoeng.* **2016**, *2*, 16043.

45. Wang, H.; Totaro, M.; Beccai, L. Toward perceptive soft robots: Progress and challenges. *Adv. Sci.* **2018**, *5*, 1800541. [CrossRef]

46. Jacob, J.; More, N.; Kalia, K.; Kapusetti, G. Piezoelectric smart biomaterials for bone and cartilage tissue engineering. *Inflammat. Regenerat.* **2018**, *38*, 2. [CrossRef]

47. Alluri, N.R.; Chandrasekhar, A.; Vivekananthan, V.; Purusothaman, Y.; Selvarajan, S.; Jeong, J.H.; Kim, S.J. Scavenging biomechanical energy using high-performance, flexible BaTiO$_3$ nanocube/PDMS composite films. *ACS Sustain. Chem. Eng.* **2017**, *5*, 4730–4738. [CrossRef]
48. Lin, Z.H.; Yang, Y.; Wu, J.M.; Liu, Y.; Zhang, F.; Wang, Z.L. BaTiO$_3$ nanotubes-based flexible and transparent nanogenerators. *J. Phys. Chem. Lett.* **2012**, *3*, 3599–3604. [CrossRef]

 *nanomaterials*

Article

Structural and Stress Properties of AlGaN Epilayers Grown on AlN-Nanopatterned Sapphire Templates by Hydride Vapor Phase Epitaxy

Chi-Tsung Tasi [1], Wei-Kai Wang [2], Sin-Liang Ou [2], Shih-Yung Huang [3], Ray-Hua Horng [4] and Dong-Sing Wuu [1,5,6,]*

[1] Department of Materials Science and Engineering, National Chung Hsing University, Taichung 40227, Taiwan; d100066018@mail.nchu.edu.tw

[2] Department of Materials Science and Engineering, Da-Yeh University, Changhua 51591, Taiwan; wk@mail.dyu.edu.tw (W.-K.W.); slo@mail.dyu.edu.tw (S.-L.O.)

[3] Department of Industrial Engineering and Management, Da-Yeh University, Changhua 51591, Taiwan; syh@mail.dyu.edu.tw

[4] Department of Electronics Engineering, National Chiao Tung University, Hsinchu 300, Taiwan; rhh@nctu.edu.tw

[5] Research Center for Sustainable Energy and Nanotechnology, National Chung Hsing University, Taichung 40227, Taiwan

[6] Innovation and Development Center of Sustainable Agriculture, National Chung Hsing University, Taichung 40227, Taiwan

* Correspondence: dsw@nchu.edu.tw; Tel.: +886-4-2284-0500 (ext. 714); Fax: +886-4-2285-5046

Received: 22 July 2018; Accepted: 8 September 2018; Published: 10 September 2018

Abstract: In this paper, we report the epitaxial growth and material characteristics of AlGaN (Al mole fraction of 10%) on an AlN/nanopatterned sapphire substrate (NPSS) template by hydride vapor phase epitaxy (HVPE). The crystalline quality, surface morphology, microstructure, and stress state of the AlGaN/AlN/NPSS epilayers were investigated using X-ray diffraction (XRD), atomic force microscopy (AFM), and transmission electron microscopy (TEM). The results indicate that the crystal quality of the AlGaN film could be improved when grown on the AlN/NPSS template. The screw threading dislocation (TD) density was reduced to 1.4×10^9 cm^{-2} for the AlGaN epilayer grown on the AlN/NPSS template, which was lower than that of the sample grown on a flat c-plane sapphire substrate (6.3×10^9 cm^{-2}). As examined by XRD measurements, the biaxial tensile stress of the AlGaN film was significantly reduced from 1,187 MPa (on AlN/NPSS) to 38.41 MPa (on flat c-plane sapphire). In particular, an increase of the Al content in the overgrown AlGaN layer was confirmed by the TEM observation. This could be due to the relaxation of the in-plane stress through the AlGaN and AlN/NPSS template interface.

Keywords: AlGaN; nanopatterned sapphire substrate; hydride vapor phase epitaxy; stress; transmission electron microscopy

---

1. Introduction

AlGaN ternary alloy templates have recently drawn increasing attention because of their potential in expanding the fabrication of optoelectronic devices operating in the ultraviolet (UV) range and high-power, high-frequency electronic devices [1–5]. Because of a critical lattice mismatch between the Al$_x$GaN$_{1-x}$ and the sapphire, heteroepitaxial growth-induced defects, such as threading dislocations (TDs), voids, and stacking faults, are usually observed [6,7] on the upper grown layer, hence destroying the performance of UV devices drastically [8–11]. Therefore, the epitaxial growth of thick, crack-free, high-quality AlGaN with a low dislocation density template plays an important role in constructing high-performance AlGaN-based optoelectronic devices. The hydride vapor phase epitaxy (HVPE) method has been shown to achieve the growth of a thick AlGaN layer serving as a template (or bulk) substrate material due to its rapid growth rate (several hundred μm/h) and relatively low cost [12,13]. However, due to the significant lattice mismatch between the AlGaN and the sapphire, the crystalline quality of the HVPE AlGaN with a low defect density is unsatisfactory. Meanwhile, epilayer cracks are induced when the critical thickness of AlGaN is exceeded during the cooling down procedure. Epitaxial lateral overgrowth (ELOG) techniques on microstripe (or honeycomb) shape-patterned sapphires have shown a promising result in reducing the defect density of the AlGaN layer [14–16]. In addition, the uses of nanopatterned sapphire substrates (NPSSs) improve the crystalline quality of the AlGaN layer by ELOG [17]. Published research using in situ AlN buffer layer below the grown $Al_{0.45}Ga_{0.55}N$ layer showed that it could not only enhance the crystallinity but also affect the surface morphology due to the misorientated crystallites [18]. The effect of various growth temperatures and V/III ratios of the AlN buffer layer on the structural properties of the subsequently grown AlGaN layer has been reported [19,20]. Another major issue is the low efficiency of Al incorporation in $Al_xGa_{1-x}N$ caused by biaxial tensile strain formation during the growing process [21]. This limited the efforts on the study of high Al content of AlGaN films and crystalline quality. It has been previously reported that high temperature growth of AlN film is considered to serve as a strain-relaxed layer to improve nitride material's structural properties [22,23]. Therefore, it is important to grow high Al content $Al_xGa_{1-x}N$ films with low defect density by the above-mentioned method. Several groups have demonstrated the AlN template/NPSS by subsequently growing UV devices by metalorganic chemical vapor deposition (MOCVD) [24–26]. Since the considerable production cost of MOCVD growth AlGaN template would be too much, HVPE method to fabricate AlGaN templates on foreign substrates are good choices for the heteroepitaxial deposition of AlGaN-based devices. In this study, the AlGaN layer was grown in a combination of ex situ MOCVD grown AlN buffer layer and NPSS surface by HVPE. In addition, the growth mechanism, crystalline quality, surface morphology, and structural properties of the AlGaN on the AlN/NPSS template were investigated.

2. Materials and Methods

A 2-inch c-plane sapphire substrate was used as a starting material for the NPSS. A SiO_2 film deposited by low-pressure chemical vapor deposition on the sapphire served as the mask layer, on which the nanoimprint resist was then spin-coated. The hexagonal hole array was transferred to the resist by nanoimprint lithography, followed by oxygen plasma descum to remove any residual resistance at the bottom of the holes. The SiO_2 film was then etched by fluorine plasma. Finally, a BCl_3/Cl_2 high-density plasma etching process was employed to etch the sapphire substrate, and the mask was removed by a dilute HF solution. Although multiple hole dimensions for nanoimprinting were attempted, the optimum NPSS used in this study was with 500 nm diameter hole arrays spaced 950 nm apart and etched to a depth of 400 nm. We deposited a 30 nm AlN buffer layer on the NPSS as an AlN/NPSS template using MOCVD, and then an AlGaN epilayer was grown on the AlN/NPSS template in an HVPE horizontal reactor as shown schematically in Figure 1a–c. For a 30 nm AlN thin film deposition, trimethylaluminum (TMAl, SAFC Hitech. Co., Ltd. Kaohsiung, Taiwan) and ammonia (NH_3, SAFC Hitech. Co., Ltd. Kaohsiung, Taiwan) were used as the precursors. H_2 was the carrier and the growth temperature at 1120 °C for 3 min. The AlGaN epilayer was also grown on a conventional sapphire substrate (CSS) as a comparison. The quartz glass reactor was covered with a furnace containing five heating zones maintained at different temperatures. Ga and Al metal chlorides serving as the group III Ga and Al precursor sources, respectively, were separately placed in the upstream region of the quartz reactor. The $AlCl_3$ and GaCl vapors were generated in the reactor by flowing HCl (APDirect Inc. Co., Ltd. Taichung, Taiwan) over the Al (10 sccm) and Ga precursor (10 sccm) sources, respectively. To avoid the formation of AlCl vapor by a reaction between the Al metals and HCl at a high temperature (which would damage the quartz reactor), the Al metal source was maintained at 500 °C. The temperature of the GaCl source was maintained between 800 °C and 900 °C. Pure N_2 gas (400 sccm) served as the carrier gas to propel the $AlCl_3$ and GaCl vapors through the two quartz tubes to the growth zone. The ammonia line consisted of NH_3 flow (2 L/min) and N_2 flow (300 sccm). During the HVPE process, the H_2 flow (Linde LienHwa Inc. Co., Ltd. Taipei, Taiwan) was kept at 2.45 L/min, N_2 flow (Linde LienHwa Inc. Co., Ltd. Taipei, Taiwan) at 200 sccm, growth pressure at 200 mbar, and growth temperature at 1080 °C.

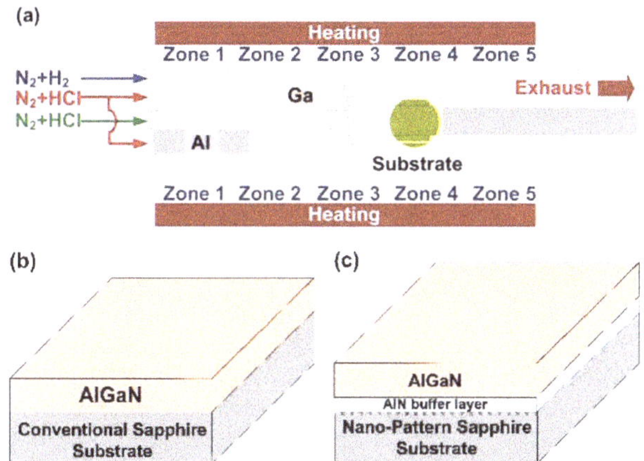

Figure 1. (**a**) A schematic diagram of the HVPE reactor used for the AlGaN grown on the (**b**) CSS and (**c**) AlN/NPSS templates.

Transmission electron microscopy (TEM; JEM-2010, JEOL, Tokyo, Japan), scanning electron microscopy (SEM; S-3000H, Hitachi, Tokyo, Japan), atomic force microscopy (AFM; 5400, Agilent, Santa Clara, CA, USA), double-crystal X-ray diffraction (DCXRD; X'Pert PRO MRD, PANalytical, Almelo, The Netherlands), and thin film stress (Toho, FLX-320-S, Nagoya, Japan) measurements were conducted to examine the microstructural properties of the AlGaN epilayers grown on the different substrate templates (e.g., CSS, AlN/NPSS, and NPSS).

3. Results and Discussion

Figure 2 shows the typical XRD scan patterns of the AlGaN grown on the CSS and AlN/NPSS templates. To evaluate the influence of strain on the Al incorporation into the AlGaN layer, two different regions (the edge and the center of the two-inch wafer) in the AlGaN grown on the CSS wafer are also displayed for comparison. In Figure 2a, the peak located at 34.53° corresponds to the diffraction from the GaN (002) plane (i.e., edge of the wafer) on the CSS template. The AlGaN (002) peak located at 34.57° (very low Al content) was observed at the center of the wafer on the CSS template (Figure 2b). This was attributed to the residual strain that occurred due to the lattice mismatch between the AlGaN and the sapphire substrate. Meanwhile, in Figure 2c, the peak located at 34.67° corresponds to the AlGaN (002) plane, whereas a weak peak around 35.98° corresponds to the AlN (002) plane on the AlN/NPSS template. Apparently, the Al composition in the AlGaN epilayer on the CSS template was lower than that on the AlN/NPSS template (Al: 10%). This is because of the strain-dependent effect on the incorporation efficiency of Al into the AlGaN layer [27]. This result indicates that the improvement on the Al incorporation might be due to a change in the surface state caused by the introduction of the AlN/NPSS template during the growth of AlGaN. Moreover, the change in the composition of $Al_xGa_{1-x}N$ alloys might be due to the lattice mismatch or strain between the AlGaN and the sapphire's rough film surface [28]. The insets in Figure 2a–c show the optical microscope morphologies of the AlGaN grown on CSS and AlN/NPSS templates, respectively. The AlGaN grown on the AlN/NPSS template exhibited the best morphology among the two other samples. It is believed that the introduction of the AlN/NPSS template was in favor of forming a smooth AlGaN film surface.

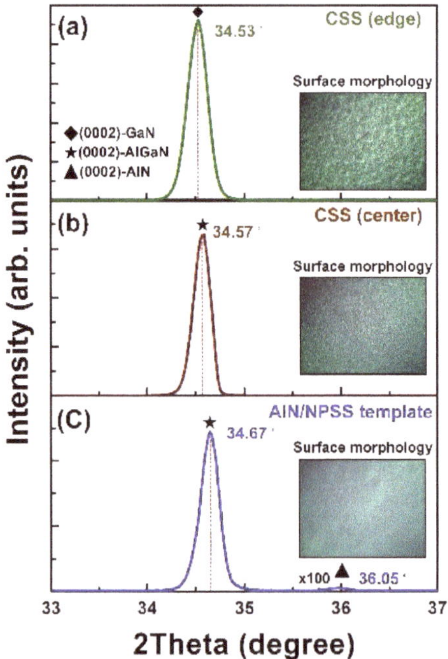

Figure 2. The typical XRD scan patterns of the AlGaN grown on (**a**) CSS (edge); (**b**) CSS (center); and (**c**) AlN/NPSS templates.

The crystal quality of these samples was also investigated using X-ray rocking curve (XRC) (plot is not shown). The XRC of the full-width at half-maximum (FWHM) value with the symmetric (002) and asymmetric (102) planes of the 3 µm thick AlGaN grown on the CSS and AlN/NPSS templates were evaluated, respectively. The FWHM values of the (002) and (102) planes of the AlGaN layer on the CSS template were estimated to be 2200 and 3600 arcsec, respectively. Meanwhile, the FWHM values of the (002) and (102) planes of the AlGaN grown on the AlN/NPSS template were 845 arcsec. These results indicate that the AlN/NPSS template improved the AlGaN layer's crystal quality by lowering the dislocation density. It is well known that the symmetric (002) and asymmetric (102) reflections can provide some information on the density of pure screw and pure edge dislocations, respectively [29]. The relationship between the dislocation density and the FWHM values of XRC can be calculated using the following equations:

$$\rho_s = \frac{\Delta\omega_s^2}{4.35c^2}, \quad \rho_e \frac{\Delta\omega_e^2}{4.35b^2}, \tag{1}$$

where ρ_s and ρ_e are the screw and edge TD densities, respectively; the quantities of ω_s and ω_e refer to the FWHM of (002) and (102), respectively; c and b are the relevant Burgers vectors of the AlGaN epilayer. The corresponding dislocation densities of (002) and (102) reflections were determined using DCXRD as shown in Figure 2b. The AlGaN film on the AlN/NPSS template exhibited a lower screw dislocation density (1.4×10^9 cm^{-2}) than that on the CSS template (6.3×10^9 cm^{-2}). Therefore, it is believed that the AlN/NPSS template could reduce the residual tensile strain, leading to fewer defects, thus improving the quality of the AlGaN layer.

Figure 3a–c shows the top-view SEM images of CSS, AlN/NPSS, and NPSS [17], respectively. It can be seen that the prepared NPSS with hole patterns in this work, and the fabrication process is described in the method section. Figure 4a–c shows the top-view SEM images of the AlGaN layer grown on CSS, AlN/NPSS, and NPSS templates [17], respectively. Because of the lattice mismatch between the AlGaN and the CSS's rough surface, incomplete 3D island coalescence with a hexagonal structure was formed (Figure 4a). On the other hand, the surface morphology of the AlGaN layer on the AlN/NPSS template was smooth and uniform (Figure 4b); the smooth surface might be due to the strain relaxation with a low defect density provided by the AlN/NPSS template. This observed result was consistent with that reported by Hagedorn et al. [18].

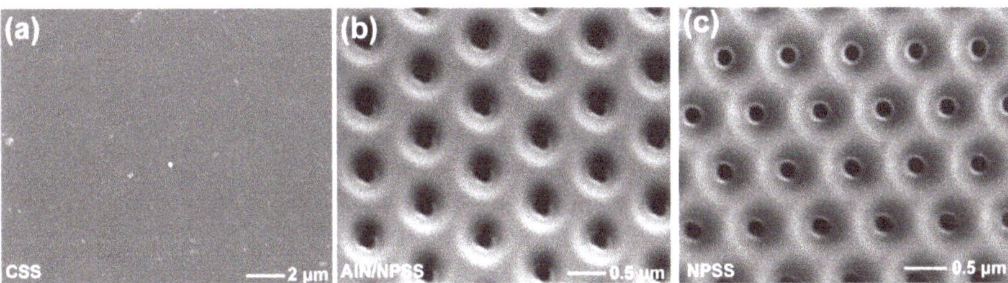

Figure 3. Top-view SEM images of the surface morphologies of the (**a**) CSS; (**b**) AlN/NPSS; and (**c**) NPSS [17].

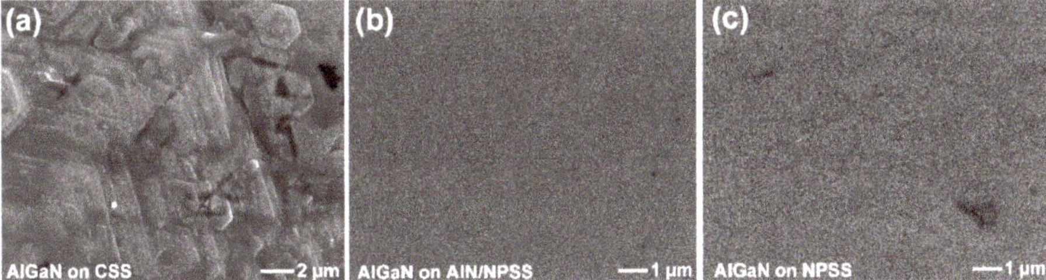

Figure 4. Top-view SEM images of the surface morphologies of the AlGaN epilayers grown on the (**a**) CSS; (**b**) AlN/NPSS; and (**c**) NPSS [17].

The corresponding surface roughness of these AlGaN samples was examined by AFM using a scan area of 10 µm^2. As shown in Figure 5, the root mean square (RMS) values of the AlGaN/CSS, AlGaN/AlN/NPSS, and NPSS [17] were 79.1, 6.66, and 14.9, respectively. The large RMS value for the surface roughness of the AlGaN film grown on CSS (i.e., AlGaN/CSS) might be due to the large lattice mismatch between the film and the substrate. The decrease in the surface roughness was related to the reduction in the dislocation density, as mentioned in the DCXRD results. These observed results conclude that the structural properties and surface morphology of the AlGaN layer were mostly defined by the substrate template.

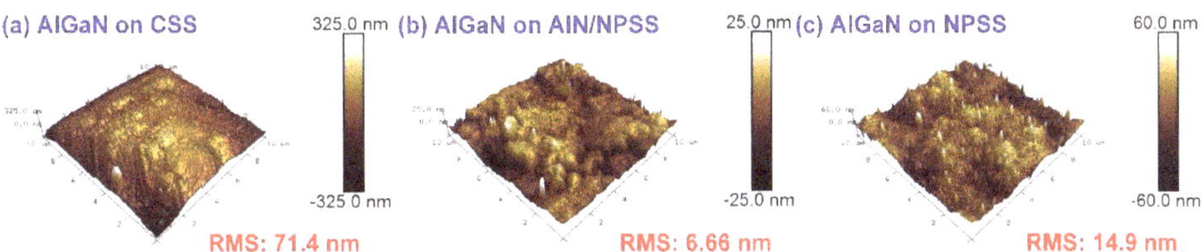

Figure 5. AFM measurements of the AlGaN grown on (**a**) CSS, (**b**) AlN/NPSS, and (**c**) NPSS [17] templates.

Since the lattice constant of the AlGaN epilayer is smaller than that of the sapphire, there exists tensile strain/stress of the AlGaN layer; thus, an AlN buffer layer is commonly used to compensate the tensile stress of the AlGaN grown on a sapphire substrate template [30]. To clearly understand the residual stress of the AlGaN layer, we estimated the strain (ε) present on the AlGaN epilayer from the FWHM of the major XRD (002) peak using the following equation [31]:

$$\varepsilon = \frac{\beta}{4tan\theta} \tag{2}$$

where β is the FWHM and θ is Bragg's diffraction angle. The calculated strain and stress are shown in Table 1. It should be noted that the stress of the AlGaN layer could be converted from tensile stress into compressive stress using the AlN/NPSS template.

Table 1. Strain (ε) and stress (σ) of the AlGaN layer grown on CSS or AlN/NPSS.

AlGaN-(002)	Substrate	2 Theta (°)	FWHM (°)	ε	σ (MPa)
	CSS	34.57	0.583	-1.6×10^{-5}	1187
	AlN/NPSS	34.65	0.235	-4.7×10^{-5}	38.41

The TEM micrographs of the AlGaN deposited on the CSS template are shown in Figure 6. Figure 6a displays the cross-sectional TEM image of AlGaN on CSS, where the thickness of the AlGaN epilayer was is approximately 250 nm. To investigate the microstructures in more detail, we chose the three regions marked I, II, and III for high-resolution (HR) TEM measurements, as shown in Figure 6a,c,d, respectively. The HRTEM image of region I was taken at the interface between the AlGaN and the CSS. In this region, the d-spacing value of the epilayer was analyzed to be 2.50 Å. However, as shown in Figure 6c,d, a larger d-spacing value of 2.59 Å appeared in both regions II and III. According to the JCPDS database, the typical d-spacing values of GaN (0002) and AlN (0002) are 2.593 Å and 2.49 Å, respectively. The d-spacing is defined as the inter-atomic spacing or the distance between adjacent planes in the crystalline materials. From the analysis of region I (Figure 6a), the d-spacing value of 2.50 Å indicates that the AlGaN (0002) phase with a very high Al content was formed in the epilayer. Meanwhile, the d-spacing value of regions II and III (2.59 Å) was extremely close to that of the typical GaN (0002), revealing that the GaN (0002) phase also appeared in the epilayer. These TEM results were in good agreement with the XRD results (Figure 2a,b). This proof confirmed that the phase separation phenomenon between the GaN (0002) and the AlGaN (0002) phases indeed occurred in the AlGaN/CSS sample. This might be attributed to the in-plane stress caused by the phase separation of the AlGaN during growth. This observed result is also consistent with those reported by Gong et al. [32]. Additionally, the dark-field TEM image observed in the two beam condition for the AlGaN epilayer deposited on CSS is shown in Figure 6e, and the screw dislocation density of this AlGaN epilayer deduced by this TEM image is 7.7×10^9 cm^{-2}. Besides, the fast Fourier transform (FFT) images for regions I and II (shown in Figure 6a) are displayed in Figure 6f,g, respectively. The result can also prove that the phase separation exists in this AlGaN epilayer.

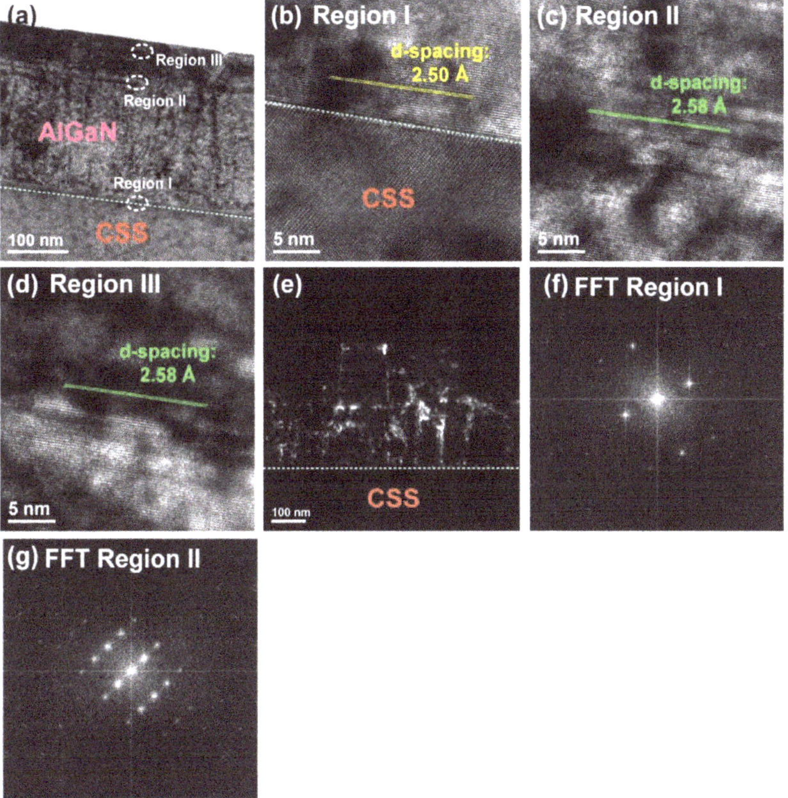

Figure 6. (**a**) A cross-sectional TEM image of the AlGaN/CSS sample. HRTEM images focused on (**b**) region I; (**c**) region II; and (**d**) region III as indicated in Figure 6a. (**e**) The dark-field TEM image observed in the two-beam condition for the AlGaN epilayer deposited on CSS. Fast Fourier transform images for regions (**f**) I and (**g**) II.

We also performed TEM measurements for the AlGaN epilayer deposited on the AlN/NPSS template, as shown in Figure 7. Figure 7a shows a cross-sectional TEM image of the AlGaN epilayer grown on the AlN/NPSS template, whereby the interface between the epilayer and the substrate was clearly observed. Although the AlN interfacial layer could not clearly been found in the present interface, it might be attributed to interdiffusion of Ga and Al during the growth process [33]. Three regions of the AlGaN epilayer (marked I, II, and III) were selected for the HRTEM measurements, as displayed in Figure 7b–d, respectively. Here, regions I and II both represented the AlGaN epilayers grown on the inclined planes (from different patterns). Meanwhile, region III represented the AlGaN epilayer grown above the top of the AlN/NPSS template. In Figure 7b, various *d*-spacing values consisting of 2.54 Å, 2.56 Å, and 2.57 Å were found in region I. Similar *d*-spacing values (2.54 Å and 2.56 Å) could also be identified in region II (Figure 7c). This reveals that the epilayer grown on the inclined planes (regions I and II) displayed the patterns belonging to the AlGaN (0002) phase. On the other hand, the *d*-spacing arrangement of the epilayer above the top of the AlN/NPSS template (region III) was more regular than that grown on the inclined planes, with one uniform *d*-spacing value of 2.56 Å. As mentioned above, the typical *d*-spacing value of GaN (0002) is 2.59 Å. Hence, the AlGaN epilayer deposited on the AlN/NPSS template indeed belonged to the AlGaN phase with no GaN phase, which agreed well with the XRD result. In addition, the dark-field TEM image observed in the two beam condition for the AlGaN epilayer deposited on AlN/NPSS template is shown in Figure 7e, and the screw dislocation density of this AlGaN epilayer deduced by this TEM image is 3.0×10^9 cm^{-2}. Based on Figures 6e and 7e, it can be found that the screw dislocation densities of these two AlGaN epilayers deduced from these TEM images are indeed similar to those evaluated from the XRD results (Figure 2). Besides, the FFT images for regions I and III (shown in Figure 7a) are displayed in Figure 7f,g, respectively. The result can also prove that only the AlGaN phase (without GaN phase) is formed in this AlGaN epilayer.

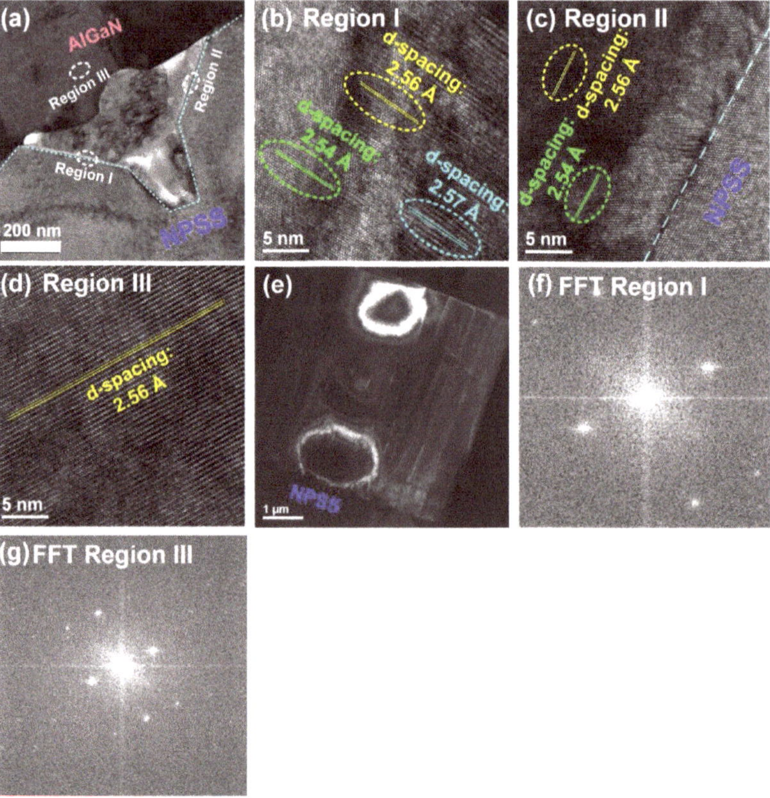

Figure 7. (**a**) A cross-sectional TEM image of the AlGaN/AlN/NPSS sample. HRTEM images focused on (**b**) region I; (**c**) region II; and (**d**) region III as indicated in Figure 7a. (**e**) The dark-field TEM image observed in the two beam condition for the AlGaN epilayer deposited on AlN/NPSS template. Fast Fourier transform images for regions (**f**) I and (**g**) III.

Based on these observations, the mechanism of Al incorporation during the AlGaN growth was proposed, as schematically illustrated in Figure 8. In Figure 8a, due to the Ga atoms with high surface mobility, Ga atoms dominate the growth mechanisms and individual islands rapidly developed for GaN growth [34]. In Figure 8b, higher Al incorporation might be due to lower strain between the AlGaN film and the AlN/NPSS template [27]. It was also assumed that the slightly misorientated NPSS substrate could provide a better opportunity for the Al and Ga atoms to interact on the surface; hence, a higher Al composition of the AlGaN film was achieved. A similar result was also previously reported by Bryan et al. [35].

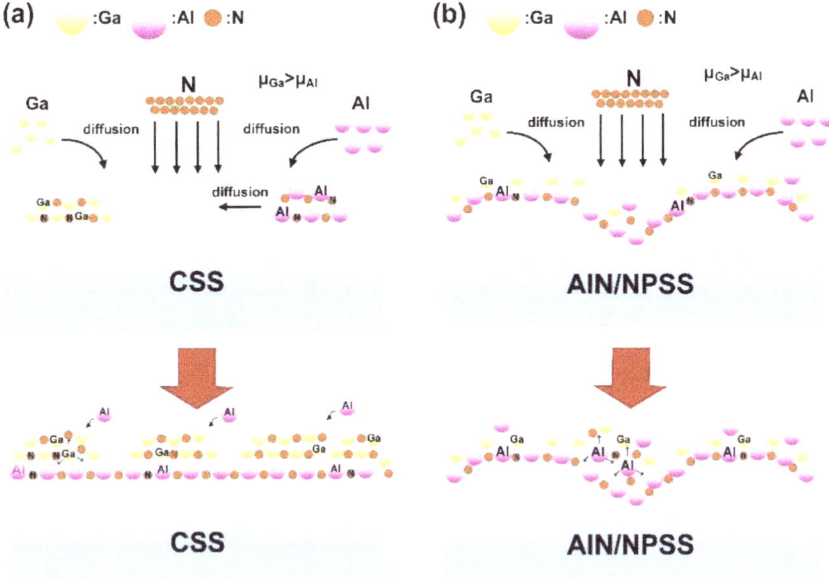

Figure 8. Schematic diagrams of the AlGaN growth mechanism on various substrates: (**a**) CSS and (**b**) AlN/NPSS.

4. Conclusions

In this study, the effects of different substrate templates on the structural and stress properties of AlGaN epilayers growth by HVPE were investigated. According to the XRD, AFM, and TEM analyses, the Al incorporation efficiency into the AlGaN epilayer could be increased using the AlN/NPSS template. The surface roughness of the layer could also be suppressed by growing the AlGaN layer on the AlN/NPSS template. As a result, we could obtain a relatively high Al content and smooth AlGaN film with a narrow XRD FWHM and low defect density. These results indicated that HVPE AlGaN/AlN/NPSS could be a promising epitaxial template for the development of high-performance AlGaN-based optoelectronics devices.

Author Contributions: D.-S.W. proposed the concept. C.-T.T., W.-K.W., and R.-H.H. conceived and designed the experiments. C.-T.T., S.-L.O., and S.-Y.H. contributed to the measurement results of the films. C.-T.T., W.-K.W., S.-L.O., and D.-S.W. wrote the manuscript. All authors have read and approved the final version of the manuscript to be submitted.

Funding: This work was supported by the Ministry of Science and Technology (Taiwan, R.O.C.) under Contract No. 104-2221-E-005-036-MY3. The authors also wish to express their sincere gratitude for the financial support from the "Innovation and Development Center of Sustainable Agriculture" from The Featured Areas Research Center Program within the framework of the Higher Education Sprout Project by the Ministry of Education (MOE) in Taiwan.

Conflicts of Interest: The authors declare no conflict of interest.

References

1. Nagamatsua, M.; Okadaa, N.; Sugimuraa, H.; Tsuzukia, H.; Moria, F.; Iidaa, K.; Bandob, A.; Iwayaa, M.; Kamiyamaa, S.; Amanoa, H.; et al. High-efficiency AlGaN-based UV light-emitting diode on laterally overgrown AlN. *J. Cryst. Growth* **2008**, *310*, 2326–2329. [CrossRef]

2. Pernot, C.; Kim, M.; Shinya Fukahori, S.; Inazu, T.; Fujita, T.; Nagasawa, Y.; Hirano, A.; Ippommatsu, M.; Iwaya, M.; Kamiyama, S.; et al. Improved efficiency of 255–280 nm AlGaN-based light-emitting diodes. *Appl. Phys. Express* **2010**, *3*, 061044. [CrossRef]

3. Yeh, C.T.; Wang, W.K.; Shen, Y.S.; Horng, R.H. 1.48-kV enhancement-mode AlGaN/GaN high-mobility transistors fabricated on 6-in. silicon by fluoride-based plasma treatment. *Jpn. J. Appl. Phys.* **2016**, *55*, 05FK06. [CrossRef]

4. Tonisch, K.; Jatal, W.; Niebelschuetz, F.; Romanus, H.; Baumann, U.; Schwierz, F.; Pezoldt, J. AlGaN/GaN-heterostructures on (111) 3C-SiC/Si pseudo substrates for high frequency applications. *Thin Solid Films* **2011**, *520*, 491–496. [CrossRef]

5. Mi, M.H.; Ma, X.H.; Yang, L.; Bin, H.; Zhu, J.J.; He, Y.L.; Zhang, M.; Wu, S.; Hao, Y. 90 nm gate length enhancement-mode AlGaN/GaN HEMTs with plasma oxidation technology for high-frequency application. *Appl. Phys. Lett.* **2017**, *111*, 173502. [CrossRef]

6. Hirayama, H.; Noguchi, N.; Kamata, N. 222 nm deep-ultraviolet AlGaN quantum well light-emitting diode with vertical emission properties. *Appl. Phys. Express* **2010**, *3*, 032102. [CrossRef]

7. Muramoto, Y.; Kimura, M.; Nouda, S. Development and future of ultraviolet light-emitting diodes: UV-LED will replace the UV lamp. *Semicond. Sci. Technol.* **2014**, *29*, 084004. [CrossRef]

8. Kneissl, M. *A Brief Review of III-Nitride UV Emitters in III-Nitride Ultraviolet Emitters: Technologies and Applications, Chap. 1*; Kneissl, M., Rass, J., Eds.; Springer: Berlin, Germany, 2016.

9. Koide, Y.; Itoh, N.; Itoh, N.; Sawaki, N.; Akasaki, I. Effect of AlN buffer layer on AlGaN/Al$_2$O$_3$ heteroepitaxial growth by metalorganic vapor phase epitaxy. *Jpn. J. Appl. Phys.* **1988**, *3*, 1156–1161. [CrossRef]

10. Wu, X.H.; Fini, P.; Tarsa, E.J.; Heying, B.; Keller, S.; Mishra, U.K.; DenBaars, S.P.; Speck, J.S. Dislocation generation in GaN heteroepitaxy. *J. Cryst. Growth* **1998**, *189/190*, 231–243. [CrossRef]

11. Amano, H.; Miyazaki, A.; Iida, K.; Kawashima, T.; Iwaya, M.; Kamiyama, S.; Akasaki, I.; Liu, R.; Bell, A.; Ponce, F.A.; et al. Defect and stress control of AlGaN for fabrication of high performance UV light emitters. *Phys. Stat. Sol.* **2004**, *12*, 2679–2685. [CrossRef]

12. Jiang, H.; Egawa, T.; Hao, M.; Liu, Y. Reduction of threading dislocations in AlGaN layers grown on AlN/sapphire templates using high-temperature GaN interlayer. *Appl. Phys. Lett.* **2005**, *87*, 241911. [CrossRef]

13. Ren, Z.; Sun, Q.; Kwon, S.Y.; Han, J.; Davitt, K.; Song, Y.K.; Nurmikko, A.V.; Cho, H.K.; Liu, W.; Smart, J.A.; et al. Heteroepitaxy of AlGaN on bulk AlN substrates for deep ultraviolet light emitting diodes. *Appl. Phys. Lett.* **2007**, *91*, 051116. [CrossRef]

14. Fleischmann, S.; Mogilatenko, A.; Hagedorn, S.; Richter, E.; Goran, D.; Schäfer, P.; Zeimer, U.; Weyers, M.; Tränkle, G. Triangular-shaped sapphire patterning for HVPE grown AlGaN layers. *Phys. Stat. Sol.* **2017**, *214*, 1600751. [CrossRef]

15. Richter, E.; Fleischmann, S.; Goran, D.; Hagedorn, S.; John, W.; Mogilatenko, A.; Prasai, D.; Zeimer, U.; Weyers, M.; Traenkle, G. Hydride vapor-phase epitaxy of c-plane AlGaN layers on patterned sapphire substrates. *J. Electron. Mater.* **2014**, *43*, 814–818. [CrossRef]

16. Mogilatenko, A.; Hagedorn, S.; Richter, E.; Zeimer, U.; Goran, D.; Weyers, M.; Trankle, G. Predominant growth of non-polar a-plane (Al,Ga)N on patterned c-plane sapphire by hydride vapor phase epitaxy. *J. Appl. Phys.* **2013**, *113*, 093505. [CrossRef]

17. Tasi, C.T.; Wang, W.K.; Tsai, T.Y.; Huang, S.Y.; Horng, R.H.; Wuu, D.S. Reduction of defects in AlGaN grown on nanoscale-patterned sapphire substrates by hydride vapor phase epitaxy. *Materials* **2017**, *10*, 605. [CrossRef] [PubMed]

18. Hagedorn, S.; Richter, E.; Zeimer, U.; Weyers, M. HVPE of Al$_x$Ga$_{1-x}$N layers on planar and trench patterned sapphire. *J. Cryst. Growth* **2012**, *353*, 129–133. [CrossRef]

19. Hagedorn, S.; Richter, E.; Zeimer, U.; Weyers, M. HVPE growth of thick Al$_{0.45}$Ga$_{0.55}$N layers on trench patterned sapphire substrates. *Phys. Status Solidi* **2013**, *10*, 355–358. [CrossRef]

20. Fleischmann, S.; Mogilatenko, A.; Hagedorn, S.; Richter, E.; Goran, D.; Schäfer, P.; Zeimer, U.; Weyers, M.; Tränkle, G. Analysis of HVPE grown AlGaN layers on honeycomb patterned sapphire. *J. Cryst. Growth* **2015**, *414*, 32–37. [CrossRef]

21. Fu, Q.M.; Peng, T.; Mei, F.; Pan, Y.; Liao, L.; Liu, C. Relaxation of compressive strain by inclining threading dislocations in Al$_{0.45}$Ga$_{0.55}$N epilayer grown on AlN/sapphire templates using graded-Al$_x$Ga$_{1-x}$N/AlN multi-buffer layers. *J. Phys. D Appl.* **2009**, *42*, 035311. [CrossRef]

22. Fleischmann, S.; Richter, E.; Mogilatenko, A.; Weyers, M.; Trankle, G. Influence of AlN buffer layer on growth of AlGaN by HVPE. *Phys. Stat. Sol.* **2017**, *254*, 1600696. [CrossRef]

23. Schowalter, L.J.; Rojo, J.C.; Slack, G.A.; Shusterman, Y.; Wang, R.; Bhat, I.; Arunmozhi, G. Epitaxial growth of AlN and Al$_{0.5}$Ca$_{0.5}$N layers on aluminum nitride substrates. *J. Cryst. Growth* **2000**, *211*, 78–81. [CrossRef]

24. Wang, T.Y.; Tasi, C.T.; Lin, K.Y.; Ou, S.L.; Horng, R.H.; Wuu, D.S. Surface evolution and effect of V/III ratio modulation on etch-pit-density improvement of thin AlN templates on nano-patterned sapphire substrates by metalorganic chemical vapor deposition. *Appl. Surf. Sci.* **2018**, *455*, 1123–1130. [CrossRef]

25. Zhang, L.; Xu, F.; Wang, J.; He, C.; Guo, W.; Wang, M.; Sheng, B.; Lu, L.; Qin, Z.; Wang, X.; et al. High-quality AlN epitaxy on nano-patterned sapphire substrates prepared by nano-imprint lithography. *Sci. Rep.* **2016**, *6*, 35934. [CrossRef] [PubMed]

26. Dong, P.; Yan, J.; Wang, J.; Zhang, Y.; Geng, C.; Wei, T.; Cong, P.; Zhang, Y.; Zeng, J.; Tian, Y.; et al. 282-nm AlGaN-based deep ultraviolet light-emitting diodes with improved performance on nano-patterned sapphire substrates. *Appl. Phys. Lett.* **2013**, *102*, 241113. [CrossRef]

27. Qin, Z.X.; Luo, H.J.; Chen, Z.Z.; Yu, T.J.; Yang, Z.J.; Xu, K.; Zhang, G.Y. Effect of AlN interlayer on incorporation efficiency of Al composition in AlGaN grown by MOVPE. *J. Cryst. Growth* **2007**, *298*, 345–356. [CrossRef]

28. Kusch, G.; Li, H.; Edwards, P.R.; Bruckbauer, J.; Sadler, T.C.; Parbrook, P.J.; Martin, R.W. Influence of substrate miscut angle on surface morphology and luminescence properties of AlGaN. *Appl. Phys. Lett.* **2014**, *104*, 092114. [CrossRef]

29. Gallinat, C.S.; Koblmüller, G.; Wu, F.; Speck, J.S. Evaluation of threading dislocation densities in In- and N-face InN. *J. Appl. Phys.* **2010**, *107*, 053517. [CrossRef]

30. Jin, R.Q.; Liu, J.P.; Zhang, J.C.; Yang, H. Growth of crack-free AlGaN film on thin AlN interlayer by MOCVD. *J. Cryst. Growth* **2004**, *268*, 35–40. [CrossRef]

31. Mallick, P.; Dash, B.N. X-ray diffraction and UV-Visible characterizations of α-Fe$_2$O$_3$ nanoparticles annealed at different temperature. *J. Nanosci. Nanotechnol.* **2013**, *3*, 130–134.

32. Gong, J.R.; Liao, W.T.; Hsieh, S.L.; Lin, P.H.; Tsai, Y.L. Strain-induced effect on the Al incorporation in AlGaN films and the properties of AlGaN/GaN heterostructures grown by metalorganic chemical vapor deposition. *J. Cryst. Growth* **2003**, *249*, 28–36. [CrossRef]

33. Chandolu, A.; Nikishin, S.; Holtz, M.; Temkin, H. X-ray diffraction study of AlN/AlGaN short period superlattices. *J. Appl. Phys.* **2007**, *102*, 114909. [CrossRef]

34. Zhao, D.G.; Liu, Z.S.; Zhu, J.J.; Zhang, S.M.; Jiang, D.S.; Yang, H.; Liang, J.W.; Li, X.Y.; Gong, H.M. Effect of Al incorporation on the AlGaN growth by metalorganic chemical vapor deposition. *Appl. Surf. Sci.* **2006**, *253*, 2452–2455. [CrossRef]

35. Bryan, I.; Bryan, Z.; Mita, S.; Rice, A.; Hussey, L.; Shelton, C.; Tweedie, J.; Mara, J.P.; Collazo, R.; Sitar, Z. The role of surface kinetics on composition and quality of AlGaN. *J. Cryst. Growth* **2016**, *451*, 65–71. [CrossRef]

Article

Aqueous Synthesis, Degradation, and Encapsulation of Copper Nanowires for Transparent Electrodes

Josef Mock [1,†], Marco Bobinger [1,*,†], Christian Bogner [1], Paolo Lugli [2] and Markus Becherer [1]

[1] Chair of Nanoelectronics, Technical University of Munich, 80333 Munich, Germany; josef.mock@tum.de (J.M.); christian.bogner@tum.de (C.B.); markus.becherer@tum.de (M.B.)

[2] Faculty of Science and Technology, Free University of Bolzano, 39100 Bolzano-Bozen, Italy; paolo.lugli@unibz.it

* Correspondence: marco.bobinger@tum.de; Tel.: +49-89-289-52738

† These authors contributed equally to this work.

Received: 17 August 2018; Accepted: 25 September 2018; Published: 28 September 2018

Abstract: Copper nanowires (CuNWs) have increasingly become subjected to academic and industrial research, which is attributed to their good performance as a transparent electrode (TE) material that competes with the one of indium tin oxide (ITO). Recently, an environmentally friendly and aqueous synthesis of CuNWs was demonstrated, without the use of hydrazine that is known for its unfavorable properties. In this work, we extend the current knowledge for the aqueous synthesis of CuNWs by studying their up-scaling potential. This potential is an important aspect for the commercialization and further development of CuNW-based devices. Due to the scalability and homogeneity of the deposition process, spray coating was selected to produce films with a low sheet resistance of 7.6 Ω/sq. and an optical transmittance of 77%, at a wavelength of 550 nm. Further, we present a comprehensive investigation of the degradation of CuNWs when subjected to different environmental stresses such as the exposure to ambient air, elevated temperatures, high electrical currents, moisture or ultraviolet (UV) light. For the oxidation process, a model is derived to describe the dependence of the breakdown time with the temperature and the initial resistance. Finally, polymer coatings made of polydimethylsiloxane (PDMS) and polymethylmethacrylate (PMMA), as well as oxide coatings composed of electron beam evaporated silicon dioxide (SiO_2) and aluminum oxide (Al_2O_3) are tested to hinder the oxidation of the CuNW films under current flow.

Keywords: copper nanowires; CuNWs; degradation; encapsulation; PDMS; PMMA; solution-based; transparent electrode

---

1. Introduction

Transparent electrodes (TEs) are commonly fabricated by materials such as carbon nanotubes (CNTs) [1], poly(3,4-ethylene dioxythiophene) polystyrene sulfonate (PEDOT:PSS) [2], graphene [3], graphene oxide [4], indium tin oxide (ITO) [5] and metal nanowires (MNWs) such as silver nanowires (AgNWs) [6] and copper nanowires (CuNWs) [7]. Hybrid systems composed of a combination of these materials such as CuNWs/graphene oxide [8], AgNWs/PEDOT:PSS [9] and AgNWs/CNTs [10] were also employed for TEs. Recently, transparent metal meshes have attracted some research interest, which is attributed to their low resistance. These meshes are made of (i) printed and electroplated silver and nickel electrodes [11], (ii) transfer printed copper grids [12] as well as (iii) copper electrodes with a micromesh structure that is patterned using ultraviolet (UV) lithography and wet etching [13].

Due to their high electro-optical performance [14], ease-of-processing in great quantities [15,16] as well as facile large-area deposition techniques at ambient conditions such as spray-coating [17], AgNWs and CuNWs are increasingly considered as next generation transparent electrodes that are able to compete with ITO [18]. It should be noted that for metal nanowires, AgNWs are the major competitor for CuNWs. Both types of metal nanowires have advantages and disadvantages. On the one hand, AgNWs possess the advantage of a high chemical robustness [19]. Further, there are rapid and recent advancements in the number and the relevance of the applications for AgNWs [20–23] as well as the improvement of the film stability [24–26], whereas CuNWs readily oxidize above a temperature of 150 °C [27], under ambient conditions. On the other hand, CuNWs are grown from low-cost copper-containing precursor salts. The extremely high abundance of copper on earth compared to silver leads to a copper price that is lowered by around a factor of 80 compared to the price for silver [28–30].

Applications for the aforementioned TE materials cover a broad spectrum that ranges from transparent heaters [31], thermoacoustic speakers [32,33], solar cells [34], microlenses [35], transparent antennas [36], photodiodes [37], touch panels [38] and organic light-emitting diodes (OLEDs) [1] to electromagnetic interference shieldings [39], as well as piezo- [40] and pyroelectric [41] energy conversion.

The first solution-based synthesis of CuNWs was reported by Chang et al. in 2005 [42]. In their work, copper(II) nitrate $Cu(NO_3)_2$ was utilized as the copper-containing precursor, ethylenediamine (EDA) as the capping agent and hydrazine (N_2H_2) was used as a reducing agent for the copper nitrate and the solvent. Since then, numerous studies that are summarized in References [27,43] have been published. A significant advancement towards an environmentally friendly synthesis has been achieved by Hwang et al. in 2016, where DI water instead of hydrazine was employed as the solvent [44]. The synthesis of the CuNWs in this contribution is based on the protocol of Hwang et al. that we recently tailored with regard to the material ratios, the process time and the process temperature, as well as the addition of alcoholic co-solvents [27]. For our synthesis, copper(II) chloride dihydrate ($CuCl_2 \cdot 2H_2O$), was used as the precursor material and *L*-Ascorbic acid (AA) was used as an environmentally friendly and mild reducing agent. Oleylamine (OM) served as the capping agent and DI water was utilized as the solvent. The self-seeding process requires neutral Cu^0 atoms, which are provided by the oxidation of AA to dehydroascorbic acid that goes along with the reduction of Cu^{2+}-ions to Cu^0-atoms. As sketched in Figure 1a that depicts a schematic for a single CuNW growth, OM has an increased adsorption rate for the (100) plane compared to the (111) plane [45]. This higher adsorption rate leads to the passivation of the wire shell and in turn, allows for the uniaxial wire growth along the (111) plane [44]. In agreement with the literature [15,44,46], the resulting CuNWs have a face-centered cubic (fcc) structure and a pentagonal cross-section [47,48], as shown in the high-resolution field-emission scanning electron microscope (FESEM) image of a single CuNW in Figure 1b.

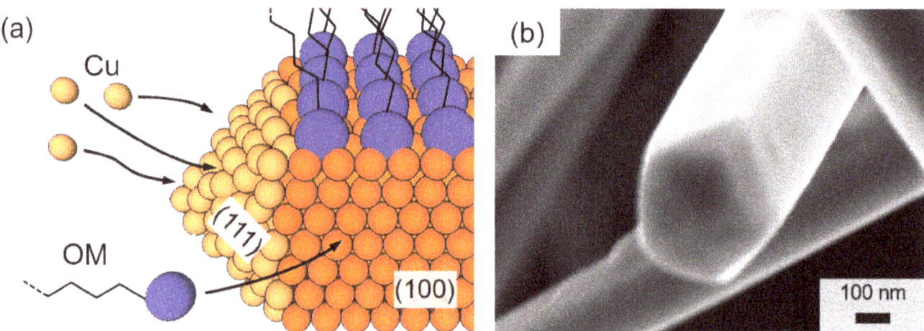

Figure 1. (**a**) Schematic for the growth of a single copper nanowire (CuNW) along the (111)-plane. The blue and orange spheres indicate oleylamine (OM) head groups and copper atoms, respectively. (**b**) High-resolution field-emission scanning electron microscope (FESEM)-image for a single copper nanowire with the characteristic pentagonal shaped cross-section.

It can be concluded that the growth mechanism of CuNWs is well understood and the aqueous synthesis protocol is tailored to an extent that allows producing TEs with properties comparable to the ones of AgNWs and ITO. However, for the commercialization of CuNWs, their up-scaling potential, that has not yet been studied for the aqueous synthesis, is crucial. Thus, in this work, we investigate the effect of the precursor-to-solvent mass ratio on the wire growth. A large drawback of CuNWs that cannot be omitted is their gradual oxidation at ambient conditions, which is readily increased for temperatures above 100 °C [49]. As tested by a few groups with success, this oxidation can be slowed by encapsulating the CuNWs with following approaches: (i) applying polymer coatings composed of polydimethylsiloxane (PDMS) [49], polymethylmethacrylate (PMMA) [49], PEDOT:PSS [50], and polyurethane acrylate resin [51] (ii) atmospheric pressure spatial atomic layer deposition (AP-SALD) of aluminum oxide (Al_2O_3) and (iii) the electrodeposition of zinc, tin and indium shells onto the nanowires, followed by their oxidation [30]. Besides encapsulation, there are also works that report on the laser-induced nanowelding of CuNWs [52] and even the photothermochemical reduction [53,54] of oxidized and non-conductive CuNW networks to highly conductive ones. Both the nanowelding of CuNW junctions and the laser-induced reduction of CuNWs improved the oxidation robustness of the films. However, to this day, a comprehensive study for the degradation of CuNWs with regards to various environmental stresses and a model for the oxidation mechanism are still missing.

Therefore, in this work, we report on a rigorous degradation study for CuNW films that are subjected to ambient air, elevated temperatures, high electrical currents, moisture, and UV light. A purposely developed and novel model that describes the time- and temperature- dependent formation of an oxide is presented. Encapsulations made of polymers or ebeam evaporated oxides are deposited to the CuNW films to increase their lifetime. Further, to the best of our knowledge, the scaling potential of the presented aqueous synthesis of CuNWs is for the first time studied by varying the precursor-to-solvent ratio. This study aids in assessing the scaling and in turn also the commercialization potential of the presented synthesis protocol.

2. Materials and Methods

2.1. Synthesis of Copper Nanowires

The protocol for the aqueous synthesis of CuNWs was adapted from one of our previous publications [27]. For a standard synthesis, 300 mg copper(II) chloride dihydrate ($CuCl_2 \cdot 2H_2O$) (Sigma Aldrich, St. Louis, MO, USA, C3279) were immersed in 25 g deionized (DI) water and bath sonicated for a duration of 5 min. Then, 900 mg oleylamine (Sigma-Aldrich, St. Louis, MO, USA, O7805) was added and the solution was horn sonicated for 90 s at a power of 200 W, using an S-450 Digital Sonifier® from Branson Ultrasonics Corporation (St. Louis, MO, USA). Subsequently, 300 mg *L*-Ascorbic acid (Sigma-Aldrich, St. Louis, MO, USA, A92902) dissolved in 5 g DI water was added. The solution was then allowed to age for 12 h in a silicone oil bath at a calibrated temperature of 79 °C. All the reagents were processed without further purification. For the study of the precursor-to-solvent weight ratio, the content of DI water was kept constant and the remaining materials were changed accordingly.

2.2. Synthesis Analysis

Scanning electron microscope (SEM)-images were recorded with an NVision40 FESEM from Carl Zeiss (Oberkochen, Germany) at an acceleration voltage of 7 kV, an extraction voltage of 5 kV and a working distance of 5–6 mm, which was optimized to achieve the best image quality. The wire lengths were evaluated manually using the software Gwyddion (v2.48). For an automated evaluation, DiameterJ (v1.018), a plugin invented for ImageJ (v1.51j8), was utilized. More details on this method are reported elsewhere [28].

2.3. Spray Ink Preparation and Deposition

The as-synthesized solution was allowed to cool down to room temperature. The top solution was decanted and the CuNW-containing precipitate was rinsed with isopropyl alcohol (IPA). To remove the OM and eventually a thin oxide shell that has formed after the synthesis, the CuNW precipitate was immersed with 5 wt% propionic acid (Sigma Aldrich, St. Louis, MO, USA, P5561) in 30 g IPA and allowed to react for 5 min. After centrifugation at a speed of 1.75 krpm for a duration of 5 min, the precipitate was immersed in 30 g IPA and sonicated for 20 min in an USC300TH bath sonicator from VWR International (Radnor, PA, USA), at frequency and power of 45 kHz and 80 W, respectively, to improve the dispersion of the CuNWs. As observed in a previous publication, bath sonication with the parameters above for a duration below 30 min does not lead to cracks or deformations of the wires [27]. For the spray deposition, a commercial and handheld airbrush (Triplex II F from Gabbert, Leipzig, Germany) with an orifice of 150 μm was utilized and operated with pressurized nitrogen (1.5 bar). To increase the evaporation of the solvent, the samples were placed on a Thermofisher RT2 hot plate and heated to a temperature of 70 °C. In a glovebox, under nitrogen atmosphere, all CuNW films were subjected to a thermal sintering treatment at a temperature of 200 °C for a duration of 30 min.

2.4. Electro-Optical Characterization

The transmittance spectra were recorded in the visible range using a 300 W xenon arc lamp, chopped at a frequency of 210 Hz. The light passes through an Oriel Cornerstone 260 $\frac{1}{4}$ monochromator and a silicon-based photodiode with a transconductance amplifier that is connected to a 70105 Oriel Merlin digital lock-in amplifier. The calibration of the photodiode was performed with a glass substrate to determine the pure transmission of the CuNW films. The sheet resistances were measured using a four-point probe head from Jandel (Linslade, UK) connected to a B2901A Keysight (Santa Rosa, CA, USA) source measuring unit (SMU). A constant current of 1 mA was sourced for all measurements.

2.5. X-ray Photoelectron Spectroscopy

X-ray photoelectron spectroscopy (XPS) measurements were performed at a base pressure of 5×10^{-10} mbar using monochromatic K_α radiation from an aluminium anode that is operated at an electrical input power of 350 W. The spectra were acquired using a SPECS Phoibos hemispherical analyzer at a pass-energy of 30 eV with an energy resolution of 0.05 eV. The raw data were processed using the software CasaXPS from Casa Software Ltd. (Teignmouth, UK). The backgrounds of the spectra were removed by Shirley background subtraction [55].

2.6. Degradation Tests

The CuNW films were sprayed to a glass substrate with the dimensions of 5.0×5.0 cm^2 and a thickness of 1.45 mm, in accordance with the process described in Section 2.3 and as reported in a previous work [27]. The films were contacted using commercial copper tape and conductive silver paint RS Pro from RS components (article number: 186-3600, Corby, UK) with a silver content of 50–75%, which resulted in an active CuNW area of 3.5×5.0 cm^2 = 17.5 cm^2.

2.6.1. Ambient Air

The degradation tests in the ambient air were performed in an air-conditioned lab environment at a temperature and a relative humidity of around 22 °C and 40%RH (=relative humidity), respectively. The samples were shielded from the light to exclude a light-induced degradation that is studied in Section 3.2.5.

2.6.2. Elevated Temperatures

For the degradation with regard to elevated temperatures, the samples were placed on an RT2 hot plate from Thermofisher, under ambient conditions, and heated from room temperature to the desired annealing temperature. The resistance-time curves were tracked in time steps of 1 s using a Keithley 2200-30-5 power supply operated at a small probe current of 5 mA that was sourced to all devices.

2.6.3. Electrical Current

The CuNW films were driven at an electrical and constant DC power of 6 W using a programmable power supply 2200-30-5 from Keithley (Cleveland, OH, USA). To keep the power at a constant level for increasing resistances, a LabVIEW (v2017) program was developed to adjust the sourced current accordingly, after each time step.

2.6.4. Moisture

The CuNW films were placed in a climatic chamber VCL 4006 from Vötsch Industrietechnik GmbH (Balingen, Germany) and the relative humidity was either set to a low value of 20%RH or a high value of 90%RH. For all measurements, the temperature was kept at a constant value of 60 °C and DI water with a resistivity of 18.2 MΩ·cm was utilized. In accordance with Sections 2.6.2 and 2.6.3, the resistance-time data was recorded using a programmable power supply, which was operated with a probe current of 5 mA.

2.6.5. UV-Light

UV light illumination of the CuNW films was performed using the illuminator box 1S from Gie-Tec GmbH (Eiterfeld, Germany) equipped with four special fluorescent tubes with a total electrical power consumption of 32 W. According to the manufacturer, the tubes emit in a wavelength range of 350 to 400 nm. The 2probe resistances of the films were measured over a duration of 1 month using a standard handheld multimeter VC830 from Voltcraft (Wollerau, Germany).

3. Results

3.1. Synthesis of Copper Nanowires

In our previous work, we have tailored the aqueous synthesis of CuNWs [27]. As the next important step that aids to assess the commercialization potential of our synthesis, we report on an up-scaling study for the aforementioned aqueous synthesis of CuNWs. For this, the precursor-to-solvent ratio is varied by around a factor of 10 and the resulting growth product is analyzed with regard to the quality of the dispersion and the nanowire diameters. A low mean diameter for the nanowires is a commonly accepted criterion for a high quality of the synthesis since it is usually accompanied by a high aspect ratio, i.e., length-to-diameter ratio, and a low haze value [56,57]. The SEM-images for the precursor-to-solvent series are depicted in Figure 2 for different mass ratios (precursor:solvent) of (a) 1:300, (b) 1:100, (c) 1:50 and (d) 1:33. Along with the precursor weight, the weights of OM and AA were adjusted accordingly. From the SEM-images in Figure 2c,d for the large precursor contents, following effects can be seen: (i) For the highest ratio of 1:33, the wire morphology is visibly degraded; (ii) the diameters are increased compared to the ones shown in the SEM-images for the lower precursor concentrations in (a) and (b); and (iii) the CuNWs tend to form clusters. After the analysis of the described aqueous CuNW synthesis, a mean diameter of 134 nm $\pm$ 4 nm and a mean length of 40 μm $\pm$ 21 μm could be obtained for the wires shown in Figure 2b. Employing the method outlined in Section 2.2 and in previous publications [28,29,58], the diameters were analyzed quantitatively and in an automated way from SEM-images (see Appendix A, Figures A1 and A2 for the high-magnification SEM-images and the diameter histograms, respectively). The mean diameters as a function of the precursor-to-solvent ratio are depicted in Figure 3.

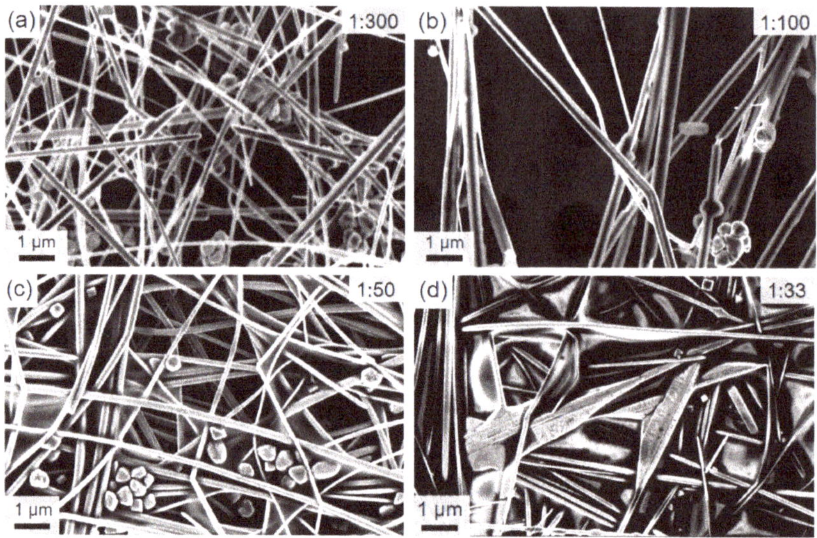

Figure 2. FESEM-images of CuNWs for the precursor-to-solvent series with different mass ratios (precursor:solvent) of (**a**) 1:300, (**b**) 1:100, (**c**) 1:50 and (**d**) 1:33.

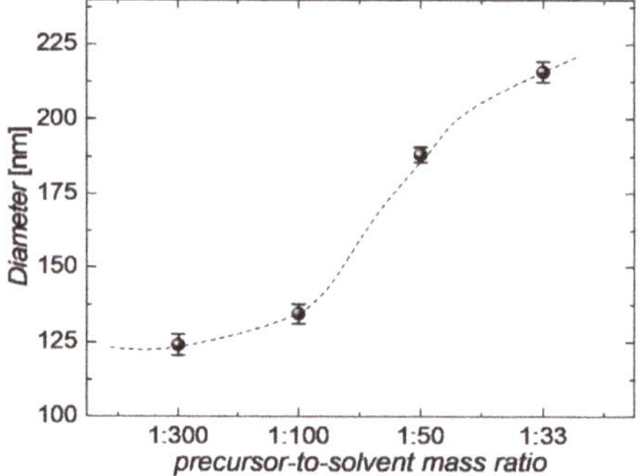

Figure 3. Mean diameters extracted using DiameterJ from the FESEM-images illustrated in Figure A1 for the precursor-to-solvent series with different mass ratios. The dashed line serves as a guide to the eye.

From Figure 3 it can be seen that the diameter decreases gradually with a reduction in the precursor concentration from a mean diameter of 215 nm to 124 nm, for a ratio of 1:33 to 1:300, respectively. We believe that there is a critical pre-cursor-to-solvent ratio of around 1:100. Below this ratio, well-defined nanowires with a small diameter grow since the formation of copper-oleylamine micelles is stable. Above a ratio of 1:100, the micelles are not stable enough since the concentrations for copper-oleylamine complexes and subsequently their interaction has increased. Similar to the case of adding alcoholic co-solvents, as previously reported [27,44], the interaction of micelles can lead to their rupture [59,60], which has a negative effect on the uniaxial wire growth since adsorption rates on the (100) and (111) planes are different. In summary, it can be concluded that a range of 10 is sufficient to study the up-scaling potential of the presented synthesis for the following reasons. For a large ratio of 1:33, the wire morphology is clearly degraded, whereas for a low ratio of 1:300, the reduction in diameter is low and already implies a saturation behavior, as can be seen from Figure 3.

3.2. Degradation of Copper Nanowires

Up to date, there are only a few studies that report on the degradation of CuNWs. So far, in agreement with the literature, the following stresses have been applied to the CuNW films: exposure to (i) sunlight [61], (ii) elevated temperatures [61], (iii) humidity [61] and (iv) electrical current [49,61]. Nevertheless, a comprehensive study that addresses the major stresses for the chemically-induced degradation of CuNWs is still missing. In the following, the degradation of CuNW films will be studied with regard to (i) exposure to ambient air at ambient temperature, (ii) exposure to elevated temperatures, (iii) prolonged electrical current flow, (iv) different moisture levels and (v) UV-visible light exposure. In Section 3.2.6, for the first time, a model for the temperature-induced oxidation of CuNWs is presented. This model describes the oxide formation on the nanowire shell and relates the breakdown time of the films to the annealing temperature and the initial resistance of the nanowire networks. Further, X-ray photoelectron spectroscopy (XPS) spectra that allow for resolving the chemical alterations induced by the environmental stresses are discussed.

3.2.1. Ambient Air

A long-term stability test under ambient atmosphere was performed for CuNW films with different initial resistance. Prior to this test and in accordance with Section 2.3, the as-synthesized and spray-deposited CuNW-films were subjected to thermal annealing at a temperature of 200 °C for a duration of 30 min, under a nitrogen atmosphere. It should be noted that this post-deposition treatment was applied for all CuNW films to lower their resistances uniformly across the film and to form a good mechanical and electrical contact at the wire-to-wire junctions, which can aid to improve the robustness of the films. The normalized increase in resistance R/R_0 for CuNW films with different transmittances that are exposed to the ambient air is depicted in Figure 4a, over a duration of 25 days. R_0 denotes the resistance value of the CuNW films at ambient conditions, before the different degradation tests are performed. The black squares indicate the data for a CuNW-film with a transmittance of around 85%, whereas the red spheres represent the data for a CuNW film with a transmittance of around 81%. It can be recognized that the film with a high transmittance has almost turned non-conductive after 25 days, whereas the resistance of the film with a low transmittance has increased by only a factor of six. The increase in resistance is attributed to the surface oxidation of the CuNWs [27], which can be recognized by the formation of metal oxide nanoparticles CuNW shell. The formation of metal oxide nanoparticles has also been observed below for the degradation with regard to electrical current and humidity (see Figure A4). This effect is illustrated in Figure 4b that shows the FESEM-image for CuNWs, which were exposed to the ambient atmosphere for more than 25 days. The fact that the film with a lower transmittance and in turn a lower initial resistance shows a reduced change in resistance over time can be understood as follows: The reduced resistance is connected with a higher wire density, which increases the probability that highly conductive paths composed of thick wires and low junction resistances are established. Since the thick wires are more resilient to oxidation, the total resistance of the film is more stable when exposed to the ambient atmosphere or a harsh environment. This behavior is in agreement with the previously published work on random percolating networks from Manning et al. [62], who observed the presence of so called *Winner Takes it All* (WTA) pathways. These WTA pathways are percolating paths with resistances that are significantly lowered compared to other paths and thus carry the major portion of the current. In their study, the lowered resistance was attributed to a lowered junction resistance, whereas for our study, we believe that the lowered resistance stems from the lowered junction resistance and pathways composed of thicker wires.

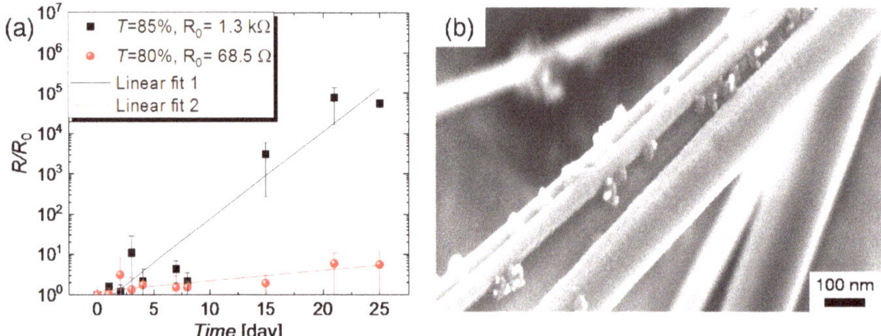

Figure 4. (**a**) Normalized increase in resistance R/R_0 of two CuNW films with different transmittances exposed to ambient conditions as a function of the time; (**b**) FESEM-image of CuNWs exposed to ambient conditions for more than 25 days.

3.2.2. Elevated Temperatures

In this section, the degradation of CuNW films that are subjected to elevated temperatures in a range of 100–200 °C are studied. For this, the films on a glass substrate are placed on a hot plate, under ambient conditions. The films are subsequently heated to various temperatures and their change in resistance with regard to the initial resistance at room temperature is tracked over time, as shown in Figure 5a. To quantify the breakdown behavior, a breakdown time t_{BD} is introduced and plotted in Figure 5b as a function of the temperatures. t_{BD} is defined as the time when the resistance of the CuNW films has doubled with respect to the initial value. For the higher temperatures, i.e., for 150, 175 and 200 °C, it can be concluded that the increase in resistance goes along with an increase in temperature. However, the change in resistance for the lower temperatures, i.e., for 100 and 125 °C, does not follow the aforementioned trend. This counter-intuitive behavior can be explained by considering the initial resistances of the films. On a qualitative level, it can be argued that a CuNW film with a lowered initial resistance shows a high wire density and thus the effect of chemical degradation induced by oxidation of the nanowire shell seems to be less. Instead, the higher number of junctions lead to a reduced influence of each junction on the complete film behavior on a denser film. To better understand this phenomenon, a model that describes the oxidation of CuNWs and captures the breakdown time, the annealing temperature, and the initial resistance of the films is derived in Section 3.2.6.

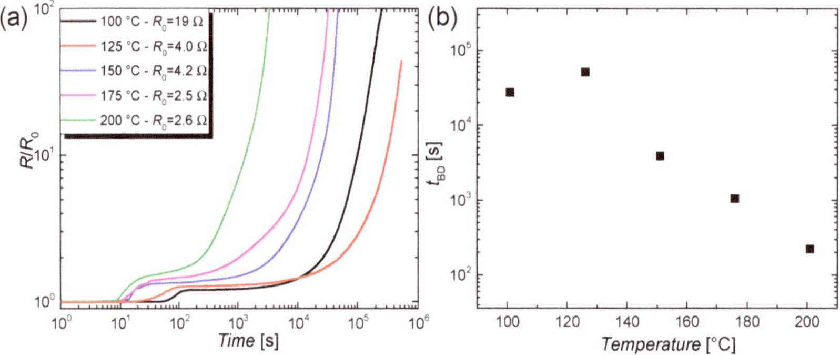

Figure 5. (**a**) Normalized increase in resistance R/R_0 over time for different temperatures. The CuNW films were placed on a hot plate in ambient air; (**b**) Breakdown time t_{BD} as a function of the annealing temperature.

3.2.3. Electrical Current

The degradation of CuNW thin films under current flow will be presented in the following. A photograph for a typical CuNW electrode under test is illustrated in Figure 6a.

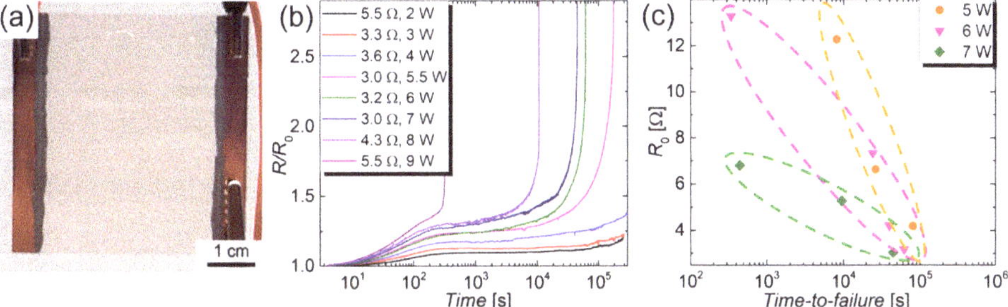

Figure 6. (**a**) Photo of a CuNW film spray-deposited to a glass substrate and electrically contacted on each side by copper tape and conductive silver ink; (**b**) Normalized resistance R/R_0 as a function of time for CuNW heaters that show resistances in a range of 3.0 Ω to 5.5 Ω and were subjected to an electrical input power ranging from 2 W to 9 W; (**c**) R_0 as a function of the time-to-failure for CuNW heaters that are subjected to powers of 5 W, 6 W and 7 W, where the ellipses indicate the regions of time-to-failure for same power values.

For this degradation study, a LabVIEW program was purposely developed to source a constant power in a range of 2 W to 9 W over an effectively heated area of 17.5 cm^2, via programmable sources, as described in Section 2.6.3. The normalized increase in resistance R/R_0 of the CuNW films is depicted in Figure 6b. It can be recognized that up to a power of 4 W the heaters are relatively stable and can be operated for more than 3 days (1 day = 0.86×10^5 s), without device failure. For higher powers, i.e., for 5.5 W, the films show a rapid increase in resistance after a duration of around 8 h, which, quickly results in a device failure. For the change in resistance behavior, three regimes can be identified: (I) a fast increase in resistance in the first 5 min attributed to the temperature-induced change in resistance, (II) a slow and gradual increase in resistance due to the gradual oxidation of the films and finally (III) a fast electrical breakdown of the films that is accompanied by fragmentation for the CuNWs, as discussed below. From the R/R_0-plots, a time-to-failure is defined as the time after which the programmable power source applies the maximum programmable voltage of 30 V to maintain the predefined input power. The time-to-failure for CuNW films is shown in Figure 6c. The dotted ellipses indicate the regions of time-to-failures for several initial resistances with the same applied power value. Independent of the applied powers and the initial resistances, the CuNW films undergo a slow and gradual oxidation followed by a rapid breakdown within around 10 s. This behavior is visualized in Figure A5, that depicts the shifted (x-axis) R/R_0-time curves for CuNW films at the moment of the breakdown. For the CuNW films it is evident that an increase in power or an increase in the initial resistance leads to a faster degradation and in turn lowers the time-to-failure from almost 1 day for $(P_1;R_{0,1}) = (5 \text{ W};4.5 \text{ Ω})$ to 7 min for $(P_2;R_{0,2}) = (7 \text{ W};7 \text{ Ω})$. The increase in power leads to an increase in temperature and in turn, enhances the oxidation and the fragmentation of the wires. A higher initial resistance and in turn lower network density also reduces the chemical robustness of the films since the probability of forming robust percolating paths with thick wires decreases with decreasing wire density. In order to verify that the degradation is enhanced for lower network densities and in turn higher initial resistances, as seen in Figure 6c, a wire density has been extracted for all CuNW-films from the microscope images shown in Figure 7 for (a) a low and (b) a high network density.

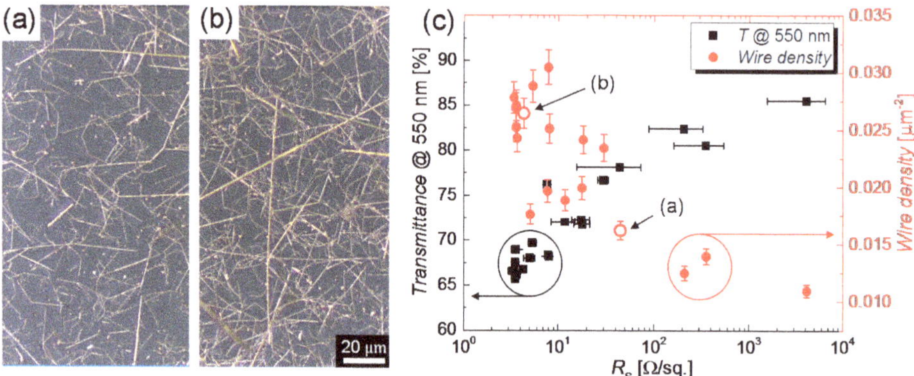

Figure 7. Light microscope images for CuNW films with (**a**) a low and (**b**) a high wire density. (**c**) Transmittance evaluated at a wavelength of 550 nm and wire density as a function of the room temperature resistance. The wire densities corresponding to the images (**a**,**b**) are indicated by red hollow symbols.

The transmittances of the films at a wavelength of 550 nm as well as their corresponding wire densities are shown in Figure 7c as a function of the sheet resistance. The data points in Figure 7c that are associated with the low and high network densities illustrated in Figure 7a,b, respectively, are drawn as hollow symbols. In accordance with the expectation, the transmittance shows a decreasing trend with decreasing initial resistance, whereas the wire density increases. The highest electro-optical performance was achieved for a film with a sheet resistance of 7.6 Ω/sq. at a transmittance of 77%. These values yield to a figure of merit ($FoM = T^{10}/R_S$) of 1000 $\times$ FoM = 9.6, which follows the commonly accepted definition from Haacke [63]. This FoM compares well to the values reported for metal nanowire-based TEs, which lie in the range of 1000 $\times$ FoM = 1–30 [14]. Next, two types of current densities that are commonly used in the literature are calculated and allow a comparison with other film heater and degradation studies. The total current I sourced to the network, the wire density n, and the total area of the heater A are used to calculate the current density per nanowire, in agreement with previous works [64]. By division with the cross-section area using the mean diameter of one nanowire d, the current density j for each individual nanowire can be calculated as follows:

$$j = \frac{I}{\left(n \cdot A \cdot \pi \cdot \frac{d^2}{4}\right)} \tag{1}$$

Depending on the wire density, for a sourced power of around 6 W, a mean current density of around 1.5 to 2.75 MA/m^2 that is sourced to the wire cross sections in the network is calculated. Another definition for the current density, denoted as $j_{Khaligh}$, was introduced by Khaligh et al., as follows [65]:

$$j_{Khaligh} = \frac{I}{A} \tag{2}$$

where I and A denote the total current and the effectively heated area, respectively. Following this definition, the current densities calculate to around 10 to 100 mA/cm^2 for different initial resistance values and sourced powers in the range of 4 W to 9 W. A CuNW-heater that has been subjected to an electrical input power of 6 W until its breakdown, after around a duration of 17 h, is shown in Figure 8a. It should be noted that we considered the current-temperature dependence of the films that heat up due to Joule heating. For this reason, the current values of the heated films and not the ones at room temperature were considered. At an electrical power of 4 W that leads to a temperature of the film of around 100 °C, the resistance increases by a factor of around 1.25, which leads to a reduction in current of 10.5%. For this calculation, we assumed a temperature coefficient of 3.2 mK^{-1} for CuNW networks, which we determined in a previous work [27].

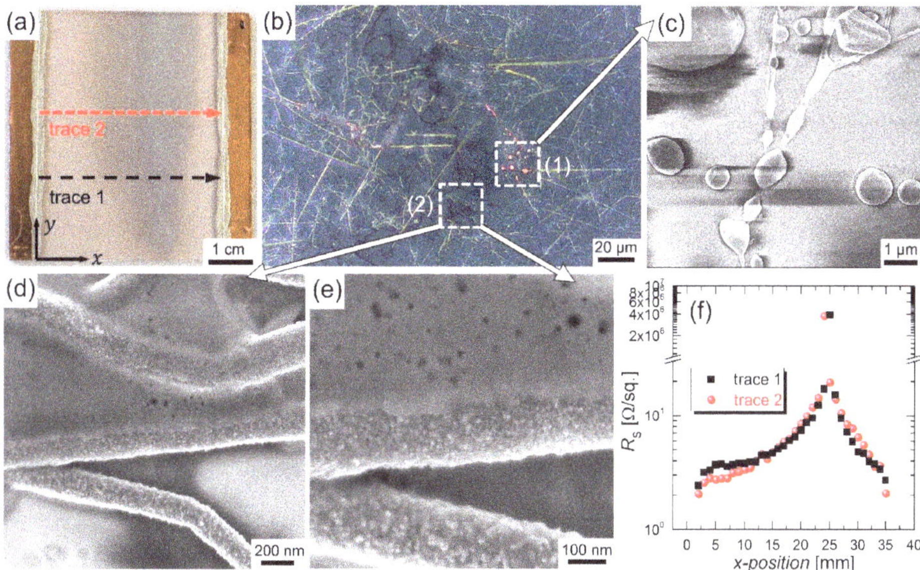

Figure 8. (**a**) Photo for a CuNW heater with a sheet resistance of around 3.2 Ω/sq. after electrical breakdown. The heater was subjected to an electrical input power of 6 W for around 17 h; (**b**) Microscope images for the CuNW film after an electrical breakdown at a position with >MΩ sheet resistance; (**c–e**) SEM-images for the CuNW heater shown in (a,b) at a position with >MΩ sheet resistance; (**f**) R_S across the CuNW heater as a function of the measurement position, as indicated in (**a**).

The reddish and greyish areas in the degraded film are associated with slightly and heavily oxidized CuNWs, respectively. The greyish appearance of oxidized CuNW films has already been observed in other studies [27]. A microscope image that provides an insight into the greyish section of the film is depicted in Figure 8b. Two main degradation mechanisms are clearly visible and further supported by SEM-images: (c) fragmentation and (d,e) oxidation that goes along with the formation of copper oxide nanoparticles, as observed in Figure 4b for the degradation under ambient air. To correlate the color of the CuNW film with the degree of oxidation, the sheet resistances were measured across the film, along the two measurement traces in x-direction that are indicated in Figure 8a (see Figure A3 for a photo of the four probe needle placement, which was aligned along the y-direction). At the position of the maximum degradation, i.e., where fragmentation can be observed, the film turns non-conductive, whereas in the remaining areas of the heater, the resistance changes from around 3 Ω in the vicinity of the contact leashes to around 20 Ω around the heavily oxidized part. The formation of a crack during the failure of metal nanowire networks under current flow, as shown in Figure 8a, has also been observed in a previous study for the case of AgNWs [66]. In this work, Sannicolo et al. [66] showed that the electrical breakdown takes the form of a global or statistical phenomenon. Further, the authors simulated the formation of a crack during the failure of the network, which propagates parallel to the bias electrodes. These results indicate that a stable operation of the CuNW heaters is possible for more than 3 days at an input power of around 4 W, which corresponds to a current density with respect to the wire cross-section of 1.2 to 2.2 MA/m^2. The temperature of the CuNW film subjected to a power of 4 W corresponds to a temperature of around 100 °C, in agreement with our previous works for CuNW-based heaters [27,28]. For higher input powers, the fragmentation and oxidation of the CuNWs are readily enhanced.

3.2.4. Moisture

For the degradation study of sprayed CuNW films with regard to elevated moisture levels, the films are placed in a climatic chamber, at a constant temperature of 60 °C, and the relative moisture is varied. The normalized increase in resistance R/R_0 is depicted in Figure 9 for two different moisture levels of (a) 90%RH and (b) 20%RH.

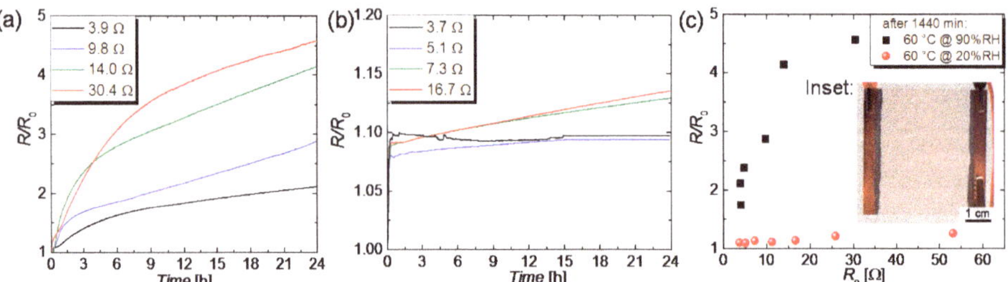

Figure 9. R/R_0 plotted as a function of time for CuNW heaters with different R_0 that were subjected to different moisture levels of (**a**) 90%RH and (**b**) 20%RH, at a temperature of 60 °C and for a duration of 24 h. (**c**) R/R_0 as a function of R_0 for the two moisture levels. The inset shows a photo for a CuNW film.

After a duration of 24 h, the resistances of the CuNW films subjected to a temperature of 60 °C and a relative moisture of 90%RH increased up to around a factor of 2 and 4.5, respectively, for the films with the lowest and the highest initial resistance. This result is in accordance with the expectation that films with a lowered initial resistance are more robust to degradation, which reflects in a reduced increase in resistance. Figure 9b shows R/R_0 as a function of the time for CuNW films subjected to a temperature of 60 °C and a humidity of 20%RH. In the beginning, the resistance of the CuNW rises fast, attributed to the resistance-temperature dependence of copper. After the climatic chamber has reached its target temperature of 60 °C, the films show a gradual increase in resistance of around 1 to 4%, for a total duration of 24 h. Similar to the behavior of the high relative moisture of 90%RH, the films with a higher initial resistance are more prone to degradation. For the films subjected to the high humidity value, the formation of metal nanoparticles could be seen, as shown in Figure A4, and observed for the case of the exposure to the ambient air (see Figure 4b) and electrical current (see Figure 8d,e). The aforementioned results are summarized in Figure 9c that shows R/R_0 as a function of the initial resistance R_0 for the two different relative moistures. It can be concluded that the degradation is greatly enhanced for the high relative moisture of 90%RH, which leads to an increase in resistance that is around a factor of 100 larger than for the low relative moisture of 20%RH.

3.2.5. UV Light

Since CuNWs are also intended to be integrated to touch panels or into aircraft and car windows, their degradation during the exposure to ultraviolet-visible (UV-vis) light has to be investigated. To study the UV-vis-light-induced degradation, CuNW films with comparable initial resistances and a transmittance of around 70% were placed in a UV-vis box that is described in Section 2.6.5, under (i) direct light exposure and (ii) shielded from the light by a thin aluminum plate. The normalized resistances over time for CuNW films subjected to both aforementioned stresses are depicted in Figure 10a. For a better separation of the degradation mechanism, the resistances for samples that are subjected to the ambient air are also characterized and the results are plotted in Figure 10a. It should be noted that, for each stress, the resistances of four different CuNW films that are exemplarily illustrated in the inset in Figure 10b for the case of exposure to UV-vis-light, were considered. From Figure 10a it can be seen that the increase in resistance is more pronounced for the films placed in the UV-vis Box than for the ones placed in the ambient air. It should be noted that the CuNW films subjected to the ambient air, as shown in Figure 10a, are more robust than the ones characterized in Figure 4a.

The difference in the degree of degradation can be attributed to the different transmittance values of the films, which were significantly lower for the UV-treated samples, i.e., around 70% compared to the 80% and 85% for the films shown in Figure 4a. The increased degradation for the CuNW films shielded from the light stems from an increased temperature in the box of around 60 °C, as measured using a Pt100 thermoresistor. The increase in resistance of the CuNW films subjected to UV-vis light is larger than the increase for the shielded samples, which indicates that UV-vis light, in fact, induces a degradation for CuNWs, as previously reported [61]. In the literature, papers that report on the UV-induced degradation of polymers [67,68] and silver nanoparticle [69] can be found. However, so far, no study gives an explanation for this effect, for the case of CuNWs.

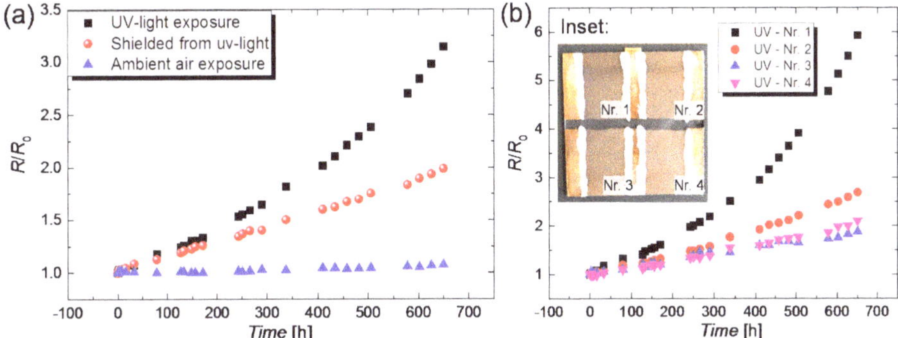

Figure 10. (**a**) Normalized increase in resistance R/R_0 over time for CuNW films with a transmittance of 70% that were (i) subjected to ultraviolet-visible (UV-vis) light, (ii) shielded from UV-vis light and (iii) exposed to the ambient air. Each symbol represents a mean resistance that was determined by averaging over four different samples; (**b**) R/R_0 over time for four different CuNW films that are subjected to prolonged UV-vis exposure. The inset depicts the four CuNW samples that were subjected to UV-vis exposure along with a labeling that is in accordance with normalized resistance curves.

We strongly believe that the degradation arises from free reactive oxygen species (ROS), generated by longwave UV light (UVA) in the range of 320–400 nm, as reported by Herrling et al. [70] and other groups [71]. A higher ROS concentration is accompanied by an increased oxidation of the metal and subsequently a faster degradation of the CuNW films. An additional degradation process can be induced by an increase in temperature. This increase in temperature can be attributed to the generation of surface and bulk plasmon polaritons that are predominantly formed by absorption of light in the UV-vis spectrum, below a wavelength of 500 nm [72,73].

3.2.6. Discussion and Modeling of the Temperature-Induced Oxidation

The degradation tests showed that the increase in resistance under ambient air is greatly increased for higher temperatures. As reported in previous works, the degradation of CuNWs in ambient air stems from the oxidation of the nanowire shell to copper oxide [49]. However, to date, there is no study for CuNWs that discusses the formation of two oxide shells with a different chemical composition on the CuNWs and presents a model for the time- and temperature-dependence of the oxide thickness, which will be discussed below. Via X-ray photoelectron spectroscopy (XPS), the oxidation of the CuNW films was studied in more detail, as shown in Figure 11, that depicts the high-resolution Cu 2p core-level spectra for CuNW films subjected to all the environmental stresses that were discussed in this paper. For the as-deposited CuNW films, no oxygen-related contribution could be identified from the Cu 2p peak, whereas for all other samples, a clear appearance of copper shake-up peaks that are centered around a binding energy of 942 eV and 962 eV can be recognized. This appearance of shake-up peaks along with a shift of the Cu $2p_{1/2}$ and Cu $2p_{3/2}$ peak to higher binding energies is a proof for the predominant formation of divalent copper species, i.e., CuO [74–76]. Further, a larger shift of the peaks to higher binding energies indicates a higher degree of oxidation.

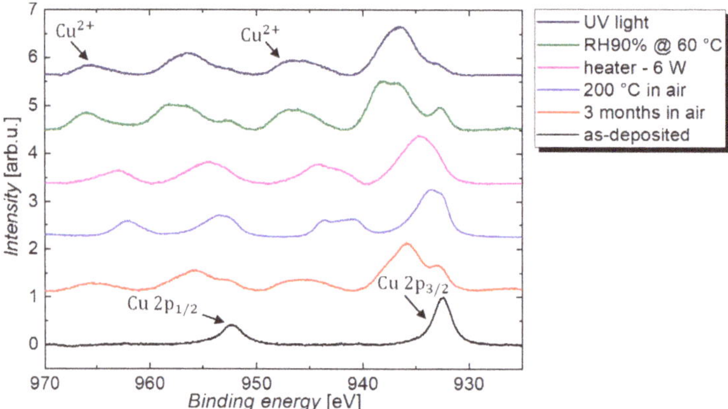

Figure 11. X-ray photoelectron spectroscopy (XPS) spectra for CuNW film that were subjected to different environmental stresses, under ambient air, i.e., (1) as-deposited, (2) exposure to ambient air for 3 months, (3) subjected to a temperature of 200 °C for a duration of 30 min, (4) heated at a power of 6 W until breakdown, (5) subjected to a relative humidity of 90%RH, at a temperature of 60 °C, and (6) UV-light exposure.

This oxidation of the CuNW film also becomes evident from its greyish appearance, as shown in Figure 12 for the photos of (a) an as-deposited CuNW film and (b) a CuNW film subjected to a temperature of 175 °C for a duration of 1 h. However, there is no study that reports on a model or a quantitative description for the oxidation mechanism of CuNWs. As sketched in Figure 12c, the corroded CuNW shell is known to consist of a thick inner shell of cuprous oxide (Cu_2O) that, on the outer face, reacts further to cupric oxide (CuO) [77,78]. The formation of CuO on the outermost shell of the CuNWs, which is in agreement with the literature, is also in accordance with the XPS spectra shown in Figure 11 that indicate that mostly CuO is present.

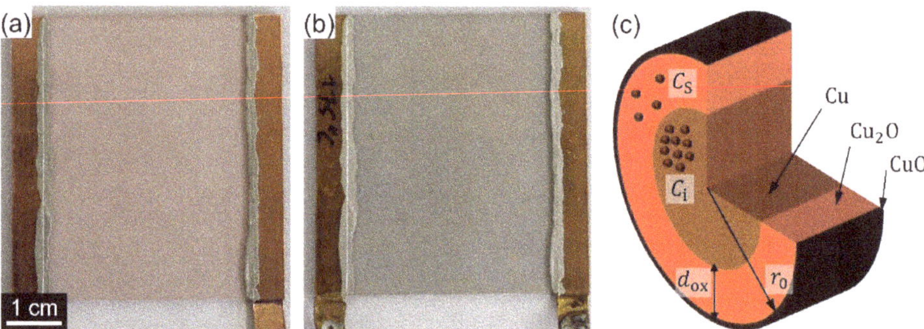

Figure 12. Photos for (**a**) an as-deposited CuNW film and (**b**) a CuNW film subjected to a temperature of 175 °C for a duration of 1 h. (**c**) Schematic of an oxidized CuNW along with the parameters that are used in the text to describe the oxidation mechanism.

The chemical reactions for the formation of Cu_2O and CuO are given in Equations (3) and (4).

$$4Cu + O_2 \rightarrow 2Cu_2O \tag{3}$$

$$2Cu_2O + O_2 \rightarrow 4CuO \tag{4}$$

For the oxidation of thin copper films, which should have chemical properties similar to the ones of CuNWs, it has been discovered that (i) the outer CuO layer with a thickness in the range of 10–50 nm is much thinner than the Cu_2O layer; (ii) the growth of the copper oxide shell is mainly due to the growth and expansion of the Cu_2O layer, and (iii) the growth mechanism is a diffusion-controlled process that will be discussed in more detail in the following [79,80]. In the case of copper, the out-diffusion of copper atoms and their subsequent reaction with oxygen dictates the oxidation process [81,82]. This mechanism is in contrast to the well-studied oxidation of silicon [83], where the diffusion of oxygen occurs in the ambient air through the silicon. The driving force for the oxidation of copper is grain-boundary-diffusion, which in turn depends on the size of the individual grains [38]. In agreement with the literature, the temperature-dependent grain-boundary-diffusion constant is described using the Arrhenius equation [84]:

$$D = D_0 \exp(-E/k_B T) \tag{5}$$

where D_0 denotes the pre-exponential factor of grain-boundary diffusion, E the activation energy, k_B the Boltzmann constant and T the temperature, respectively. Next, an expression for the time-dependent oxide thickness $d_{ox}(t)$ is derived. As a starting point, Fick's law is employed to describe the diffusion F_d of copper atoms, which is dictated by the concentration gradient of the copper atoms C_i in the non-oxidized and C_s the oxidized region of the CuNW, as sketched in Figure 12c:

$$F_d = D/d_{ox}(t) \cdot \tag{6}$$

After some math and physical assumptions that are described in more detail in Appendix C, a parabolic rate law is derived for the time-dependent oxide thickness, as follows:

$$dox(t) = \sqrt{B \cdot t} \tag{7}$$

where B denotes a reaction-specific constant, in accordance with the Deal-Grove model [83]. After deriving the parabolic law for the oxide thickness in Equation (7) and having in mind that the diffusion of copper atoms is described by the Arrhenius law in Equation (5), a relation between the breakdown-time t_{BD}, as shown in Figure 5b, and the annealing temperature T can be derived. For simplicity, it is assumed that the resistance of the different CuNW films $R_{0,i}$ can be modeled by the resistance of a metallic cylinder with the formula:

$$R_{0,i} = \rho \cdot \frac{l}{A} \tag{8}$$

where ρ denotes the bulk resistivity of copper and l and A denotes the length and the cross-section of the conductor. It has to be noted that Equation (8) represents a drastic simplification for the modeling of the film resistance and is only employed for a qualitative description of the oxidation process. During the oxidation, the CuNW shell is consumed to copper oxide and the cross section that can effectively contribute to the electrical percolation is reduced. Hence, a time-dependent resistance $R(t)$ that depends on the time-dependent oxide thickness $d_{ox}(t)$ is given by:

$$R(t) = \rho \cdot l / \pi (r_0 - d_{ox}(t))^2 \tag{9}$$

where r_0 denotes the initial radius of the pristine nanowire, as sketched in Figure 12c. After plugging Equation (7) in Equation (9) and substituting the reaction specific constant B with the Arrhenius Equation (5) (see Appendix C for more details), an analytical dependence can be given for the temperature-breakdown time dependence, as follows:

$$\frac{1}{T} = a \cdot \ln\left(\sqrt{b \cdot R_{0,i} \cdot t_{BD}}\right) + c \tag{10}$$

where a, b and c denote constants. To verify the validity of this equation for the degradation of CuNWs, the annealing temperatures and breakdown times from Figure 5b were plotted in Figure 13. The graph shows a clear linear dependence, which proves that the derived temperature dependence for the breakdown time agrees well with our model.

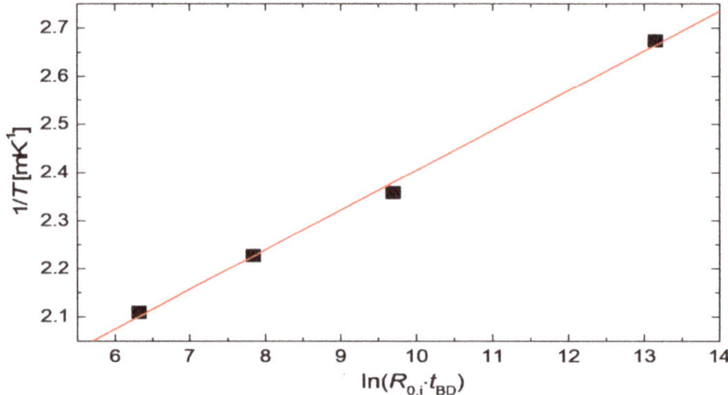

Figure 13. Temperature-breakdown time dependence for the data from Figure 5b. The solid line represents a linear fit to the experimental data, in agreement with Equation (10).

The presented model can help to predict the breakdown of CuNW-based devices if the process temperature and the initial film resistance are known.

3.3. Encapsulation of Copper Nanowires

Despite the high relevance of this aspect, there are only a very few studies that report on the encapsulation of CuNW films [30,49,50], as highlighted in the introduction section. In this work, we test various coating materials such as PDMS, PMMA and electron beam evaporated oxides as coating materials. The aim of this study is to increase the lifetime of CuNW-based devices under current flow or at elevated temperatures and to compare the effectiveness of a low-cost polymer-based coating with an expensive ebeam evaporated oxide coating. The normalized increase R/R_0 as a function of the time for CuNW films subjected to an electrical input power of 6 W is depicted in Figure 14 for the coating materials (a) PDMS, (b) PMMA, and (c) SiO_2 and Al_2O_3.

The resistance transient for an uncoated CuNW film is plotted in the graphs (a–c) and is denoted as the reference. So far, PDMS has been used as a substrate material for CuNWs [85] but little is known about its performance as an encapsulation material. In Figure 14 it can be seen that the PDMS coating leads to a reduction in lifetime from around 17 h for the reference film to around 4–6 h for the films coated with a thickness of either 4 or 22 μm. On the one hand, this behavior can be attributed to the high permeation rate of PDMS for gases in the ambient air [86]. On the other hand, since the reduction in lifetime is drastic, the degradation could also be induced by a chemical interaction between the PDMS and the CuNW film. From the R/R_0 response of CuNW films coated with PMMA, it can be concluded that a thin film with a thickness of 200 nm, applied by spin coating, has no significant effect as an encapsulation material. For thicker films with a thickness of around 9 μm, the lifetimes of the CuNW networks were 31 h and 44 h, respectively, which corresponds to an increase by 82% and 160%, respectively, compared to the lifetime of the reference film. This result is in agreement with the work of other groups who successfully tested PMMA as a coating material for CuNW films [49,61]. As the last coating materials, electron beam evaporated SiO_2 and Al_2O_3 were, to the best of our knowledge, for the first time tested in combination with CuNWs. From Figure 14c it can be seen that the oxide coatings lead to a lifetime of 21 h, 34 h (both encapsulations are composed of 50 nm Al_2O_3 & 250 nm SiO_2) and 40 h (1000 nm SiO_2), respectively, which corresponds to an increase in lifetime of 24%, 100% and

135%, respectively, with regard to the reference sample. A combination of SiO_2 and Al_2O_3 was tested since stacks composed of different oxide layers were reported to show a lowered gas permeability. This effect is in accordance with the work from Dameron et al. [87] who found that the evaporation of SiO_2 onto Al_2O_3 can heal out defects in the Al_2O_3 film that could otherwise serve as diffusion sides for the permeation of ambient gases.

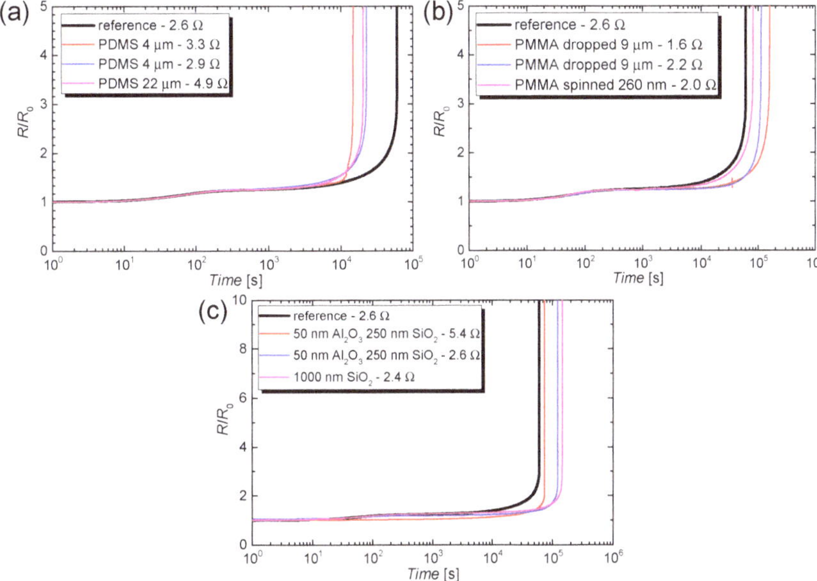

Figure 14. Normalized increase in resistance R/R_0 as a function of the time for CuNW films coated with (**a**) PDMS, (**b**) PMMA and (**c**) SiO_2 and Al_2O_3. The CuNW films were heated over an effective area of 3.5×5 cm^2 by applying an electrical input power of 6 W.

4. Conclusions

In summary, we investigated the up-scaling potential of an aqueous synthesis of CuNWs, and studied the degradation and the encapsulation of sprayed CuNW films. For the synthesis, we have observed that the precursor-to-solvent weight ratio should be kept below 1:100 to allow a stable micelle formation of the copper-oleylamine complex during growth and to improve the dispersibility of the nanowires. To the best of our knowledge, our work is the first report on the effect of the precursor-to-solvent ratio, which is an important factor for the upscaling potential of the presented synthesis. For the commercialization of CuNWs, the upscaling potential is a key factor, since there are various different applications requiring TEs such as heaters, touch panels, solar cells or OLEDs. To substitute ITO in all mentioned devices, large amounts of CuNWs are needed. Further, we rigorously studied the degradation mechanism for CuNWs, which will serve as a reference in future to explain and predict the failure of CuNW-based devices in follow-up studies. For the CuNW film degradation, we observed that chemical degradation via oxidation is the main degradation mechanism, as observed in other studies. We believe that the UV-driven degradation of CuNWs, that has not yet been understood in the literature stems from the UV-induced generation of ozone that leads to rapid oxidation of the CuNWs. In addition to these degradation studies, a purposely-developed model was presented to correlate the breakdown time of the CuNW film with the temperature and the initial resistance of the films. The encapsulation of the CuNW films using polymers such as PDMS proved to be difficult. In contrast to the expectation, the PDMS coating lowered the lifetime of the films, whereas a positive trend could be observed for PMMA, in agreement with the literature, as well as for electron beam evaporated oxide coatings that have been tested for the first time. We found that the maximum improvement in the lifetime is comparable for PMMA and ebeam evaporated oxides. Therefore, for economic reasons, PMMA represents a promising coating material since it can be applied at a large scale, under ambient conditions, and the material costs are also low. To further increase the lifetime of CuNW-based devices, we propose to use a combination of two techniques that have already been tested with success: The laser-induced nanowelding of CuNW junctions, followed by the sealing of the film with a PMMA encapsulation.

Author Contributions: Conceptualization, M.B. (Marco Bobinger) and M.B. (Markus Becherer); Investigation, J.M., M.B. (Marco Bobinger) and C.B.; Methodology, M.B. (Marco Bobinger) and J.M.; Software, J.M.; Validation, M.B. (Markus Becherer), M.B. (Marco Bobinger) and J.M.; Data Curation, J.M. and M.B. (Marco Bobinger); Writing-Original Draft Preparation, J.M. and M.B. (Marco Bobinger); Writing-Review & Editing, M.B. (Markus Becherer); Supervision, M.B. (Markus Becherer); Project Administration, M.B. (Markus Becherer) and P.L.; Funding Acquisition, M.B. (Markus Becherer) and P.L., please turn to the CRediT taxonomy for the term explanation.

Funding: This work was supported by the German Research Foundation (DFG) and the Technical University of Munich (TUM) within the funding program Open Access Publishing. The authors also thank the DFG and the NSERC for financial support of the Alberta/Technische Universität München Graduate School for Functional Hybrid Materials ATUMS (IRTG2022, NSERC CREATE) as well as the TUM Graduate School and IGSSE.

Conflicts of Interest: The authors declare no conflict of interest.

Appendix A Synthesis of CuNWs

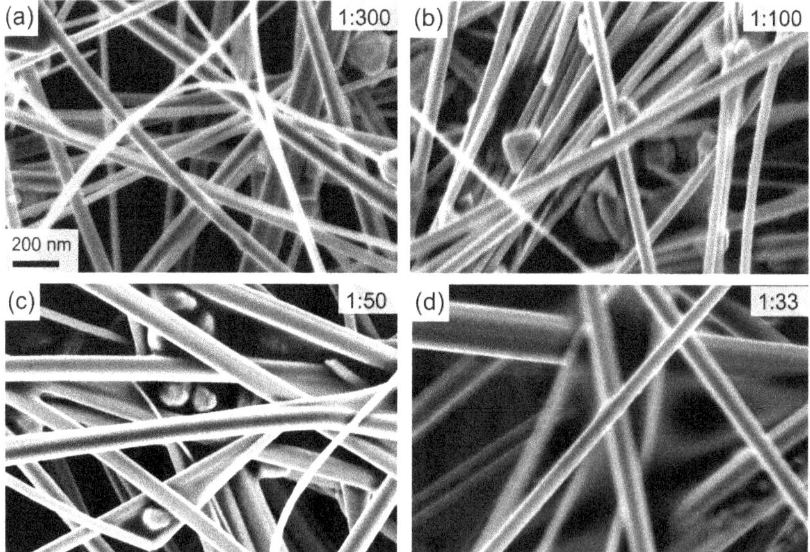

Figure A1. High-resolution FESEM-images of CuNWs for the precursor-to-solvent series with different mass ratios (precursor:solvent) of (**a**) 1:300, (**b**) 1:100, (**c**) 1:50 and (**d**) 1:33. The scale bar in (**a**) applies to all images.

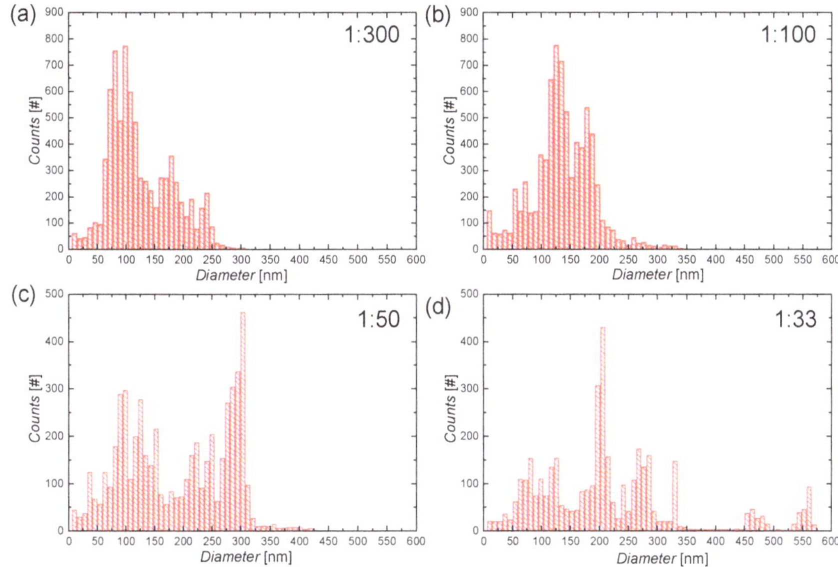

Figure A2. Diameter histograms for the precursor-to-solvent series with different mass ratios (precursor:solvent) of (**a**) 1:300, (**b**) 1:100, (**c**) 1:50 and (**d**) 1:33.

Appendix B Degradation of CuNWs

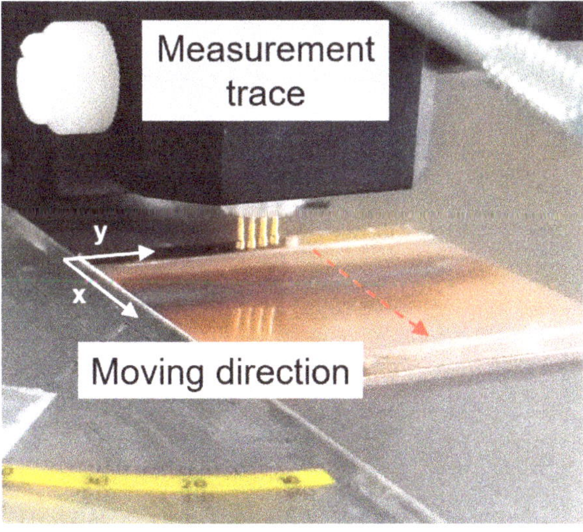

Figure A3. Photo for the four-probe measurement across a CuNW film after electrical breakdown.

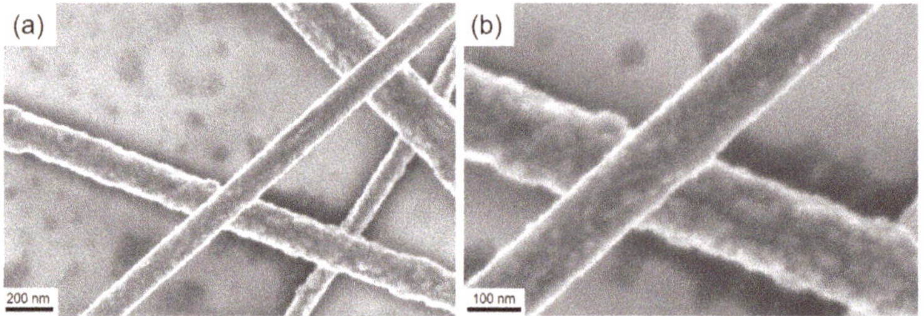

Figure A4. SEM-images for a CuNW film subjected to a relative humidity and temperature of 90%RH and 60 °C, for a duration of 24 h, under (**a**) a low and (**b**) a high magnification.

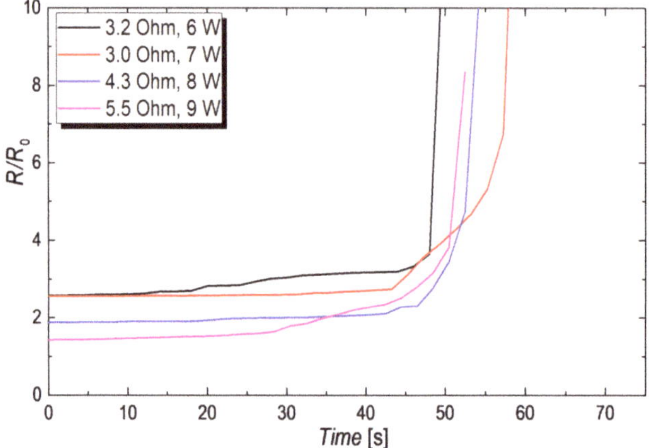

Figure A5. Normalized increase in resistance R/R_0 over time for CuNW films with a different initial resistance that were subjected to different electrical input powers. The x-axes for all films have been shifted to allow comparing the breakdown regions of each device.

Appendix C Oxidation Model for CuNWs

The physical assumption applied during the derivation of the time dependence of the oxide thickness is a variant of the so-called Deal-Grove Model. The underlying principle of the model is the equality of the diffusion flux of copper atoms and the reaction rate of the oxidation process. The diffusion flux F_d is described by Equation (A1), as follows:

$$F_d = \frac{D}{d_{ox}(t)} \cdot (C_s - C_i) \tag{A1}$$

where D denotes the diffusion constant, $d_{ox}(t)$ the time-dependent oxide thickness and C_s and C_i the concentration of copper atoms in the copper oxide shell and in the pristine copper, respectively, in agreement with the schematic for the oxidation of a single nanowire in Figure 12. The reaction-rate F_r is defined by Equation (2).

$$F_r = M \cdot \frac{d}{dt} d_{ox}(t) = k_s \cdot C_i \tag{A2}$$

where M is the oxygen density in the oxide layer and k_s is the reaction-rate coefficient. Equations (A1) and (A2) together yield to the differential expression for the oxide thickness of:

$$\frac{d}{dt} d_{ox}(t) = k_s \cdot \frac{C_i}{M} = k_s \cdot \frac{C_s}{M \cdot \left(1 + \frac{k_s}{D} \cdot d_{ox}(t)\right)} \tag{A3}$$

This differential equation can be solved under the assumption that $d_{ox}(t)$ is much larger than D/k_s, which yields to the parabolic rate law that is also derived in the main text.

$$d_{ox}(t) = \sqrt{B \cdot t}; B = 2 \cdot \frac{D \cdot C_s}{M} \tag{A4}$$

In the following, the equation that relates the annealing temperature with the initial resistance and the breakdown time, i.e., Equation (10) in the main text, is derived in more detail. In Equation (9), the nanowire radius r_0 can be substituted by:

$$r_0 = \sqrt{\pi \cdot \frac{R_0}{\rho \cdot L}} \tag{A5}$$

where R_0 and L denote by the effective length and resistance of the nanowire network and ρ the resistivity of copper, respectively. For the time-dependent oxide thickness $d_{ox}(t)$ in Equation (9), the parabolic rate law in Equation (A4) is utilized, which gives the following equation:

$$\frac{R_t}{R_0} = \frac{1}{\left(1 - \sqrt{\pi \frac{R_0}{\rho L} B t}\right)^2} \tag{A6}$$

As described in Equation (A4) in the supporting information and Equation (5) in the main text, the following expression is derived for B:

$$B = D \cdot e^{\frac{-E}{kT}} \rightarrow \sqrt{\frac{\rho}{R_0 t_{bd}}} e^{\frac{-E}{2kT}} \tag{A7}$$

when the expression for B provided in Equation (A7) is plugged into Equation (A6), the equation for the initial resistance-breakdown time dependence of the annealing temperature in Equation (10) in the main text can be derived.

References

1. Li, J.; Hu, L.; Wang, L.; Zhou, Y.; Grüner, G.; Marks, T.J. Organic light-emitting diodes having carbon nanotube anodes. *Nano Lett.* **2006**, *6*, 2472–2477. [CrossRef] [PubMed]
2. Vosgueritchian, M.; Lipomi, D.J.; Bao, Z. Highly conductive and transparent PEDOT:PSS films with a fluorosurfactant for stretchable and flexible transparent electrodes. *Adv. Funct. Mater.* **2012**, *22*, 421–428. [CrossRef]
3. Li, X.; Zhu, Y.; Cai, W.; Borysiak, M.; Han, B.; Chen, D.; Piner, R.D.; Colombo, L.; Ruoff, R.S. Transfer of large-area graphene films for high-performance transparent conductive electrodes. *Nano Lett.* **2009**, *9*, 4359–4363. [CrossRef] [PubMed]
4. Yin, Z.; Sun, S.; Salim, T.; Wu, S.; Huang, X.; He, Q.; Lam, Y.M.; Zhang, H. Organic photovoltaic devices using highly flexible reduced graphene oxide films as transparent electrodes. *ACS Nano* **2010**, *4*, 5263–5268. [CrossRef] [PubMed]
5. Meng, L.; dos Santos, M. Properties of indium tin oxide films prepared by rf reactive magnetron sputtering at different substrate temperature. *Thin Solid Films* **1998**, *322*, 56–62. [CrossRef]
6. Sun, Y.; Gates, B.; Mayers, B.; Xia, Y. Crystalline silver nanowires by soft solution processing. *Nano Lett.* **2002**, *2*, 165–168. [CrossRef]
7. Rathmell, A.R.; Bergin, S.M.; Hua, Y.L.; Li, Z.Y.; Wiley, B.J. The growth mechanism of copper nanowires and their properties in flexible, transparent conducting films. *Adv. Mater.* **2010**, *22*, 3558–3563. [CrossRef] [PubMed]
8. Dou, L.; Cui, F.; Yu, Y.; Khanarian, G.; Eaton, S.W.; Yang, Q.; Resasco, J.; Schildknecht, C.; Schierle-Arndt, K.; Yang, P. Solution-processed copper/reduced-graphene-oxide core/shell nanowire transparent conductors. *ACS Nano* **2016**, *10*, 2600–2606. [CrossRef] [PubMed]

9. Tokuno, T.; Nogi, M.; Jiu, J.; Suganuma, K. Hybrid transparent electrodes of silver nanowires and carbon nanotubes: A low-temperature solution process. *Nanoscale Res. Lett.* **2012**, *7*, 281. [CrossRef] [PubMed]

10. Stapleton, A.J.; Afre, R.A.; Ellis, A.V.; Shapter, J.G.; Andersson, G.G.; Quinton, J.S.; Lewis, D.A. Highly conductive interwoven carbon nanotube and silver nanowire transparent electrodes. *Sci. Technol. Adv. Mater.* **2013**, *14*, 035004. [CrossRef] [PubMed]

11. Chen, X.; Guo, W.; Xie, L.; Wei, C.; Zhuang, J.; Su, W.; Cui, Z. Embedded Ag/Ni metal-mesh with low surface roughness as transparent conductive electrode for optoelectronic applications. *ACS Appl. Mater. Interfaces* **2017**, *9*, 37048–37054. [CrossRef] [PubMed]

12. Kang, M.G.; Joon Park, H.; Hyun Ahn, S.; Jay Guo, L. Transparent Cu nanowire mesh electrode on flexible substrates fabricated by transfer printing and its application in organic solar cells. *Sol. Energy Mater. Sol. Cells* **2010**, *94*, 1179–1184. [CrossRef]

13. Khan, A.; Lee, S.; Jang, T.; Xiong, Z.; Zhang, C.; Tang, J.; Guo, L.J.; Wen-Di, L. High-performance flexible transparent electrode with an embedded metal mesh fabricated by cost-effective solution process. *Small* **2016**, *12*, 3021–3030. [CrossRef] [PubMed]

14. Langley, D.; Giusti, G.; Mayousse, C.; Celle, C.; Bellet, D.; Simonato, J.-P.P. Flexible transparent conductive materials based on silver nanowire networks: A review. *Nanotechnology* **2013**, *24*, 452001. [CrossRef] [PubMed]

15. Li, S.; Chen, Y.; Huang, L.; Pan, D. Large-scale synthesis of well-dispersed copper nanowires in an electric pressure cooker and their application in transparent and conductive networks. *Inorg. Chem.* **2014**, *53*, 4440–4444. [CrossRef] [PubMed]

16. Lee, J.H.; Lee, P.; Lee, D.; Lee, S.S.; Ko, S.H. Large-scale synthesis and characterization of very long silver nanowires via successive multistep growth. *Cryst. Growth Des.* **2012**, *12*, 5598–5605. [CrossRef]

17. Scardaci, V.; Coull, R.; Lyons, P.E.; Rickard, D.; Coleman, J.N. Spray deposition of highly transparent, low-resistance networks of silver nanowires over large areas. *Small* **2011**, *7*, 2621–2628. [CrossRef] [PubMed]

18. Ye, S.; Rathmell, A.R.; Chen, Z.; Stewart, I.E.; Wiley, B.J. Metal nanowire networks: The next generation of transparent conductors. *Adv. Mater.* **2014**, *26*, 6670–6687. [CrossRef] [PubMed]

19. Bobinger, M.; Dergianlis, V.; Becerer, M.; Lugli, P. Comprehensive synthesis study of well-dispersed and solution-processed metal nanowires for transparent heaters. *J. Nanomater.* **2018**, *2018*, 1–13. [CrossRef]

20. Kwon, J.; Suh, Y.D.; Lee, J.; Lee, P.; Han, S.; Hong, S.; Yeo, J.; Lee, H.; Ko, S.H. Recent progress in silver nanowire based flexible/wearable optoelectronics. *J. Mater. Chem. C* **2018**, *6*, 7445. [CrossRef]

21. Jung, J.; Lee, H.; Ha, I.; Cho, H.; Kim, K.K.; Kwon, J.; Won, P.; Hong, S.; Ko, S.H. Highly stretchable and transparent electromagnetic interference shielding film based on silver nanowire percolation network for wearable electronics applications. *ACS Appl. Mater. Interfaces* **2017**, *9*, 44609–44616. [CrossRef] [PubMed]

22. Jeong, S.; Cho, H.; Han, S.; Won, P.; Lee, H.; Hong, S.; Yeo, J.; Kwon, J.; Ko, S.H. High efficiency, transparent, reusable, and active PM2.5 filters by hierarchical Ag nanowire percolation network. *Nano Lett.* **2017**, *17*, 4339–4346. [CrossRef] [PubMed]

23. Lee, J.; An, K.; Won, P.; Ka, Y.; Hwang, H.; Moon, H.; Kwon, Y.; Hong, S.; Kim, C.; Lee, C.; et al. A dual-scale metal nanowire network transparent conductor for highly efficient and flexible organic light emitting diodes. *Nanoscale* **2017**, *9*, 1978–1985. [CrossRef] [PubMed]

24. Oh, J.Y.; Lee, D.; Jun, G.H.; Ryu, H.J.; Hong, S.H. High conductivity and stretchability of 3D welded silver nanowire filled graphene aerogel hybrid nanocomposites. *J. Mater. Chem. C* **2017**, *5*, 8211–8218. [CrossRef]

25. Lee, H.; Hong, S.; Lee, J.; Suh, Y.D.; Kwon, J.; Moon, H.; Kim, H.; Yeo, J.; Ko, S.H. Highly stretchable and transparent supercapacitor by ag-au core-shell nanowire network with high electrochemical stability. *ACS Appl. Mater. Interfaces* **2016**, *8*, 15449–15458. [CrossRef] [PubMed]

26. Hong, S.; Lee, H.; Lee, J.; Kwon, J.; Han, S.; Suh, Y.D.; Cho, H.; Shin, J.; Yeo, J.; Ko, S.H. Highly stretchable and transparent metal nanowire heater for wearable electronics applications. *Adv. Mater.* **2015**, *27*, 4744–4751. [CrossRef] [PubMed]

27. Bobinger, M.; Mock, J.; La Torraca, P.; Becherer, M.; Lugli, P.; Larcher, L. Tailoring the aqueous synthesis and deposition of copper nanowires for transparent electrodes and heaters. *Adv. Mater. Interfaces* **2017**, *4*, 1700568. [CrossRef]

28. Bobinger, M.R.R.; La Torraca, P.; Mock, J.; Becherer, M.; Cattani, L.; Angeli, D.; Larcher, L.; Lugli, P. Solution-processing of copper nanowires for transparent heaters and thermo-acoustic loudspeakers. *IEEE Trans. Nanotechnol.* **2018**, *17*, 940–947. [CrossRef]

29. Bobinger, M.; Mock, J.; Becherer, M.; Torraca, P.L.; Angeli, D.; Larcher, L.; Lugli, P. Characterization and modelling of transparent heaters based on solution-processed copper nanowires. In Proceedings of the 2017 IEEE 17th International Conference on Nanotechnology, NANO 2017, Pittsburgh, PA, USA, 25–28 July 2017; IEEE: Piscataway, NJ, USA, 2017; pp. 151–154.

30. Chen, Z.; Ye, S.; Stewart, I.E.; Wiley, B.J. Copper nanowire networks with transparent oxide shells that prevent oxidation without reducing transmittance. *ACS Nano* **2014**, *8*, 9673–9679. [CrossRef] [PubMed]

31. Bobinger, M.; Angeli, D.; Colasanti, S.; La Torraca, P.; Larcher, L.; Lugli, P. Infrared, transient thermal, and electrical properties of silver nanowire thin films for transparent heaters and energy-efficient coatings. *Phys. Status Solidi* **2017**, *214*, 1600466. [CrossRef]

32. Xiao, L.; Chen, Z.; Feng, C.; Liu, L.; Bai, Z.Q.; Wang, Y.; Qian, L.; Zhang, Y.; Li, Q.; Jiang, K.; et al. Flexible, stretchable, transparent carbon nanotube thin film loudspeakers. *Nano Lett.* **2008**, *8*, 4539–4545. [CrossRef] [PubMed]

33. La Torraca, P.; Bobinger, M.; Pavan, P.; Becherer, M.; Zhao, S.; Koebel, M.; Cattani, L.; Lugli, P.; Larcher, L. High efficiency thermoacoustic loudspeaker made with a silica aerogel substrate. *Adv. Mater. Technol.* **2018**, 1800139. [CrossRef]

34. Yu, Z.; Li, L.; Zhang, Q.; Hu, W.; Pei, Q. Silver nanowire-polymer composite electrodes for efficient polymer solar cells. *Adv. Mater.* **2011**, *23*, 4453–4457. [CrossRef] [PubMed]

35. Zaiba, S.; Kouriba, T.; Ziane, O.; Stéphan, O.; Bosson, J.; Vitrant, G.; Baldeck, P.L. Metallic nanowires can lead to wavelength-scale microlenses and microlens arrays. *Opt. Express* **2012**, *20*, 15516–15521. [CrossRef] [PubMed]

36. Kirsch, N.J.; Vacirca, N.A.; Plowman, E.E.; Kurzweg, T.P.; Fontecchio, A.K.; Dandekar, K.R. Optically transparent conductive polymer rfid meandering dipole antenna. In Proceedings of the 2009 IEEE International Conference on RFID, Orlando, FL, USA, 27–28 April 2009; pp. 278–282. [CrossRef]

37. Falco, A.; Cinà, L.; Scarpa, G.; Lugli, P.; Abdellah, A. Fully-sprayed and flexible organic photodiodes with transparent carbon nanotube electrodes. *ACS Appl. Mater. Interfaces* **2014**, *6*, 10593–10601. [CrossRef] [PubMed]

38. Lee, J.; Lee, P.; Lee, H.; Lee, D.; Lee, S.S.; Ko, S.H. Very long Ag nanowire synthesis and its application in a highly transparent, conductive and flexible metal electrode touch panel. *Nanoscale* **2012**, *4*, 6408–6414. [CrossRef] [PubMed]

39. Lee, T.-W.; Lee, S.-E.; Jeong, Y.G. Highly effective electromagnetic interference shielding materials based on silver nanowire/cellulose papers. *ACS Appl. Mater. Interfaces* **2016**, *8*, 13123–13132. [CrossRef] [PubMed]

40. Park, T.; Kim, B.; Kim, Y.; Kim, E. Highly conductive PEDOT electrodes for harvesting dynamic energy through piezoelectric conversion. *J. Mater. Chem. A* **2014**, *2*, 5462–5469. [CrossRef]

41. Park, T.; Na, J.; Kim, B.; Kim, Y.; Shin, H.; Kim, E. Photothermally activated pyroelectric polymer films for harvesting of solar heat with a hybrid energy cell structure. *ACS Nano* **2015**, *9*, 11830–11839. [CrossRef] [PubMed]

42. Chang, Y.; Lye, M.L.; Zeng, H.C. Large-scale synthesis of high-quality ultralong copper nanowires. *Langmuir* **2005**, *21*, 3746–3748. [CrossRef] [PubMed]

43. Zhao, S.; Han, F.; Li, J.; Meng, X.; Huang, W.; Cao, D.; Zhang, G.; Sun, R.; Wong, C.P. Advancements in copper nanowires: synthesis, purification, assemblies, surface modification, and applications. *Small* **2018**, *14*, 1800047. [CrossRef] [PubMed]

44. Hwang, C.; An, J.; Choi, B.D.; Kim, K.; Jung, S.-W.; Baeg, K.-J.; Kim, M.-G.; Ok, K.M.; Hong, J. Controlled aqueous synthesis of ultra-long copper nanowires for stretchable transparent conducting electrode. *J. Mater. Chem. C* **2016**, *4*, 1441–1447. [CrossRef]

45. Yang, H.-J.; He, S.-Y.; Tuan, H.-Y. Self-seeded growth of five-fold twinned copper nanowires: Mechanistic study, characterization, and SERS applications. *Langmuir* **2014**, *30*, 602–610. [CrossRef] [PubMed]

46. Jin, M.; He, G.; Zhang, H.; Zeng, J.; Xie, Z.; Xia, Y. Shape-controlled synthesis of copper nanocrystals in an aqueous solution with glucose as a reducing agent and hexadecylamine as a capping agent. *Angew. Chem. Int. Ed.* **2011**, *50*, 10560–10564. [CrossRef] [PubMed]

47. Ye, S.; Stewart, I.E.; Chen, Z.; Li, B.; Rathmell, A.R.; Wiley, B.J. How copper nanowires grow and how to control their properties. *Acc. Chem. Res.* **2016**, *49*, 442–451. [CrossRef] [PubMed]

48. Ye, S.; Rathmell, A.R.; Ha, Y.-C.; Wilson, A.R.; Wiley, B.J.; Ye, S.; Rathmell, A.R.; Wilson, A.R.; Wiley, B.J.; Ha, Y. The role of cuprous oxide seeds in the one-pot and seeded syntheses of copper nanowires. *Small* **2014**, *10*, 1771–1778. [CrossRef] [PubMed]

49. Zhai, H.; Wang, R.; Wang, X.; Cheng, Y.; Shi, L.; Sun, J. Transparent heaters based on highly stable Cu nanowire. *Nano Res.* **2016**, *9*, 3924–3936. [CrossRef]

50. Chen, J.; Zhou, W.; Chen, J.; Fan, Y.; Zhang, Z.; Huang, Z.; Feng, X.; Mi, B.; Ma, Y.; Huang, W. Solution-processed copper nanowire flexible transparent electrodes with PEDOT:PSS as binder, protector and oxide-layer scavenger for polymer solar cells. *Nano Res.* **2015**, *8*, 1017–1025. [CrossRef]

51. Kim, D.; Kwon, J.; Jung, J.; Kim, K.; Lee, H. A Transparent and flexible capacitive-force touch pad from high-aspect-ratio copper nanowires with enhanced oxidation resistance for applications in wearable electronics. *Small Methods* **2018**, *2*, 1800077. [CrossRef]

52. Han, S.; Hong, S.; Ham, J.; Yeo, J.; Lee, J.; Kang, B.; Lee, P.; Kwon, J.; Lee, S.S.; Yang, M.Y.; et al. Fast plasmonic laser nanowelding for a Cu-nanowire percolation network for flexible transparent conductors and stretchable electronics. *Adv. Mater.* **2014**, *26*, 5808–5814. [CrossRef] [PubMed]

53. Han, S.; Hong, S.; Yeo, J.; Kim, D.; Kang, B.; Yang, M.Y.; Ko, S.H. Nanorecycling: Monolithic integration of copper and copper oxide nanowire network electrode through selective reversible photothermochemical reduction. *Adv. Mater.* **2015**, *27*, 6397–6403. [CrossRef] [PubMed]

54. Park, J.H.; Han, S.; Kim, D.; You, B.K.; Joe, D.J.; Hong, S.; Seo, J.; Kwon, J.; Jeong, C.K.; Park, H.J.; et al. Plasmonic-tuned flash Cu nanowelding with ultrafast photochemical-reducing and interlocking on flexible plastics. *Adv. Funct. Mater.* **2017**, *27*, 1701138. [CrossRef]

55. Herrera-Gomez, A.; Bravo-Sanchez, M.; Ceballos-Sanchez, O.; Vazquez-Lepe, M.O. Practical methods for background subtraction in photoemission spectra. *Surf. Interface Anal.* **2014**, *46*, 897–905. [CrossRef]

56. Lee, E.-J.; Kim, Y.-H.; Hwang, D.K.; Choi, W.K.; Kim, J.-Y. Synthesis and optoelectronic characteristics of 20 nm diameter silver nanowires for highly transparent electrode films. *RSC Adv.* **2016**, *6*, 11702–11710. [CrossRef]

57. Araki, T.; Jiu, J.; Nogi, M.; Koga, H.; Nagao, S.; Sugahara, T.; Suganuma, K. Low haze transparent electrodes and highly conducting air dried films with ultra-long silver nanowires synthesized by one-step polyol method. *Nano Res.* **2014**, *7*, 236–245. [CrossRef]

58. Hotaling, N.A.; Bharti, K.; Kriel, H.; Simon, C.G., Jr.; Simon, C.G. DiameterJ: A validated open source nanofiber diameter measurement tool. *HHS Public Access* **2015**, *61*, 327–338. [CrossRef] [PubMed]

59. Maibaum, L.; Dinner, A.R.R.; Chandler, D. Micelle formation and the hydrophobic effect. *J. Phys. Chem. B* **2004**, *108*, 6778–6781. [CrossRef]

60. Lang, J.; Zana, R. Effect of alcohols and oils on the kinetics of micelle formation-breakdown in aqueous solutions of ionic surfactants. *J. Phys. Chem.* **1986**, *90*, 5258–5265. [CrossRef]

61. Celle, C.; Cabos, A.; Fontecave, T.; Laguitton, B.; Benayad, A.; Guettaz, L.; Pélissier, N.; Nguyen, V.H.; Bellet, D.; Muñoz-Rojas, D.; Simonato, J.-P. Oxidation of copper nanowire based transparent electrodes in ambient conditions and their stabilization by encapsulation: Application to transparent film heaters. *Nanotechnology* **2018**, *29*, 085701. [CrossRef] [PubMed]

62. Manning, H.G.; Niosi, F.; da Rocha, C.G.; Bellew, A.T.; O'Callaghan, C.; Biswas, S.; Flowers, P.F.; Wiley, B.J.; Holmes, J.D.; Ferreira, M.S.; et al. Emergence of winner-takes-all connectivity paths in random nanowire networks. *Nat. Commun.* **2018**, *9*, 3219. [CrossRef] [PubMed]

63. Haacke, G. New figure of merit for transparent conductors. *J. Appl. Phys.* **1976**, *47*, 31901–91106. [CrossRef]

64. Khaligh, H.H.; Xu, L.; Khosropour, A.; Madeira, A.; Romano, M.; Pradére, C.; Tréguer-Delapierre, M.; Servant, L.; Pope, M.A.; Goldthorpe, I.A. The Joule heating problem in silver nanowire transparent electrodes. *Nanotechnology* **2017**, *28*, 425703. [CrossRef] [PubMed]

65. Khaligh, H.H.; Goldthorpe, I.A. Failure of silver nanowire transparent electrodes under current flow. *Nanoscale Res. Lett.* **2013**, *8*, 235. [CrossRef] [PubMed]

66. Sannicolo, T.; Charvin, N.; Flandin, L.; Kraus, S.; Papanastasiou, D.T.; Celle, C.; Simonato, J.-P.; Muñoz-Rojas, D.; Jiménez, C.; Bellet, D. Electrical mapping of silver nanowire networks: A versatile tool for imaging network homogeneity and degradation dynamics during failure. *ACS Nano* **2018**, *12*, 4648–4659. [CrossRef] [PubMed]

67. Copinet, A.; Bertrand, C.; Govindin, S.; Coma, V.; Couturier, Y. Effects of ultraviolet light (315 nm), temperature and relative humidity on the degradation of polylactic acid plastic films. *Chemosphere* **2004**, *55*, 763–773. [CrossRef] [PubMed]

68. Rivaton, A.; Chambon, S.; Manceau, M.; Gardette, J.L.; Lemaître, N.; Guillerez, S. Light-induced degradation of the active layer of polymer-based solar cells. *Polym. Degrad. Stab.* **2010**, *95*, 278–284. [CrossRef]

69. Gorham, J.M.; MacCuspie, R.I.; Klein, K.L.; Fairbrother, D.H.; Holbrook, R.D. UV-induced photochemical transformations of citrate-capped silver nanoparticle suspensions. *J. Nanoparticle Res.* **2012**, *14*, 1139. [CrossRef]

70. Herrling, T.; Jung, K.; Fuchs, J. Measurements of UV-generated free radicals/reactive oxygen species (ROS) in skin. *Spectrochim. Acta Part A Mol. Biomol. Spectrosc.* **2006**, *63*, 840–845. [CrossRef] [PubMed]

71. Rittié, L.; Fisher, G.J. UV-light-induced signal cascades and skin aging. *Ageing Res. Rev.* **2002**, *1*, 705–720. [CrossRef]

72. Takagi, K.; Nair, S.V.; Watanabe, R.; Seto, K.; Kobayashi, T.; Tokunaga, E. Surface plasmon polariton resonance of gold, silver, and copper studied in the kretschmann geometry: Dependence on wavelength, angle of incidence, and film thickness. *J. Phys. Soc. Jpn.* **2017**, *86*, 124721. [CrossRef]

73. Duan, J.L.; Cornelius, T.W.; Liu, J.; Karim, S.; Yao, H.J.; Picht, O.; Rauber, M.; Müller, S.; Neumann, R. Surface plasmon resonances of Cu Nanowire Arrays. *J. Phys. Chem. C* **2009**, *113*, 13583–13587. [CrossRef]

74. Seifert, M.; Vargas, J.E.B.; Bobinger, M.; Sachenhauser, M.; Cummings, A.W.; Roche, S.; Garrido, J.A.; Sachsenhauser, M.; Cummings, A.W.; Roche, S.; et al. Role of grain boundaries in tailoring electronic properties of polycrystalline graphene by chemical functionalization. *2D Mater.* **2015**, *2*, 024008. [CrossRef]

75. Platzman, I.; Brener, R. Oxidation of polycrystalline copper thin films at ambient conditions. *J. Phys. Chem. C* **2008**, *112*, 1101–1108. [CrossRef]

76. Fleisch, T.H.; Mains, G.J. Reduction of copper oxides by UV radiation and atomic hydrogen studied by XPS. *Appl. Surf. Sci.* **1982**, *10*, 51–62. [CrossRef]

77. Park, J.-H.; Natesan, K. Oxidation of copper and electronic transport in copper oxides. *Oxid. Met.* **1993**, *39*, 411–435. [CrossRef]

78. Wan, Y.; Wang, X.; Sun, H.; Li, Y.; Zhang, K.; Wu, Y. Corrosion behavior of copper at elevated temperature. *Int. J. Electrochem. Sci.* **2012**, *7*, 7902–7914. [CrossRef]

79. Lee, S.-K.; Hsu, H.-C.; Tuan, W.-H. Oxidation behavior of copper at a temperature below 300 °C and the methodology for passivation. *Mater. Res.* **2016**, *19*, 51–56. [CrossRef]

80. Papadimitropoulos, G.; Vourdas, N.; Vamvakas, V.E.; Davazoglou, D. Deposition and characterization of copper oxide thin films. *J. Phys. Conf. Ser.* **2005**, *10*, 182–185. [CrossRef]

81. Nerle, U. Thermal oxidation of copper for favorable formation of cupric oxide (CuO) semiconductor. *IOSR J. Appl. Phys.* **2013**, *5*, 1–7. [CrossRef]

82. Ramanandan, G.K.P.; Ramakrishnan, G.; Planken, P.C.M. Oxidation kinetics of nanoscale copper films studied by terahertz transmission spectroscopy. *J. Appl. Phys.* **2012**, *111*, 123517. [CrossRef]

83. Deal, B.E.; Grove, A.S. General relationship for the thermal oxidation of silicon. *J. Appl. Phys.* **1965**, *36*, 3770. [CrossRef]

84. Mehrer, H. *Diffusion in Solids: Fundamentals, Methods, Materials, Diffusion-Controlled Processes*; Springer-Verlag: Berlin, Germany, 2007; ISBN 978-3-540-71486-6.

85. Won, Y.; Kim, A.; Yang, W.; Jeong, S.; Moon, J. A highly stretchable, helical copper nanowire conductor exhibiting a stretchability of 700%. *NPG Asia Mater.* **2014**, *6*, e132. [CrossRef]

86. Berean, K.; Ou, J.Z.; Nour, M.; Latham, K.; McSweeney, C.; Paull, D.; Halim, A.; Kentish, S.; Doherty, C.M.; Hill, A.J.; et al. The effect of crosslinking temperature on the permeability of PDMS membranes: Evidence of extraordinary CO_2 and CH_4 gas permeation. *Sep. Purif. Technol.* **2014**, *122*, 96–104. [CrossRef]

87. Dameron, A.; Davidson, S.; Burton, B.; Carcia, P.; McLean, R.; George, S. Gas diffusion barriers on polymers using multilayers fabricated by Al_2O_3 and rapid SiO_2 atomic layer deposition. *J. Phys. Chem. C* **2008**, *112*, 4573–4580. [CrossRef]

 *nanomaterials*

Article

Investigation of Optimum Mg Doping Content and Annealing Parameters of $Cu_2Mg_xZn_{1-x}SnS_4$ Thin Films for Solar Cells

Yingrui Sui [1], Yu Zhang [1], Dongyue Jiang [1], Wenjie He [1], Zhanwu Wang [1], Fengyou Wang [1], Bin Yao [2] and Lili Yang [1,*]

[1] Key Laboratory of Functional Materials Physics and Chemistry of the Ministry of Education, Jilin Normal University, Siping 136000, China
[2] State Key Laboratory of Superhard Materials and College of Physics, Jilin University, Changchun 130012, China
* Correspondence: llyang@jlnu.edu.cn; Tel.: +86-434-329-4566

Received: 18 May 2019; Accepted: 7 June 2019; Published: 30 June 2019

Abstract: $Cu_2Mg_xZn_{1-x}SnS_4$ ($0 \leq x \leq 0.6$) thin films were prepared by a simple, low-temperature (300 °C) and low-cost sol–gel spin coating method followed by post-annealing at optimum conditions. We optimized the annealing conditions and investigated the effect of Mg content on the crystalline quality, electrical and optical performances of the $Cu_2Mg_xZn_{1-x}SnS_4$ thin films. It was found that the $Cu_2Mg_xZn_{1-x}SnS_4$ film annealed at 580 °C for 60 min contained large grain, less grain boundaries and high carrier concentration. Pure phase kesterite $Cu_2Mg_xZn_{1-x}SnS_4$ ($0 \leq x \leq 0.6$) thin films were obtained by using optimal annealing conditions; notably, the smaller Zn^{2+} ions in the Cu_2ZnSnS_4 lattice were replaced by larger Mg^{2+} ions. With an increase in x from 0 to 0.6, the band gap energy of the films decreased from 1.43 to 1.29 eV. When the ratio of Mg/Mg + Zn is 0.2 ($x = 0.2$), the grain size of $Cu_2Mg_xZn_{1-x}SnS_4$ reaches a maximum value of 1.5 μm and the surface morphology is smooth and dense. Simultaneously, the electrical performance of $Cu_2Mg_xZn_{1-x}SnS_4$ thin film is optimized at $x = 0.2$, the carrier concentration reaches a maximum value of 3.29×10^{18} cm^{-3}.

Keywords: $Cu_2Mg_xZn_{1-x}SnS_4$; thin films; photoelectric performance; sol–gel; sulfuration treatment; solar cell

---

1. Introduction

In recent years, the semiconductor Cu_2ZnSnS_4 (CZTS) has attracted enormous attention as an ideal absorber material for low-cost thin film solar cells. For thin film $CuInGaSe_2$ (CIGS) and CdTe solar cells, reliable efficiencies of more than 20% have been achieved [1,2]. However, the limited resources and extremely high costs of In and Ga, and toxicity of Se and Cd significantly limit further development of CIGS and CdTe solar cells. CZTS is regarded as a substitute for CIGS, wherein the high-cost and rare In and Ga, and toxic Se are replaced by low-cost and earth-abundant Zn, Sn and S, respectively. In addition to being composed of abundant and non-toxic elements, CZTS exhibits remarkable photoelectric properties as an absorbing layer, including a high absorption coefficient ($>10^4$ cm^{-1}) and a suitable band gap (1.40–1.50 eV) [3]. To date the best efficiency of pure CZTS has broken through 11% [4,5], but it is still far below than that of CIGS (21.7%) [6]. In order to realize the industrialization of low-cost and environmental protection CZTS solar cells, it is necessary to further improve the efficiency of CZTS based thin film solar cells. The low efficiencies of CZTS solar cells are attributed to factors such as low crystallinity, large open circuit voltage (V_{oc}) deficit as well as poor band alignment at the CdS/CZTS interface [7–9]. As we all know, V_{oc} is linearly related to the band gap of CZTS. Band gap engineering has emerged as an effective method to adjust the band alignment at the CdS/CZTS heterojunction interface [10,11]. Therefore, it is extremely crucial to find the absorption layer that is conducive to tuning the band gap of CZTS.

The band gap of CZTS can be decreased from 1.5 eV to 1.0 eV by doping Se and tuning the S/Se ratio [12], which is a widely used method to tune the band gap of CZTS. However, it is not easy to precisely control the S/Se ratio during the annealing process, because of the difference in the volatility of S and Se [13–15]. In addition, Se is extremely hazardous to human health and the environment. Cation substitution is considered a highly effective means to tune the band gap of CZTS, particularly the substitution of Zn with Cd, the latest efficiency has exceeded 11% [5]. Several studies have demonstrated that via Cd doping, the band gap of CZTS can be adjusted and the crystalline quality of the CZTS films can be effectively improved, leading to a significant improvement in the V_{oc} of CZTS [16,17]. However, Cd toxicity is a serious issue. In addition, several other elements have been introduced to CZTS, such as Sb and alkali metals (Na, K) [18–23]. However, it is found that these doping elements have little influence on the tuning band gap, because the ionic radii of these elements are mismatched, leading their incorporation only on the grain boundaries or surfaces rather than in the CZTS lattice. Therefore, the development of a facile and environmentally friendly method to tune the band gap of CZTS is highly imperative.

Tuning the band gap of CZTS by substituting Zn with Mg is more advantageous compared to that with other elements. Firstly, compared to Na, K and Sb ions, Mg^{2+} ions occupy Zn^{2+} sites in the CZTS lattice rather than segregate on the CZTS grain boundaries and surfaces, because the radii of Mg^{2+} and Zn^{2+} are very similar. In addition, the introduction of Mg in the CZTS absorber layer has unique strengths because of the low price, high reserves and being environmentally friendly. Mg is more abundant than Zn, less costly than Ge and environmentally friendly compared to Cd [24]. Lastly, some possible impurity phases due to the existence of ZnS during the synthesis process of the CZTS precursor solution may be eliminated or reduced. Since the ZnS binary phase exists stably in the solution condition but the MgS binary phase is unstable [24]. These advantages make the substitution of Zn with Mg an effective approach to adjust the band gap of the CZTS. So far, the research about Cu_2MgSnS_4 nanoparticles prepared by hot-injection have been reported [24], in our previous studies, the $Cu_2Mg_xZn_xSn(S,Se)_4$ thin films with different Mg concentration have been successfully synthesized by the sol–gel method [25], but the study of $Cu_2Mg_xZn_{1-x}SnS_4$ thin films as the absorption layer has not been investigated.

In this work, it is the first time $Cu_2Mg_xZn_{1-x}SnS_4$ thin films with different Mg content were fabricated by the sol–gel method. The crystal structure and electro-optic performance of the $Cu_2Mg_xZn_{1-x}SnS_4$ films were systematically characterized. It is found that Mg^{2+} ions were successfully incorporated into the CZTS lattice, which occupied the Zn^{2+} ion sites, the prepared $Cu_2Mg_xZn_{1-x}SnS_4$ thin films had kesterite structures, indicating that Mg doping did not affect the crystal structure. Moreover, Mg doping resulted in an increase in the particle size and enhancement in the crystallinity and electrical properties of the CZTS film. With an increase in x from 0 to 0.6, the band gap of the $Cu_2Mg_xZn_{1-x}SnS_4$ film decreased from 1.43 to 1.29 eV.

2. Experimental Method

2.1. Sample Preparation

Soda lime glass (SLG) substrates were used to deposit $Cu_2Mg_xZn_{1-x}SnS_4$ thin films. Firstly, $Cu_2Mg_xZn_{1-x}SnS_4$ precursor solutions with different Mg contents were prepared by the sol-gel method. The copper (II) acetate monohydrate (0.8086 g), tin (II) chloride hydrate (0.5077 g), zinc acetate and magnesium chloride hexahydrate (0.4794 g) were dissolved in 2-methoxyethanol (10 mL) and stirred for 10 min, then the solution was evenly mixed with thiourea (1.3702 g). For the sake of obtaining precursor solutions with different Mg contents, the mole ratios of Mg/(Mg + Zn) were set to 0, 0.1, 0.2, 0.4 and 0.6 in the solution. After the complete dissolution of the metal compounds, monoethanolamine was added as a stabilizer, and the stirring process was continued until the solution became clear and transparent.

$Cu_2Mg_xZn_{1-x}SnS_4$ thin films were deposited onto SLG substrates by spin coating at 3000 rpm for 30 s, followed by sintering for 5 min on a hot plate at 300 °C in air. To obtain micrometer thick $Cu_2Mg_xZn_{1-x}SnS_4$ films, the coating and sintering processes were repeated. Next, the precursor films were rapidly annealed under a sulfur atmosphere at different annealing temperatures and times.

2.2. Materials Characterization

The crystal structures of the prepared $Cu_2Mg_xZn_{1-x}SnS_4$ thin films were characterized by power X-ray diffraction (XRD; λ = 0.15406 nm/max-ga, Rigaku Corporation, Tokyo, Japan) and Raman spectroscopy with an excitation wavelength of 514 nm. The chemical composition of the $Cu_2Mg_xZn_{1-x}SnS_4$ films and the chemical bonding states of the constituents were characterized by X-ray photoelectron spectroscopy (XPS; Thermo Fisher Scientific, Waltham, MA, USA) using Al Kα as the X-ray source. Scanning electron microscopy (SEM) was performed using Hitachi S-4800 (JEOL Ltd., Tokyo, Japan). The electrical properties of the $Cu_2Mg_xZn_{1-x}SnS_4$ films were measured in the van der Pauw configuration. The optical properties of the films were measured using an ultraviolet-visible-near-infrared (UV-Vis-NIR) spectrophotometer (UV-3101PC, Tokyo, Japan).

3. Results and Discussion

It has been widely reported that annealing conditions significantly affect the properties of CZTS films. To optimize the annealing conditions for $Cu_2Mg_xZn_{1-x}SnS_4$ films, they were annealed under different conditions. Figure 1a–f show the SEM images of the $Cu_2Mg_xZn_{1-x}SnS_4$ (x = 0.2) films annealed at different conditions. Samples A1, A2 and A3 were annealed for 60 min under a sulfur atmosphere at 540, 580 and 600 °C, respectively. Figure 1a–c shows the SEM images of samples A1, A2 and A3, respectively. As shown in the surface SEM image in Figure 1a, sample A1 contained small nanoparticles (30–100 nm); in addition, a small hole was observed on the surface of the film. Figure 1b shows the SEM image of sample A2; as observed, with an increase in the annealing temperature, the crystalline quality of the $Cu_2Mg_xZn_{1-x}SnS_4$ film significantly improved; grain size increased up to 1.4 µm; and the surface became smooth, dense and crack free. However, with a further increase in the annealing temperature to 600 °C, the grain size decreased to 500–900 nm, and more voids and nanoparticles were observed on the surface, as shown in Figure 1c. As shown in Figure 1a–c, sample A2 exhibited optimal crystalline quality, indicating that the optimum annealing temperature was 580 °C. Samples B1, B2 and B3 were annealed at 580 °C under a sulfur atmosphere for 30, 45 and 75 min respectively. Figure 1d–f show the SEM images of samples B1, B2 and B3. Compared to that of sample A2 annealed at 580 °C for 60 min, the crystalline quality of samples B1, B2 and B3 was inferior. Moreover, the surfaces of B1, B2 and B3 were uneven and porous, as shown in Figure 1d–f. As shown in Figure 1b,d,e, with an increase in the annealing time from 30 min and 45 min, the grain size increased from 70–200 nm to 100–500 nm, and then, with a further increase in the annealing time to 60 min, the grain size increased to 100–1400 nm. However, as the film was annealed for a longer time (75 min), the grain size of the $Cu_2Mg_xZn_{1-x}SnS_4$ film decreased to 200–700 nm. The surface morphological examination indicated that the $Cu_2Mg_xZn_{1-x}SnS_4$ grain growth gradually occurred, and an optimal crystallization quality was achieved by annealing at 580 °C for 60 min.

Figure 1. Scanning electron microscopy (SEM) images of $Cu_2Mg_xZn_{1-x}SnS_4$ ($0 \leq x \leq 0.6$) thin films annealed at a different annealing condition: (**a**) 540 °C, 60 min; (**b**) 580 °C, 60 min; (**c**) 600 °C, 60 min; (**d**) 580 °C, 30 min; (**e**) 580 °C, 45 min and (**f**) 580 °C, 75 min.

Table 1 lists the electrical transport parameters of the $Cu_2Mg_xZn_{1-x}SnS_4$ ($x = 0.2$) film annealed at different annealing conditions. As observed, the film invariably exhibited p-type conductivity. With an increase in the annealing temperature from 540 °C to 600 °C, the carrier concentration first sharply increased from 4.12×10^{15} cm^{-3} (sample A1) to 3.29×10^{18} cm^{-3} (sample A2), and then decreased to 3.79×10^{17} cm^{-3} (sample A3); notably, the resistivity decreased from 9.43×10^0 Ωcm to 1.16×10^{-1} Ωcm, and then increased to 1.53×10^0 Ωcm. The mobility decreased from 3.70×10^0 cm^2 V^{-1} S^{-1} to 1.01×10^{-1} cm^2 V^{-1} S^{-1} and then increased to 7.87×10^{-1} cm^2 V^{-1} S^{-1}. Similarly, with an increase in the annealing time from 30 min to 75 min, the carrier concentration first increased from 3.21×10^{14} cm^{-3} (sample B1) to 3.79×10^{15} cm^{-3} (sample B2), reached the maximum value of 3.29×10^{18} cm^{-3} (sample A2) and finally reduced to 4.62×10^{16} cm^{-3} (sample B3). Simultaneously, the mobility decreased from 6.02×10^0 cm^2 V^{-1} S^{-1} for sample B1 to 2.02×10^0 cm^2 V^{-1} S^{-1} for sample B2, and then reached the minimum 1.01×10^{-1} cm^2 V^{-1} S^{-1} for A2 and finally slightly elevated to 9.32×10^{-1} cm^2 V^{-1} S^{-1} for sample B3. According to the SEM and Hall results, the carrier concentration gradually increases with the annealing time changes from 30 to 60 min, but it starts to decrease when the annealing time increases from 60 to 75 min. It can be explained that when the annealing time increases from 30 to 60 min, the crystallinity of $Cu_2Mg_xZn_{1-x}SnS_4$ films is improved, the defects at the grain boundaries are passivated, resulting in the increase of carrier concentration. When the annealing time increases from 60 to 75 min, the crystallization quality is slightly deteriorated, as shown in the previous SEM results, therefore, the carrier concentration decrease. The change of mobility is opposite to that of the carrier concentration. When the annealing time changes from 30 to 60 min, the mobility decreases with the increasing of the carrier concentration, and when the annealing time increases from 60 to 75 min, the mobility starts to increase with the decreasing of the carrier concentration. Finally, it was found when the film was annealed at 580 °C for 60 min, the $Cu_2Mg_xZn_{1-x}SnS_4$ film has the optimum crystallization quality and the best electrical performance with the carrier concentration of 3.29×10^{18} cm^{-3} and the mobility of 1.01×10^{-1} cm^2 V^{-1} s^{-1}. It is concluded that the change of the electrical properties may have great relevance to the defects passivated in the grain boundaries by improving the crystallinity properties.

Table 1. Electrical properties of the $Cu_2Mg_xZn_{1-x}SnS_4$ (x = 0.2) thin films annealed at different annealing conditions.

Sample	ρ (Ω.cm)	n (cm^{-3})	μ (cm^2 V^{-1} S^{-1})	Type
A1	9.43×10^0	4.12×10^{15}	3.70×10^0	*p*
A2	1.16×10^{-1}	3.29×10^{18}	1.01×10^{-1}	*p*
A3	1.53×10^0	3.79×10^{17}	7.87×10^{-1}	*p*
B1	7.99×10^1	3.21×10^{14}	6.02×10^0	*p*
B2	8.97×10^0	3.79×10^{15}	2.02×10^0	*P*
B3	4.53×10^0	4.62×10^{16}	9.32×10^{-1}	*P*

To evaluate the crystalline quality and investigate the existence of impurity phases, the films were subjected to XRD analysis. Figure 2 illustrates the XRD patterns of the $Cu_2Mg_xZn_{1-x}SnS_4$ ($0 \leq x \leq 0.6$) thin films. As shown in Figure 2a, strong diffraction peaks at 2θ = 28.53, 32.99, 47.33 and 56.17° were observed for all films, which were assigned to the (112), (200), (220) and (312) diffraction planes of kesterite CZTS (JCPDS card no. 26-0575) [26,27]. In addition, two weak peaks were observed at 2θ = 69.27° and 76.44°, which were ascribed to the (008) and (332) planes of kesterite CZTS [28], suggesting that the crystalline quality of the $Cu_2Mg_xZn_{1-x}SnS_4$ films was satisfactory. Apart from the diffraction peaks of CZTS, no secondary phase peaks were detected, indicating that Mg doping did not affect the crystal structure of the CZTS film. As observed, with an increase in x from 0 to 0.1, the intensity of the (112) peak slightly increased, and then with a further increase in x to 0.2, the peak intensity reached the maximum, implying that the crystalline quality of the $Cu_2Mg_xZn_{1-x}SnS_4$ thin film with x = 0.2 is the best. However, with increasing x from 0.2 to 0.6, the intensity of the (112) peak gradually decreased and became the lowest at x = 0.6; this gradual deterioration in the crystallinity of the $Cu_2Mg_xZn_{1-x}SnS_4$ thin films with increasing x was attributed to excessive Mg doping. Figure 2b shows the enlarged view of the (112) peaks. As observed, with an increasing Mg content, the (112) peak unidirectionally shifted to smaller 2θ values, suggesting an increase in the lattice constant of $Cu_2Mg_xZn_{1-x}SnS_4$. It is well known that the change of the ion radius in CZTS usually results in the change of lattice parameters [29–31]. The occupation of Zn^{2+} sites in the CZTS host lattice by Mg^{2+} ions results in an increase in the $Cu_2Mg_xZn_{1-x}SnS_4$ lattice parameters, because the covalent radius of Mg^{2+} (1.36 Å) is larger than that of Zn^{2+} (1.25 Å). Thus, the XRD results indicated that with Mg doping, the phase structure of CZTS did not change, and the Zn^{2+} sites in the CZTS host lattice were occupied by Mg^{2+}.

The formation of pure kesterite $Cu_2Mg_xZn_{1-x}SnS_4$ cannot be properly confirmed by XRD, because the lattice parameters of CZTS and some possible impurity phases such as tetragonal Cu_2SnS_3, cubic ZnS and Cu_xS are similar [32,33]. Therefore, to confirm the formation of pure kesterite $Cu_2Mg_xZn_{1-x}SnS_4$, the samples were subjected to Raman spectroscopy analysis.

Figure 3 shows the Raman spectra of $Cu_2Mg_xZn_{1-x}SnS_4$ ($0 \leq x \leq 0.6$) films. As shown, the spectra contained the dominant characteristic peak at 333 cm^{-1} and two relatively weak peaks at 288 cm^{-1} and 375 cm^{-1}. These Raman peaks were attributed to the A1, A2 and E vibration modes of the S atom in kesterite CZTS, respectively; these results agreed well with those previously reported [34,35]. Notably, no other ternary or binary phase (Cu_2SnS_3, SnS_2, SnS, ZnS) peaks were observed in the Raman spectra. In addition, as shown in Figure 3, with an increase in x from 0 to 0.6, the Raman peak, particularly for the peak of A_1 vibration mode, was slightly red shift systematically. Figure 3 displays the A1 mode peak position variation as a function of the Mg content; as observed, the peak shifted from 336.79 cm^{-1} to 332.13 cm^{-1} with an increase in the Mg content. Combining with the XRD results, the change in the A1 peak position could be ascribed to lattice expansion due to the substitution of the smaller Zn ions by the larger Mg ions in $Cu_2Mg_xZn_{1-x}SnS_4$. The redshift in the lattice vibrations were attributed to the lower bonding force of Mg–S than that of Zn–S, resulting from the larger covalent radius of Mg than that of Zn. A similar Raman peak shift caused by ion replacement has been reported in previous studies [36]. Combined with XRD results to analyze the result of Raman spectra, it was found that no other impurity compounds were detected in $Cu_2Mg_xZn_{1-x}SnS_4$ films when the x was in the range of 0 to 0.6. The pure kesterite $Cu_2Mg_xZn_{1-x}SnS_4$ thin films were successfully prepared.

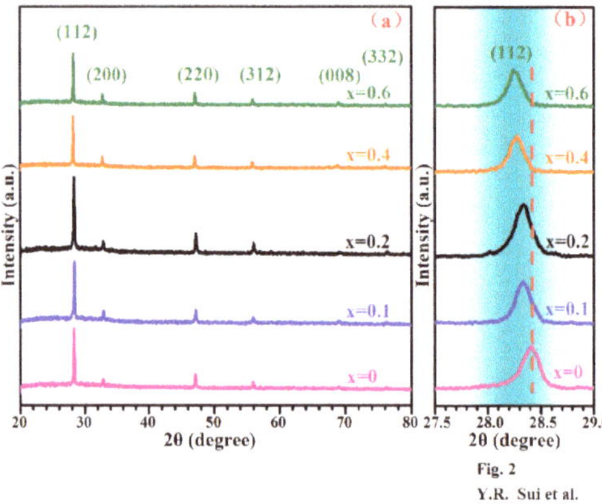

Figure 2. (**a**) XRD spectra of $Cu_2Mg_xZn_{1-x}SnS_4$ ($0 \leq x \leq 0.6$) thin films. (**b**) Enlarged view of the corresponding (112) diffraction peaks of the $Cu_2Mg_xZn_{1-x}SnS_4$ ($0 \leq x \leq 0.6$) thin films.

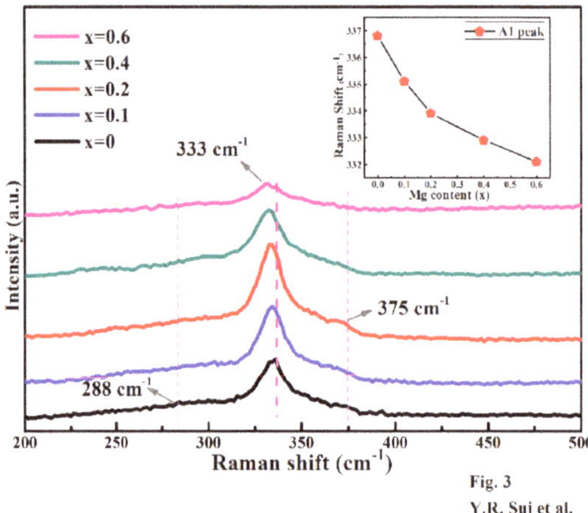

Figure 3. Raman spectra of $Cu_2Mg_xZn_{1-x}SnS_4$ ($0 \leq x \leq 0.6$) thin films. Inset: The main Raman peaks of A_1 mode as a function of the Mg content.

Notably, the chemical composition of the $Cu_2Mg_xZn_{1-x}SnS_4$ films and the chemical bonding states of the constituents significantly affect the solar cell performance. Hence, the $Cu_2Mg_xZn_{1-x}SnS_4$ films were characterized by XPS. Figure 4a–d show the XPS profiles of the constituent metals (Cu, Zn, Sn and Mg) of the representative $Cu_2Mg_xZn_{1-x}SnS_4$ ($x = 0.2$) sample. Figure 4a displays the Cu 2p XPS profile. The two peaks at 952.4 eV and 931.7 eV were attributed to Cu 2p1/2 and Cu 2p3/2. In addition, the peak separation value agreed well with the standard value of 20.7 eV, indicating that Cu was present in the +1 combined-state [37]. Figure 4b illustrates the XPS spectrum of Zn 2p. The two peaks located at

1044.6 eV and 1022.1 eV were attributed to Zn 2p1/2 and Zn 2p3/2, respectively, the splitting energy was 22.5 eV. The splitting value is consistent with the standard value of 22.97 eV, which confirms that Zn exists in a +1 state [38]. The Sn 3d XPS profile is displayed in Figure 4c. As observed, two peaks of Sn $3d_{3/2}$ and Sn $3d_{5/2}$, situated at 494.3 and 485.9 eV were detected; the peak separation value was 8.4 eV, which agreed with the standard value, implying that Sn was in the Sn^{4+} oxidation state [39]. Figure 4d presents the Mg 1s XPS profile; the peak at 1303.7 eV was assigned to the Mg 1s core level, indicating the presence of divalent Mg^{2+} [24]. According to the results of XPS, the valence states of Cu, Zn, Mg and Sn were +1, +2, +4 and +2 respectively. This further confirmed the substitution of Zn in CZTS by Mg, agreeing well with the XRD and Raman results.

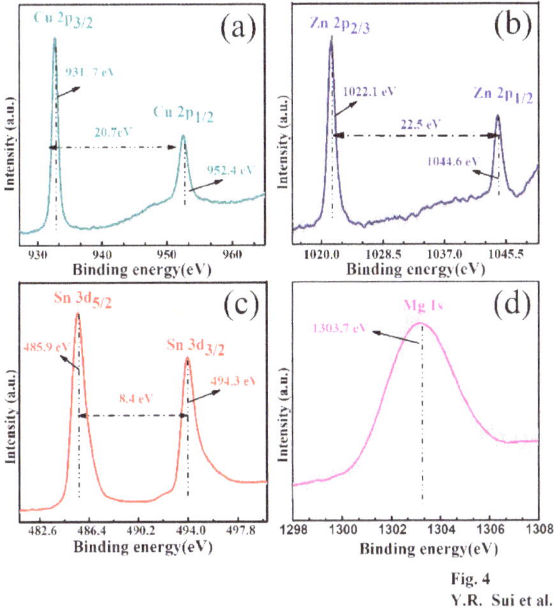

Figure 4. X-ray photoelectron spectroscopy (XPS) spectrum of $Cu_2Mg_xZn_{1-x}SnS_4$ ($x = 0.2$) thin films: (**a**) Cu, (**b**) Zn, (**c**) Sn and (**d**) Mg.

The atomic contents of Cu, Zn, Sn, S and Mg in the $Cu_2Mg_xZn_{1-x}SnS_4$ ($0 \leq x \leq 0.6$) films are listed in Table 2. When the percentages of Mg/(Mg + Zn) for the precursor solution of $Cu_2Mg_xZn_{1-x}SnS_4$ ($0 \leq x \leq 0.6$) were 0, 10, 20, 40 and 60, the percentages of Mg/(Mg + Zn) in $Cu_2Mg_xZn_{1-x}SnS_4$ films were 0, 7.79, 14.22, 34.72 and 54.61, respectively. Notably, the elemental loss during annealing and the preparation process cannot be neglected; nonetheless, the Mg/(Mg + Zn) ratio in the $Cu_2Mg_xZn_{1-x}SnS_4$ films increased with an increase in the Mg content in the precursor solution. Figure 5 summarizes the atomic contents of the constituent elements of the $Cu_2Mg_xZn_{1-x}SnS_4$ ($0 \leq x \leq 0.6$) films, according to the energy dispersive X-ray spectroscopy (EDS) results presented in Table 2. As observed, the atomic content of Mg gradually increased with a decrease in the atomic content of Zn from 17.95 to 7.39; moreover, the Mg/(Mg + Zn) ratio also increased. This indicated that Mg was incorporated into the CZTS lattice, replacing Zn. Furthermore, the changes in the atomic contents of other elements in the $Cu_2Mg_xZn_{1-x}SnS_4$ films were negligible. The result is in good agreement with the conclusion that Mg will substitute the site of Zn obtained from the analysis result of XRD and Raman.

Table 2. EDS composition analyses of the $Cu_2Mg_xZn_{1-x}SnS_4$ ($0 \leq x \leq 0.6$) thin films.

Sample	Cu (at %)	Zn (at %)	Mg (at %)	Sn (at %)	S (at %)	Mg/(Mg + Zn) (at %)
$x = 0$	25.07	17.95	0	10.30	46.98	0
$x = 0.1$	25.34	15.51	1.31	10.03	47.81	7.79
$x = 0.2$	25.52	14.66	2.43	10.61	46.78	14.22
$x = 0.4$	25.31	11.32	6.02	10.32	47.03	34.72
$x = 0.6$	25.08	7.39	8.89	10.77	47.87	54.61

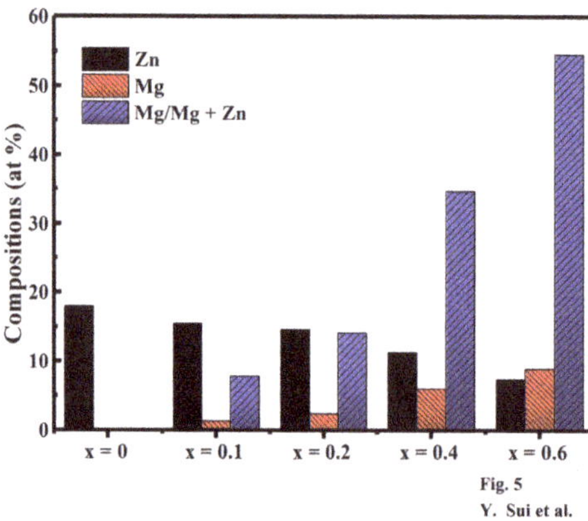

Figure 5. Energy dispersive X-ray spectroscopy (EDS) composition analyses of $Cu_2Mg_xZn_{1-x}SnS_4$ ($0 \leq x \leq 0.6$) thin films.

To determine the effect of the Mg content on the crystalline quality of the $Cu_2Mg_xZn_{1-x}SnS_4$ ($0 \leq x \leq 0.6$) films, the films were detected by SEM as shown in Figure 6a–e. Figure 6a displays the surface SEM images of the $Cu_2Mg_xZn_{1-x}SnS_4$ film with $x = 0$. As observed, the film consisted of irregular nanoscale grains (40–500 nm). Moreover, the surface of the film was relatively rough, but compact. Obviously, the irregular grain boundaries and small particles are not conducive to the improvement of the efficiency for the CZTS solar cells. As shown in Figure 6b, with an increase in x to 0.1, the film crystallinity enhanced and the grain size increased to 400–1200 nm. Furthermore, the surface morphology was significantly improved and become smooth and compact. With a further increase in the value of x to 0.2, the film surface became very flat and dense, as displayed in Figure 6c; in addition, the grain size further increased to 0.7–1.5 μm, which was conducive to achieving high efficiencies for CZTS solar cells. Figure 6d shows the SEM image of the $Cu_2Mg_xZn_{1-x}SnS_4$ film with $x = 0.4$. As observed, the grain size sharply decreased to 300–900 nm, but the grains were larger than those of $Cu_2Mg_xZn_{1-x}SnS_4$ with $x = 0$, and densely stacked. The crystalline quality of the $Cu_2Mg_xZn_{1-x}SnS_4$ film continued to deteriorate with further increase in x to 0.6. As shown in Figure 6e, the grain size of $Cu_2Mg_xZn_{1-x}SnS_4$ film reduced to 200–400 nm, occasionally, a few larger grains were observed on the film surface. As seen, the surface morphology of the film with $x = 0.6$ was uneven and irregular. The bar chart in Figure 6 shows the average particle diameter as a function of the Mg content. As seen, the average size gradually increased with an increase in the value of x from 0 to 0.2 and reached the maximum at $x = 0.2$, then with further increase in x from 0.2 to 0.6, the size sharply decreased. It is well known that the good grain growth and smooth surface is of great significance to the fabrication of high power conversion efficiency (PCE) CZTS solar cells. Because the absorption layer with larger particle size can reduce the grain boundaries area, which is conducive to decrease the recombination of photon-generated carrier and increase the efficiency of CZTS solar cells. Based on the results of SEM, when the value of x was 0.2, it was concluded that the crystallization quality of $Cu_2Mg_xZn_{1-x}SnS_4$ films achieved the best results, the grain size was the largest and the surface was the smoothest and denser, which is most suitable for the absorber layer of the CZTS solar cells.

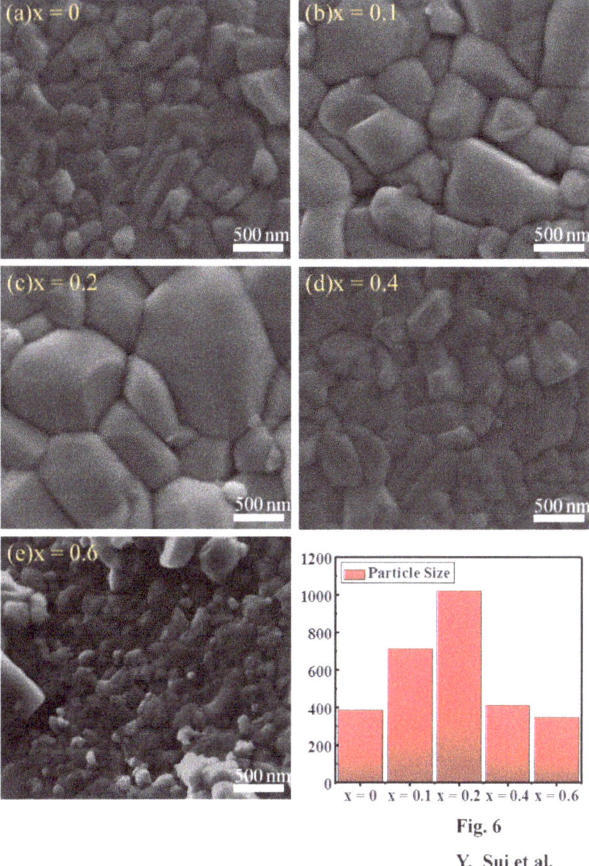

Fig. 6

Y. Sui et al.

Figure 6. SEM images of $Cu_2Mg_xZn_{1-x}SnS_4$ ($0 \leq x \leq 0.6$) thin films with different Mg content: (**a**) $x = 0$, (**b**) $x = 0.1$, (**c**) $x = 0.2$, (**d**) $x = 0.4$ and (**e**) $x = 0.6$. Inset: The average diameter of particles as a function of the Mg content.

UV-Vis-NIR spectroscopy was carried out to investigate the influence of Mg content on the band gap (E_g) values of the $Cu_2Mg_xZn_{1-x}SnS_4$ ($0 \leq x \leq 0.6$) thin films. Figure 7a illustrates the plots of $(\alpha h\upsilon)^2$ against $h\upsilon$ for the films, where α and $h\upsilon$ are the absorption coefficient and photon energy, respectively. The E_g values for the $Cu_2Mg_xZn_{1-x}SnS_4$ ($0 \leq x \leq 0.6$) films can be obtained by optical absorption measurements, according to Tauc's relation [40]:

$$(\alpha h\upsilon) = A(h\upsilon - E_g)^n, \tag{1}$$

where A is a constant, n = 1/2, 3/2, 2 and 3 for the allowed direct, forbidden direct, allowed indirect and forbidden indirect transitions, respectively [41]. In general, $Cu_2Mg_xZn_{1-x}SnS_4$ is regarded as a direct band gap semiconductor, therefore, n = 1/2. The values of E_g for the $Cu_2Mg_xZn_{1-x}SnS_4$ thin films with $x = 0, 0.1, 0.2, 0.4$ and 0.6 calculated according to Tauc's relation were 1.43, 1.36, 1.35, 1.33 and 1.29 eV, respectively. The inset of Figure 7a shows the UV–vis absorption spectra of the representative $Cu_2Mg_xZn_{1-x}SnS_4$ with $x = 0.2$. It was found that the $Cu_2Mg_xZn_{1-x}SnS_4$ film had a stronger absorption intensity in the short wavelength range, which is suitable as the absorber layer of the CZTS solar cells. Figure 7b shows the variation in the band gap energy as a function of the Mg content. As observed, the

E_g value reduced from 1.43 to 1.29 eV with an increase in x from 0 to 0.6, which can be ascribed to the change in the lattice parameter, resulting from the occupation of Zn sites by Mg. According to the first principles calculation results for the CZTS semiconductor, the minimum of the conduction band depends on the Sn 3d and S 3p antibonding orbitals and the maximum of the valence band is primarily related to p–d hybridization between Cu and S [42–44]. In the present work, the Mg element will take the site of Zn in CZTS, which will not affect the band gap of CZTS based on the theoretical analysis mentioned above. However, the band gap of $Cu_2Mg_xZn_{1-x}SnS_4$ linearly varied as x increased from 0 to 0.6. Similar phenomena that the band gap of CZTS changes regularly because Zn is replaced by other elements (Cd, Ge) have been mentioned in previous studies [33,44], they ascribed the change to an increase in the unit cell volume, which led to a reduction in the antibonding component of the s–p and s–s hybridization between S^{2-} and Sn^{4+}, resulting in a decrease in the minimum of the conduction band. In the present work, the substitution of Zn by Mg increased the volume of the unit cell and reduced the antibonding component of s–p and s–s hybridization between S^{2-} and Sn^{4+}. Hence, the minimum of the conduction band and the band gap of $Cu_2Mg_xZn_{1-x}SnS_4$ gradually decreased with increasing Mg content.

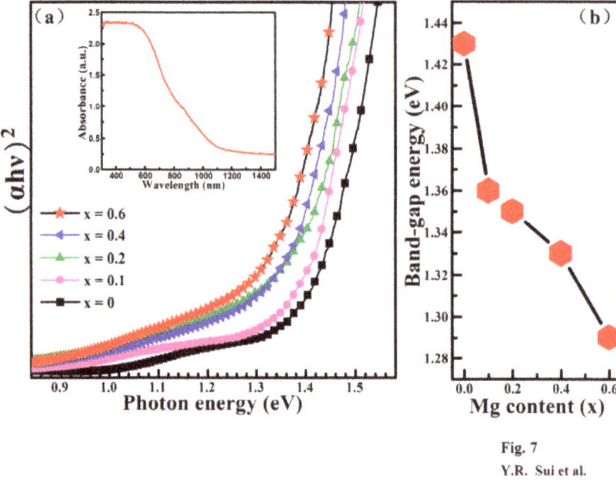

Figure 7. (a) The plot of $(\alpha h\nu)^2$ vs $h\nu$ for the absorption spectra. The inset shows the UV–vis absorption spectra of the representative $Cu_2Mg_xZn_{1-x}SnS_4$ with $x = 0.2$; (b) Band gap variation as a function of the Mg content.

The conductivity (ρ), carrier concentration (n) and mobility (μ) for the $Cu_2Mg_xZn_{1-x}SnS_4$ ($0 \leq x \leq 0.6$) thin films were determined by the van der Pauw method at room temperature and the results are presented in Table 3. Tests were repeated on the same sample to ensure precision and reliability of the electrical performances of the $Cu_2Mg_xZn_{1-x}SnS_4$ films. All samples with different Mg contents exhibited p-type semiconductor characteristics. In addition, as the value of x increased from 0 to 0.2, the carrier concentration of the $Cu_2Mg_xZn_{1-x}SnS_4$ films increased from 6.95×10^{16} cm^{-3} to 3.29×10^{18} cm^{-3}. However, with a further increase in x from 0.2 to 0.6, the carrier concentration gradually decreased to 2.02×10^{17} cm^{-3}. Simultaneously, the mobility decreased from 2.63×10^0 cm^2 V^{-1} s^{-1} to 1.01×10^{-1} cm^2 V^{-1} s^{-1} with an increase in x from 0 to 0.2, and then, the mobility increased to 1.43×10^0 cm^2 V^{-1} s^{-1} as x increased to 0.6. The increase in the carrier concentration with an increase in x from 0 to 0.2 was attributed to the passivation of grain boundary defects, resulting from an improvement in the crystalline quality of the $Cu_2Mg_xZn_{1-x}SnS_4$ thin film. As previously reported for K-doped and Na-doped CZTSSe and CIGS solar cells [45,46], the carrier concentration markedly improved because of passivation of grain boundary defects, resulting from an improvement in the crystallization properties. Moreover, the decrease in the carrier concentration observed in this study with an increase in x from 0.2 to 0.6 was attributed to the deterioration of crystalline quality due to the small grain size, multi-hole and irregular surface morphology. Notably, the $Cu_2Mg_xZn_{1-x}SnS_4$ film with $x = 0.2$ exhibited an optimal electrical conductivity.

Table 3. Electrical properties of the $Cu_2Mg_xZn_{1-x}SnS_4$ ($0 \leq x \leq 0.6$) thin films.

Sample	ρ (Ω.cm)	n (cm^{-3})	μ (cm^2 V^{-1} S^{-1})	Type
$x = 0$	3.73×10^1	6.95×10^{16}	2.63×10^0	p
$x = 0.1$	3.21×10^1	2.46×10^{17}	3.23×10^{-1}	p
$x = 0.2$	1.16×10^{-1}	3.29×10^{18}	1.01×10^{-1}	p
$x = 0.4$	1.92×10^{-1}	1.21×10^{18}	1.22×10^0	p
$x = 0.6$	1.12×10^0	2.02×10^{17}	1.43×10^0	p

4. Conclusions

In conclusion, we have prepared $Cu_2Mg_xZn_{1-x}SnS_4$ films with different Mg contents and investigated the influence of the annealing temperature and time on the performance of the films. The optimal annealing temperature and time was found to be 580 °C and 60 min, respectively. Moreover, under the optimal annealing conditions, we investigated the effect of Mg content on the performance of the $Cu_2Mg_xZn_{1-x}SnS_4$ films in detail. It was found that the $Cu_2Mg_xZn_{1-x}SnS_4$ films were ideal for use as absorption layers in solar cells because of their continuous tunable band gaps, favorable photoelectric performance and high crystallinity. With an increase in x from 0 to 0.6, the band gap increased from 1.43 to 1.29 eV. Notably, the continuous tunable band gap could facilitate the tuning of band alignment at the $Cu_2Mg_xZn_{1-x}SnS_4$/CdS heterojunction by changing the Mg content. Furthermore, the $Cu_2Mg_xZn_{1-x}SnS_4$ film with $x = 0.2$ exhibited superior crystallinity and surface morphology compared to other $Cu_2Mg_xZn_{1-x}SnS_4$ films. Meanwhile, at $x = 0.2$, the electrical conductivity of $Cu_2Mg_xZn_{1-x}SnS_4$ film reached the optimal level, with a carrier concentration of 3.29×10^{18} cm^{-3} and a mobility of 1.01×10^{-1} cm^2 V^{-1} s^{-1}. Notably, after being annealed at 580 °C for 60 min, the $Cu_2Mg_xZn_{1-x}SnS_4$ film with the optimal Mg/(Mg + Zn) ratio of 0.2 exhibited favorable photoelectric performance and enhanced crystalline quality, making it a promising candidate for the preparation of high-efficiency solar cells with tunable band gap absorption layers.

Author Contributions: Conceptualization, L.Y.; Writing-Original Draft Preparation, Y.S. and Y.Z.; Software, D.J.; Formal analysis, Y.S., W.H. and Z.W.; Investigation, Y.S., Y.Z. and F.W.; Writing-Review & Editing, Y.S. and B.Y.

Funding: This research was funded by the National Natural Science Foundation of China under Grant Nos. 61505067, 61605059, 61475063, 61775081, 61705079.

Conflicts of Interest: The authors declare no conflict of interest.

References

1. Lokhande, A.C.; Chalapathy, R.B.V.; He, M.; Jo, E.; Gang, M.; Pawar, S.A.; Lokhande, C.D.; Kim, J.H. Development of Cu_2SnS_3 (CTS) thin film solar cells by physical techniques: A status review. *Sol. Energy Mater. Sol. Cells* **2016**, *153*, 84–107. [CrossRef]
2. Khadka, D.B.; Kim, J.H. Band gap engineering of alloyed $Cu_2ZnGe_xSn_{1-x}Q_4$ (Q = S,Se) films for solar cell. *J. Phys. Chem. C* **2015**, *119*, 1706–1713. [CrossRef]
3. Friedlmeier, T.M.; Wieser, N.; Walter, T.; Dittrich, H.; Schock, H.W. Hetero-junctions based on Cu_2ZnSnS_4 and $Cu_2ZnSnSe_4$ thin films. In Proceedings of the 14th European PVSEC and Exhibition, Barcelona, Spain, 30 June–4 July 1997; pp. 1242–1245.
4. Yan, C.; Huang, J.; Sun, K.; Johnston, S.; Zhang, Y.; Sun, H.; Pu, A.; He, M.; Liu, F.; Eder, K.; et al. Cu_2ZnSnS_4 solar cells with over 10% power conversion efficiency enabled by heterojunction heat treatment. *Nat. Energy* **2018**, *3*, 764–772. [CrossRef]
5. Yan, C.; Sun, K.W.; Huang, J.L.; Johnston, S.; Liu, F.Y.; Veettil, B.P.; Sun, K.; Pu, A.; Zhou, F.Z.; Stride, J.A.; et al. Xiaojing hao beyond 11% efficient sulfide kesterite $Cu_2Zn_xCd_{1-x}SnS_4$ Solar cell: Effects of cadmium alloying. *ACS Energy Lett.* **2017**, *2*, 930–936. [CrossRef]

6. Jackson, P.; Hariskos, D.; Wuerz, R.; Kiowski, O.; Bauer, A.; Friedlmeier, T.M.; Powalla, M. Properties of Cu(In,Ga)Se$_2$ solar cells with new record efficiencies up to 21.7%. *Phys. Status Solidi* **2015**, *9*, 28–31.

7. Mitzi, D.B.; Gunawan, O.; Todorov, T.K.; Barkhouse, D.A.R. Prospects and performance limitations for Cu–Zn–Sn–S–Se photovoltaic technology. *Philos. Trans. R. Soc.* **2013**, *371*, 1–22. [CrossRef] [PubMed]

8. Yang, Y.; Wang, G.; Zhao, W.; Tian, Q.; Huang, L.; Pan, D. Solution-Processed Highly Efficient Cu$_2$ZnSnSe$_4$ Thin Film Solar Cells by Dissolution of Elemental Cu, Zn, Sn, and Se Powders. *ACS Appl. Mater. Interfaces* **2015**, *7*, 460–464. [CrossRef] [PubMed]

9. Miskin, C.; Yang, W.; Hages, C.; Carter, N.; Joglekar, C.; Stach, E.; Agrawal, R. 9.0% efficient Cu$_2$ZnSn(S,Se)$_4$ solar cells from selenized nanopartcle inks. *Prog. Photovolt.* **2015**, *23*, 654–659. [CrossRef]

10. Hages, C.; Levcenco, S.; Miskin, C.; Alsmeier, J.; Abou-Ras, D.; Wilks, R.; Bär, M.; Unold, T.; Agrawal, R. Improved performance of ge-alloyed CZTGeSSe thin-film solar cells through control of elemental losses. *Prog. Photovolt.* **2015**, *23*, 376–384. [CrossRef]

11. Lokhande, A.C.; Gurav, K.V.; Jo, E.J.; Lokhande, C.D.; Kim, J.H. Chemical synthesis of Cu$_2$SnS$_3$ (CTS) nanoparticles: A status review. *J. Alloys Compd.* **2016**, *656*, 295–310. [CrossRef]

12. Woo, K.; Kim, Y.; Yang, W.; Kim, K.; Kim, I.; Oh, Y.; Kim, J.; Moon, J. band-gap-graded Cu$_2$ZnSn(S$_{1-x}$, Se$_x$)$_4$ solar cells fabricated by an ethanol-based, particulate precursor ink route. *Sci. Rep.* **2013**, *3*, 3069. [CrossRef] [PubMed]

13. Xin, H.; Katahara, J.K.; Braly, I.L.; Hillhouse, H.W. 8% efficient Cu$_2$ZnSn(S,Se)$_4$ solar cells from redox equilibrated simple precursors in DMSO. *Adv. Energy Mater.* **2014**, *4*, 1301823. [CrossRef]

14. Duan, H.; Yang, W.; Bob, B.; Hsu, C.; Lei, B.; Yang, Y. The Role of Sulfur in Solution-Processed Cu$_2$ZnSn(S,Se)$_4$ and its effect on defect properties. *Adv. Funct. Mater.* **2013**, *23*, 1466–1471. [CrossRef]

15. Todorov, T.; Reuter, K.; Mitzi, D. High-efficiency solar cell with earth-abundant liquid-processed absorber. *Adv. Mater.* **2010**, *22*, E156–E159. [CrossRef]

16. Fu, J.; Tian, Q.W.; Zhou, Z.J.; Kou, D.X.; Meng, Y.N.; Zhou, W.H.; Wu, S.X. Improving the performance of solution-processed Cu$_2$ZnSn(S,Se)$_4$ photovoltaic materials by Cd^{2+} substitution. *Chem. Mater.* **2016**, *28*, 16. [CrossRef]

17. Xiao, Z.Y.; Li, Y.F.; Yao, B.; Deng, R.; Ding, Z.H.; Wu, T. Bandgap engineering of Cu$_2$Cd$_x$Zn$_{1-x}$SnS$_4$ alloy for photovoltaic applications: A complementary experimental and first-principles study. *J. Appl. Phys.* **2013**, *114*, 18. [CrossRef]

18. Hsieh, Y.T.; Han, Q.; Jiang, C.; Song, T.B.; Chen, H.; Meng, L.; Zhou, H.; Yang, Y. Efficiency enhancement of Cu$_2$ZnSn(S,Se)$_4$ solar cells via alkali metals doping. *Adv. Energy Mater.* **2016**, *6*, 1502386. [CrossRef]

19. Sutter-Fella, C.M.; Stuckelberger, J.A.; Hagendorfer, H.; La Mattina, F.; Kranz, L.; Nishiwaki, S.; Uhl, A.R.; Romanyuk, Y.E.; Tiwari, A.N. Sodium assisted sintering of chalcogenides and its application to solution process Cu$_2$ZnSn(S,Se)$_4$ thin film solar cells. *Chem. Mater.* **2014**, *26*, 1420–1425. [CrossRef]

20. Yuan, M.; Mitzi, D.B.; Liu, W.; Kellock, A.J.; Chey, S.J.; Deline, V.R. Optimization of CIGS-based PV device through antimony doping. *Chem. Mater.* **2010**, *22*, 285–287. [CrossRef]

21. Johnson, M.; Baryshev, S.V.; Thimsen, E.; Manno, M.; Zhang, X.; Veryovkin, I.V.; Leighton, C.; Aydil, E.S. Alkali metal-enhanced grain growth in Cu$_2$ZnSnS$_4$ thin films. *Energy Environ. Sci.* **2014**, *7*, 1931–1938. [CrossRef]

22. Yuan, M.; Mitzi, D.B.; Gunawan, O.; Kellock, A.J.; Chey, S.J.; Deline, V.R. Antimony assisted low-temperature processing of CuIn$_{1-x}$Ga$_x$Se$_{2-y}$Sy solar cells. *Thin Solid Films* **2010**, *519*, 852–856. [CrossRef]

23. Tai, K.F.; Fu, D.C.; Chiam, S.Y.; Huan, C.H.A.; Batabyal, S.K.; Wong, L.H. Antimony doping in solution-processed Cu$_2$ZnSn(S,Se)$_4$ solar cells. *ChemSusChem* **2015**, *8*, 3504–3511. [CrossRef] [PubMed]

24. Wei, M.; Du, Q.; Wang, R.; Jiang, G.; Liu, W.; Zhu, C. Synthesis of new earth-abundant kesterite cu$_2$mgsns$_4$ nanoparticles by hot-injection method. *Chem. Lett.* **2014**, *43*, 1149–1151. [CrossRef]

25. Zhang, Y.; Sui, Y.R.; Wu, Y.J.; Jiang, D.Y.; Wang, Z.W.; Wang, F.Y.; Lv, S.Q.; Yao, B.; Yang, L.L. Synthesis and investigation of environmental protection and earth-abundant kesterite Cu$_2$Mg$_x$Zn$_{1-x}$Sn(S,Se)$_4$ thin films for Solar Cells. *Ceram. Int.* **2018**, *44*, 15249–15255. [CrossRef]

26. Kishor Kumar, Y.B.; Suresh Babu, G.; Uday Bhaskar, P.; Sundar Raja, V. Preparation and characterization of spray-deposited Cu$_2$ZnSnS$_4$ thin films. *Sol. Energy Mater. Sol. Cells* **2009**, *93*, 1230–1237. [CrossRef]

27. Sui, Y.R.; Wu, Y.J.; Zhang, Y.; Wang, F.Y.; Gao, Y.B.; Lv, S.Q.; Wang, Z.W.; Sun, Y.F.; Wei, M.B.; Yao, B.; et al. Synthesis of simple, low cost and benign sol–gel Cu$_2$In$_x$Zn$_{1-x}$SnS$_4$ alloy thin films: Influence of different rapid thermal annealing conditions and their photovoltaic solar cells. *RSC Adv.* **2018**, *8*, 9038–9048. [CrossRef]

28. Sui, Y.R.; Wu, Y.J.; Zhang, Y.; Wang, Z.W.; Wei, M.B.; Yao, B. Indium effect on structural, optical and electrical properties of $Cu_2In_xZn_{1-x}SnS_4$ alloy thin films for Solar Cell. *Superlattices Microstruct.* **2017**, *111*, 579–590. [CrossRef]

29. Kuo, D.H.; Wubet, W. Mg dopant in $Cu_2ZnSnSe_4$: An n-type former and a promoter of electrical mobility up to120 cm^2 V^{-1} s^{-1}. *J. Solid State Chem.* **2014**, *215*, 122–127. [CrossRef]

30. Ibraheam, A.S.; Al-Douri, Y.; Hashim, U.; Ghezzar, M.R.; Addou, A.; Ahmed, W.K. Cadmium effect on optical properties of $Cu_2Zn_{1-x}Cd_xSnS_4$ quinternary alloys nanostructures. *Sol. Energy* **2015**, *114*, 39–50. [CrossRef]

31. Kevin, P.; Malik, M.A.; Mcadams, S.; O'Brien, P. Synthesis of nanoparticulate alloys of the composition $Cu_2Zn_{1-x}Fe_xSnS_4$: Structural, optical, and magnetic properties. *J. Am. Chem. Soc.* **2015**, *137*, 15086–15089. [CrossRef]

32. Walsh, A.; Chen, S.; Wei, S.H.; Gong, X.G. Kesterite thin-film solar cells: Advances in materials modelling of Cu_2ZnSnS_4. *Adv. Energy Mater.* **2012**, *2*, 400–409. [CrossRef]

33. Jiang, Y.H.; Yao, B.; Li, Y.F.; Ding, Z.H.; Xiao, Z.Y.; Yang, G.; Liu, R.J.; Liu, K.S.; Sun, Y.M. Effect of Cd content and sulfurization on structures and properties of Cd doped Cu_2SnS_3 thin films. *J. Alloys Compd.* **2017**, *721*, 92–99. [CrossRef]

34. Khare, A.; Himmetoglu, B.; Johnson, M.; Norris, D.J.; Cococcioni, M.; Aydil, E.S. Calculation of the lattice dynamics and Raman spectra of copper zinc tin chalcogenides and comparison to experiments. *J. Appl. Phys.* **2012**, *118*, 083707. [CrossRef]

35. Dumcenco, D.; Huang, Y.S. The vibrational properties study of kesterite Cu_2ZnSnS_4 single crystals by using polarization dependent Raman spectroscopy. *Opt. Mater.* **2013**, *35*, 419–425. [CrossRef]

36. Tong, Z.; Yan, C.; Su, Z.; Zeng, F.; Yang, J.; Li, Y.; Jiang, L.; Lai, Y.; Liu, F. Effects of potassium doping on solution processed kesterite Cu_2ZnSnS_4 thin film solar cells. *Appl. Phys. Lett.* **2014**, *105*, 223903. [CrossRef]

37. Tsega, M.; Dejene, F.B.; Kuo, D.H. Morphological evolution and structural properties of $Cu_2ZnSn(S,Se)_4$ thin films deposited from single ceramic target by a one-step sputtering process and selenization without H_2Se. *J. Alloys Compd.* **2015**, *642*, 140–147. [CrossRef]

38. Chen, H.Y.; Yu, S.M.; Shin, D.W.; Yoo, J.B. Solvothermal synthesis and characterization of chalcopyrite $CuInSe_2$ nanoparticles. *Nanoscale Res. Lett.* **2010**, *5*, 217–223. [CrossRef]

39. Rondiya, S.; Wadnerkar, N.; Jadhav, Y.; Jadkar, S.; Haram, S.; Kabir, M. Structural, electronic, and optical properties of Cu_2NiSnS_4: A combined experimental and theoretical study toward photovoltaic applications. *Chem. Mater.* **2017**, *29*, 3133–3142. [CrossRef]

40. Li, Y.F.; Yao, B.; Lu, Y.M.; Wei, Z.P.; Gai, Y.Q.; Zheng, C.J.; Li, B.H.; Shen, D.Z.; Fan, X.W.; Tang, Z.K. Realization of p-type conduction in undoped MgxZn1-xO thin films by controlling Mg content. *Appl. Phys. Lett.* **2007**, *91*, 2115. [CrossRef]

41. Chen, S.; Walsh, A.; Luo, Y.; Yang, J.H.; Gong, X.G.; Wei, S.H. Wurtzite-derived poly types of kesterite and stannite quaternary chalcogenide semiconductors. *Phys. Rev. B* **2010**, *82*, 195203. [CrossRef]

42. Zhang, Y.; Sun, X.; Zhang, P.; Yuan, X.; Huang, F.; Zhang, W. Structural properties and quasiparticle band structures of cu-based quaternary semiconductors for photovoltaic applications. *J. Appl. Phys.* **2012**, *111*, 0637091–0637096. [CrossRef]

43. Son, D.H.; Kim, D.H.; Park, S.N.; Yang, K.J.; Nam, D.; Cheong, H.; Kang, J.K. Growth and device characteristics of CZTSSe thin-film solar cells with 8.03% efficiency. *Chem. Mater.* **2015**, *27*, 5180–5188. [CrossRef]

44. Shu, Q.; Yang, J.H.; Chen, S.; Huang, B.; Xiang, H.; Gong, X.G.; Wei, S.H. $Cu_2Zn(Sn,Ge)Se_4$ and $Cu_2Zn(Sn,Si)Se_4$ alloys as photovoltaic materials: Structural and electronic properties. *Phys. Rev. B* **2013**, *87*, 1152081–1152085. [CrossRef]

45. Zhou, H.; Song, T.; Hsu, W.; Luo, S.; Ye, S.; Duan, H.; Hsu, C.; Yang, W.; Yang, Y. Rational defect passivation of $Cu_2ZnSn(S,Se)_4$ photovoltaics with solution-processed Cu_2ZnSnS_4:Na nanocrystals. *J. Am. Chem. Soc.* **2013**, *135*, 15998–16001. [CrossRef] [PubMed]

46. Laemmle, A.; Wuerz, R.; Powalla, M. Efficiency enhancement of $Cu(In,Ga)Se_2$ thin-film solar cells by a post-deposition treatment with potassium fluoride. *Phys. Status Solidi RRL* **2013**, *7*, 631–634. [CrossRef]

Article

Long Low-Loss-Litium Niobate on Insulator Waveguides with Sub-Nanometer Surface Roughness

Rongbo Wu [1,2,†], Min Wang [3,4,†], Jian Xu [3,4], Jia Qi [1,2], Wei Chu [1,4], Zhiwei Fang [3,4], Jianhao Zhang [1,2], Junxia Zhou [3,4], Lingling Qiao [1], Zhifang Chai [3,4], Jintian Lin [1] and Ya Cheng [1,3,4,5,*]

[1] State Key Laboratory of High Field Laser Physics, Shanghai Institute of Optics and Fine Mechanics, Chinese Academy of Sciences, Shanghai 201800, China; rbwu@siom.ac.cn (R.W.); qijia@siom.ac.cn (J.Q.); chuwei@siom.ac.cn (W.C.); jhzhang@siom.ac.cn (J.Z.); qiaolingling@siom.ac.cn (L.Q.); jintianlin@siom.ac.cn (J.L.)

[2] University of Chinese Academy of Sciences, Beijing 100049, China

[3] State Key Laboratory of Precision Spectroscopy, East China Normal University, Shanghai 200062, China; mwang@phy.ecnu.edu.cn (M.W.); jxu@phy.ecnu.edu.cn (J.X.); zwfang@phy.ecnu.edu.cn (Z.F.); 52180920026@stu.ecnu.edu.cn (J.Z.); zfchai@phy.ecnu.edu.cn (Z.C.)

[4] XXL—The Extreme Optoelectromechanics Laboratory, School of Physics and Materials Science, East China Normal University, Shanghai 200241, China

[5] Collaborative Innovation Center of Extreme Optics, Shanxi University, Taiyuan 030006, Shanxi, China

* Correspondence: ya.cheng@siom.ac.cn

† These authors contributed equally to this paper.

Received: 5 October 2018; Accepted: 2 November 2018; Published: 6 November 2018

Abstract: In this paper, we develop a technique for realizing multi-centimeter-long lithium niobate on insulator (LNOI) waveguides with a propagation loss as low as 0.027 dB/cm. Our technique relies on patterning a chromium thin film coated on the top surface of LNOI into a hard mask with a femtosecond laser followed by chemo-mechanical polishing for structuring the LNOI into the waveguides. The surface roughness on the waveguides was determined with an atomic force microscope to be 0.452 nm. The approach is compatible with other surface patterning technologies, such as optical and electron beam lithographies or laser direct writing, enabling high-throughput manufacturing of large-scale LNOI-based photonic integrated circuits.

Keywords: lithium niobate; waveguide; photonic integrated circuit; propagation loss; optical lithography; chemo-mechanical polishing

1. Introduction

Photonic integrated circuits (PICs) have shown the potential for use in complex information processing systems employing both quantum and classical light sources [1,2]. To increase computational efficiency and reconfigurability, PIC-based optical computers/calculators must have low propagation loss, fast tunability, and efficient optical interfacing. Currently, several materials have been utilized to construct large-scale PICs, including silicon and some semiconductor materials [3–6], fused silica [7,8], and bulk lithium niobate (LN) [9,10]. The advantage of silicon-based PICs is the high refractive index of silicon that enables the fabrication of compact light circuits with strong confinement and tight bends. In addition, the lithographic technology for high precision patterning of silicon and semiconductors is mature. However, silicon-based PICs intrinsically suffer from a relatively high propagation loss and a transmission window prohibitive for visible and shorter wavelengths. PICs can be built on fused silica and bulk LN crystals by local modification of the refractive index via either illumination of light or ion doping. Unfortunately, the refractive index increases achieved using these approaches are usually in the order of 10^{-3} to 10^{-2}, resulting in large PICs footprints being required for minimizing the bending loss. Most importantly, the typical propagation losses of waveguides in state-of-the-art PICs are typically in the order of 10^{-1} dB/cm or higher, which ultimately limits the performance of PIC-based optical computers.

A revolutionary approach for building high-performance PICs has been emerging, enabled by the successful application of high quality lithium niobate on insulator (LNOI) nanophotonic structures. The first experimental proof of this approach was provided by first patterning the LNOI into the designated geometries using a femtosecond laser. The draft structures obtained after the femtosecond laser patterning, which has a relatively high sidewall roughness in the order of tens of nanometers, were then polished with a focused ion beam (FIB) milling to smoothen the sidewall [11]. This concept was soon incorporated into other lithographic technologies, such as optical lithography and electron beam writing (EBW), for defining the planar patterns on LNOI substrates followed by reactive ion etching to complete the nanostructuring of the LNOI [12,13]. The initial focus was mainly on optical microresonators [11–21], and other devices (such as waveguides and photonic crystals [22–28]) appeared shortly, taking advantage of the high surface smoothness of the sidewalls as a result of the ion dry etching. So far, the propagation loss in the LNOI waveguides has reached 0.04 dB/cm, highlighting their potential for use in large-scale PIC applications [29].

Notably, the ion etching step, which is necessary for achieving high quality sidewalls on LNOI nanophotonic structures, leaves a low but non-negligible surface roughness that is difficult to completely remove [29]. Moreover, the use of FIB or EBW in the patterning of LNOI makes the approach impractical for fabricating large-scale PICs due to their low throughputs and limited range of motion. Recently, we developed a technique for fabricating high-quality optical microresonators on LNOI with a quality factor above 10^7 [30]. Since this technique does not involve an ion beam etching process, surface smoothness beyond that allowed by ion beam etching can be readily achieved, and the footprint sizes of PICs can be increased by patterning the LNOI photonic structures with either laser direct writing or optical lithography. Here, we experimentally show that we are able to realize 10-cm-long LNOI waveguides with a propagation loss of 0.027 dB/cm, which benefited from the low surface roughness of 0.452 nm measured with an atomic force microscope (AFM). The low loss waveguides can be essential building blocks for light modulation, beam delivery and manipulation, nonlinear optics, and optical signal processing.

2. Materials and Methods

A commercially available X-cut LNOI wafer fabricated by ion slicing (NANOLN, Jinan Jingzheng Electronics Co., Ltd., Jinan, Shandong, China) was chosen in our experiment as the material upon which the LNOI waveguides were produced [31]. The LN thin film had a thickness of 400 nm and was bonded to a SiO_2 layer 2-μm-thick, which was grown on a LN substrate. The fabrication procedures are schematically illustrated in Figure 1, including (1) deposition of a thin layer of chromium (Cr) with a thickness of 600 nm on LNOI by magnetron sputtering and (2) patterning of the Cr film using femtosecond laser ablation. It is critical to carefully choose the pulse energy of the femtosecond laser to ensure the complete removal of the Cr film without damaging the LNOI underneath, which is enabled by the unique characteristics of the interaction of femtosecond laser pulses with various types of materials [32]. More details on the laser parameters can be found in Wu, R., et al. [30]. To produce a tightly focused spot ~1 μm in diameter, a $100\times$ objective lens (M Plan Apo NIR, Mitutoyo Corporation, Kawasaki, Kanagawa, Japan) with a numerical aperture (NA) of 0.7 was used in our experiment. Femtosecond laser direct writing was conducted by translating the LNOI sample with a computer-controlled XY motion stage (ABL15020WB and ABL15020, Aerotech Inc., Pittsburgh, PA, USA, translation resolution ~100 nm). The focus of the laser beam was controlled in the Z direction using another one-dimensional stage with a translation resolution of 100 nm (ANT130-110-L-ZS, Aerotech Inc., Pittsburgh, PA, USA) on which the objective lens was installed. A charged coupled device (CCD) was installed above the objective lens for monitoring the fabrication process. The laser power was chosen to ensure the complete removal of the Cr thin film while keeping the LNOI underneath the Cr film intact. The patterned Cr disk served as a hard mask in the subsequent CM polishing.

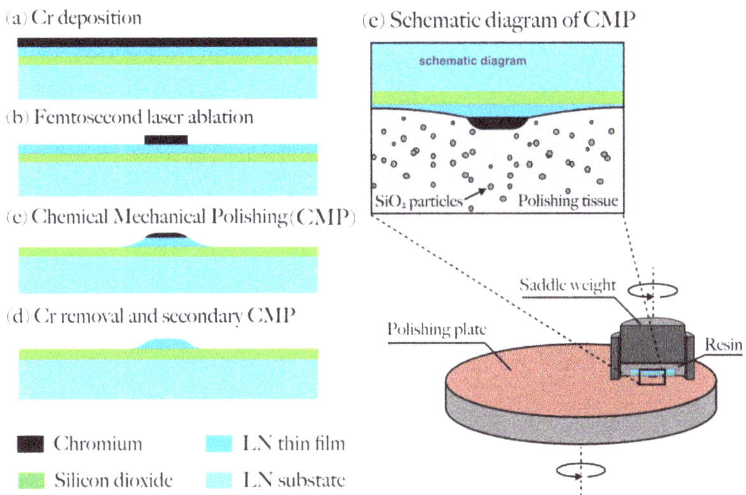

Figure 1. (**a**–**d**) Flow-chart of fabrication of lithium niobate on insulator (LNOI) waveguide and (**e**) schematic diagram of chemo-mechanical polishing (CMP).

The CM polishing was carried out using a wafer polishing machine (NUIPOL802, Hefei Kejing Materials Technology Co., Ltd., Hefei, Anhui, China). More details on CM polishing can also be found in Wu, R, et al. [30]. Note that the Cr film has a higher hardness than the LNOI, so LNOI would be completely removed by CM polishing, unless it is protected by a Cr mask. Finally, the Cr mask was removed by immersing the fabricated sample in a Cr etching solution (Chromium etchant, Alfa Aesar GmbH, Haverhill, MA, USA) for 10 min. All the procedures up to this stage were the same as those used for fabricating a high-Q microresonator, as described by Wu, R., et al. [30], For fabricating low-loss waveguides, the sample underwent an additional CM polishing at a relatively low pressure with a shorter polishing duration to improve the smoothness of the upper surface of LN waveguide, as illustrated in Figure 1d.

To characterize the propagation loss in the LNOI waveguide, we constructed a whispering gallery ring resonator through which the propagation loss of the LNOI waveguide was determined using $\alpha = 2\pi n_{eff}/(Q\lambda)$, where α is the attenuation coefficient, n_{eff} the effective refractive index, Q the quality factor of the ring resonator, and λ is the wavelength of the light beam. Both n_{eff} and the Q-factor were determined from the transmission spectrum of microring resonator. The experimental setup for measuring the Q factor of the ring resonator is schematically shown in Figure 2. The light produced by a tunable laser (LTB-6728, Newport Corporation, Santa Clara, CA, USA) was coupled to a curved LNOI waveguide whose geometric parameters, including the thickness and width, were the same as that of the LNOI ring, determined with the use of a fiber lens. The exiting light was collected using a 20× objective lens (MPlanFL N, Olympus Corporation, Tokyo, Japan) into a detector (Model 1811, Newport Corporation, Santa Clara, CA, USA). The bend of the upper waveguide was intentionally introduced for preventing the stray light from the fiber laser from entering the objective lens located in front of the detector. The polarization direction of the light was adjusted with a fiber polarization controller. The curved waveguide was coupled to the LNOI ring resonator via evanescent coupling. Specifically, by carefully adjusting the distance between the coupling waveguide and the microring resonator, we were able to achieve the coupling condition crucial for obtaining an accurate intrinsic Q factor of the ring resonator. We fabricated five microrings and the measured optical losses were close to each other (with Q factors between 10^6 and 10^7), indicating that the fabrication technique is reliable and reproducible.

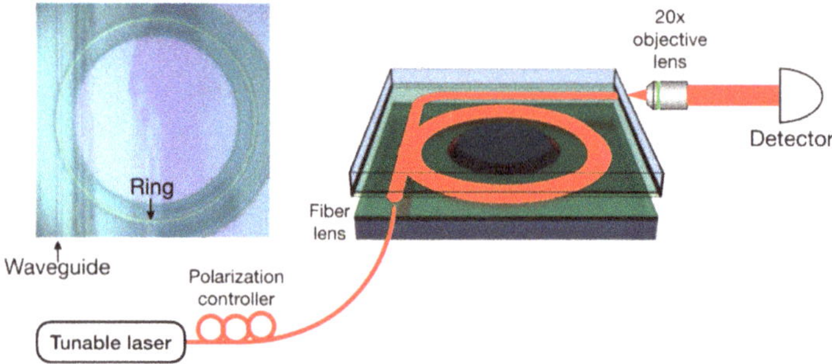

Figure 2. Schematic of the experimental setup for measuring the Q factor of the microring resonator. Left inset: Optical micrograph image of the microring resonator coupling with the waveguide, as indicated by the black arrows.

3. Results

Figure 3a shows the top-view scanning electron microscope (SEM) images of a LN micro ring resonator with a diameter of 160 μm. The width of the waveguide was ~3 μm. The close-up-view SEM of an arc of the ring highlights the high surface smoothness of the CM polished sample. A further atomic force microscope (AFM) inspection, as illustrated in Figure 3c, confirmed that a surface roughness Rq as low as 0.452 nm was achieved. The same fabrication technique was also used to fabricate a continuous 11-cm-long optical waveguide, as shown by the digital-camera-captured picture in Figure 3d, with details provided in the zoomed optical micrographs in Figure 3e–f. The total time of femtosecond laser ablation for fabricating the 11-cm-long waveguide was 90 min. At this moment, the footprint size of the PICs was only limited by the LNOI wafer size. The LNOI wafer can be made larger without much difficulty [33].

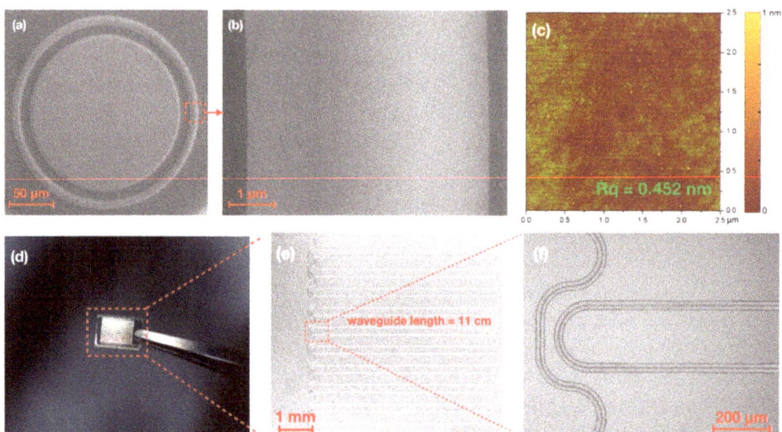

Figure 3. (**a**) Top-view scanning electron microscope (SEM) image of a lithium niobate (LN) microring resonator; (**b**) Zoomed view of the ridge of the microring resonator in (**a**); (**c**) Atomic force microscope (AFM) image of the ridge; (**d**) Picture of a chip consisting of an 11-cm-long waveguide captured by digital camera; (**e**,**f**) Zoomed images of the waveguides on the chip captured with an optical microscope.

The optical loss characterization was performed using whispering-gallery-resonator-loss measurements. The propagation loss α is related to the Q factor of the ring resonator. Figure 4a shows the measured transmission spectrum for the wavelength range between 1546 and 1564 nm. The free spectral range (FSR) of the microring resonator was determined to be 2.71 nm, which is consistent with the 160 µm diameter of our ring resonator. The resonant lines appeared regularly spaced, indicating that mostly low-order modes exist in the ring resonator. One of the whispering-gallery modes at the resonant wavelength of 1560.48 nm was chosen for the measurement of the loaded Q factor by fitting with a Lorentz function, which reached 5.70×10^6, corresponding to an intrinsic Q factor of 1.14×10^7 in the critical coupling regime as evidenced in Figure 4b. The effective refractive index, $n_{eff} = \lambda^2/(2\pi R \cdot FSR)$, with a ring radius R of 80 µm and wavelength of 1560.48 nm, was calculated to be 1.79, which is in good agreement with our finite-difference time-domain (FDTD) simulation result given in the inset of Figure 4b. Combining the effective refractive index and the Q-factor obtained from the transmission spectrum, the propagation loss in the microring resonator was calculated to be 0.027 dB/cm using the aforementioned expression $\alpha = 2\pi n_{eff}/(Q\lambda)$. This result represents the upper limit of the propagation loss in the LNOI waveguide fabricated using our method.

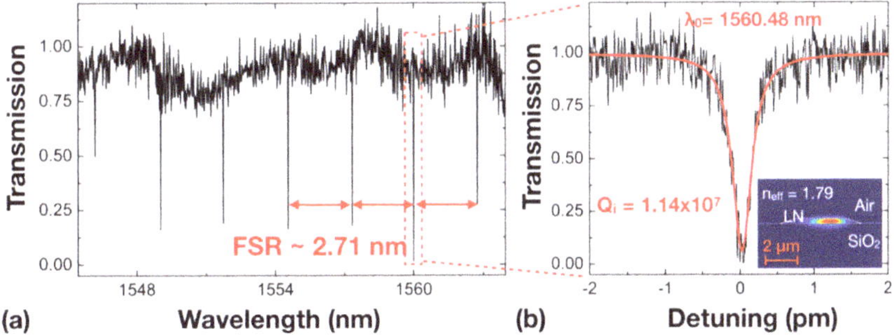

Figure 4. (**a**) Transmission spectrum of the LN microring resonator; (**b**) The Lorentz fitting (red curve) reveals a loaded Q factor of 5.70×10^6, corresponding to an intrinsic Q factor of 1.14×10^7. Inset: The optical mode distribution and n_{eff} in the ring waveguide calculated using finite-difference time-domain (FDTD) simulation.

4. Discussion

The low propagation loss of 0.027 dB/cm was a result of the low surface roughness Rq of 0.452 nm on the fabricated LNOI waveguides. The surface roughness was improved using our technique in which the LNOI is purely patterned by the chemo-mechanical polishing without any use of ion beam etching [30]. The ion beam etching inherently leaves a small amount of surface roughness on the nearly vertical sidewalls, which is difficult to completely remove by top surface polishing [34,35]. This is the major reason that we were able to obtain a propagation loss lower than the waveguides fabricated by FIB or reactive ion etching. In the current experiment, we used a femtosecond laser to pattern the Cr hard mask. Generally speaking, the sidewall roughness on the Cr mask patterned by femtosecond laser ablation should be higher than the surface roughness of the LNOI photonic structures produced by ion beam etching. However, the sidewall roughness on the Cr mask only transfers to the underneath LNOI near the top surface, thus it can be completely suppressed with an additional polishing process for thinning the LNOI substrate after the removal of the Cr mask (Figure 1d).

Notably, the propagation loss obtained by measuring the Q factor of the ring resonator may have been underestimated for the straight segments in the LNOI waveguides, as presented in Figure 3d–f, due to a higher radiative loss in the ring resonator. Ultimately, the propagation loss of the LNOI waveguides is limited by the absorption in crystalline LN, which is well known to be in the order of $\sim 10^{-3}$ dB/cm. Our measured loss was still one order of magnitude away from the theoretical limit, which could be attributed to several factors, including the radiative loss in the ring resonator and some unknown contamination on the surface of the LNOI waveguide, as the measurements were all carried out in a low-class clean room, existence of absorptive defects in the LNOI substrate owing to the imperfect crystal growth, and the remaining surface roughness left by the chemo-mechanical polishing process. Thus, to realize LNOI waveguides with propagation losses in the order of 1×10^{-3} dB/cm, many refinements should be systematically investigated in the future.

The fabrication resolution of femtosecond laser direct writing is typically in the order of 1 µm for inorganic materials such as glass, semiconductors, and metals. However, today's optical lithography can easily achieve sub-micron or even 100-nm-level patterning resolutions. This should be sufficient for fabricating single-mode LNOI waveguides of narrower widths by which PICs, such as Mach-Zender interferometers and polarization convertors, can be built. The mode field of the LNOI waveguides can be tuned by coating them with fused silica for suppressing higher order modes as well as scattering loss due to the reduced contrast of the refractive index between the LNOI waveguide and the cladding environment [36]. With these improvements, the LNOI waveguides will become a major building block for PIC applications.

5. Conclusions

In this study, we achieved a propagation loss of 0.027 dB/cm in LNOI waveguides fabricated by combining femtosecond laser micromachining for patterning the Cr mask and chemo-mechanical polishing for transferring the laser-written patterns to the LNOI beneath the Cr mask. By eliminating the low-throughput FIB or EBW process, our technique enables the rapid fabrication of longer low-loss optical waveguides, which are only limited by the range of motion of the air bearing stage and the size of the LNOI wafers. Thus, our approach promotes the fabrication efficiency and reduces the cost of manufacturing LNOI PICs. The waveguides and ring resonators can be adopted for constructing complex PICs, enabling the mass production of LNOI PICs for optical communication and computation applications.

Author Contributions: Conceptualization, Y.C.; methodology, Y.C.; software, R.W.; validation, R.W. and M.W.; formal analysis, R.W.; investigation, R.W., M.W., J.Q. and J.X.; resources, R.W., M.W., W.C., Z.F., J.Z. (Jianhao Zhang), J.Z. (Junxia Zhou), L.Q., Z.C. and J.L.; data curation, R.W. and M.W.; writing—original draft preparation, Y.C. and R.W.; writing—review and editing, Y.C., R.W., M.W., J.L. and J.X.; visualization, R.W., J.Q., J.X. and M.W.; supervision, Y.C.; project administration, L.Q. and Z.C.; funding acquisition, Y.C., J.L. and J.X.

Funding: This research was funded by the National Natural Science Foundation of China (Grant Nos. 11734009, 11874375, 61590934, 61635009, 61327902, 61505231, 11604351, 11674340, 61575211, 61675220, 61761136006), the Strategic Priority Research Program of Chinese Academy of Sciences (Grant No. XDB16000000), the Key Research Program of Frontier Sciences, Chinese Academy of Sciences (Grant No. QYZDJ-SSW-SLH010), the Project of Shanghai Committee of Science and Technology (Grant No. 17JC1400400), the Shanghai Rising-Star Program (Grant No. 17QA1404600), and the Shanghai Pujiang Program (Grant No.18PJ1403300).

References

1. Ladd, T.D.; Jelezko, F.; Laflamme, R.; Nakamura, Y.; Monroe, C.; O'Brien, J.L. Quantum computers. *Nature* **2010**, *464*, 45–53. [CrossRef] [PubMed]
2. Shen, Y.; Harris, N.C.; Skirlo, S.; Prabhu, M.; Baehr-Jones, T.; Hochberg, M.; Sun, X.; Zhao, S.; Larochelle, H.; Englund, D.; et al. Deep learning with coherent nanophotonic circuits. *Nat. Photonics* **2017**, *11*, 441. [CrossRef]
3. Harris Nicholas, C.; Bunandar, D.; Pant, M.; Steinbrecher Greg, R.; Mower, J.; Prabhu, M.; Baehr-Jones, T.; Hochberg, M.; Englund, D. Large-scale quantum photonic circuits in silion. *Nacnophotonics* **2016**, *5*, 456–468. [CrossRef]
4. Dietrich, C.P.; Fiore, A.; Thompson, M.G.; Kamp, M.; Höfling, S. GaAs integrated quantum photonics: Towards compact and multi-functional quantum photonic integrated circuits. *Laser Photonics Rev.* **2016**, *10*, 870–894. [CrossRef]

5.   Najafi, F.; Mower, J.; Harris, N.C.; Bellei, F.; Dane, A.; Lee, C.; Hu, X.; Kharel, P.; Marsili, F.; Assefa, S.; et al. On-chip detection of non-classical light by scalable integration of single-photon detectors. *Nat. Commun.* **2015**, *6*, 5873. [CrossRef] [PubMed]

6.   Yang, K.Y.; Oh, D.Y.; Lee, S.H.; Yang, Q.-F.; Yi, X.; Shen, B.; Wang, H.; Vahala, K. Bridging ultrahigh-Q devices and photonic circuits. *Nat. Photonics* **2018**, *12*, 297–302. [CrossRef]

7.   Qiang, X.; Zhou, X.; Wang, J.; Wilkes, C.M.; Loke, T.; O'Gara, S.; Kling, L.; Marshall, G.D.; Santagati, R.; Ralph, T.C.; et al. Large-scale silicon quantum photonics implementing arbitrary two-qubit processing. *Nat. Photonics* **2018**, *12*, 534–539. [CrossRef]

8.   Marshall, G.D.; Politi, A.; Matthews, J.C.F.; Dekker, P.; Ams, M.; Withford, M.J.; O'Brien, J.L. Laser written waveguide photonic quantum circuits. *Opt. Express* **2009**, *17*, 12546–12554. [CrossRef] [PubMed]

9.   Sohler, W.; Hu, H.; Ricken, R.; Quiring, V.; Vannahme, C.; Herrmann, H.; Büchter, D.; Reza, S.; Grundkötter, W.; Orlov, S.; et al. Integrated optical devices in lithium niobate. *Opt. Photonics News* **2008**, *19*, 24–31. [CrossRef]

10.   Jin, H.; Liu, F.M.; Xu, P.; Xia, J.L.; Zhong, M.L.; Yuan, Y.; Zhou, J.W.; Gong, Y.X.; Wang, W.; Zhu, S.N. On-chip generation and manipulation of entangled photons based on reconfigurable lithium-niobate waveguide circuits. *Phys. Rev. Lett.* **2014**, *113*, 103601. [CrossRef] [PubMed]

11.   Lin, J.; Xu, Y.; Fang, Z.; Song, J.; Wang, N.; Qiao, L.; Fang, W.; Cheng, Y. Second harmonic generation in a high-Q lithium niobate microresonator fabricated by femtosecond laser micromachining. *arXiv*, 2014; arXiv:1405.6473.

12.   Wang, J.; Bo, F.; Wan, S.; Li, W.; Gao, F.; Li, J.; Zhang, G.; Xu, J. High-Q lithium niobate microdisk resonators on a chip for efficient electro-optic modulation. *Opt. Express* **2015**, *23*, 23072–23078. [CrossRef] [PubMed]

13.   Wang, C.; Burek, M.J.; Lin, Z.; Atikian, H.A.; Venkataraman, V.; Huang, I.-C.; Stark, P.; Lončar, M. Integrated high quality factor lithium niobate microdisk resonators. *Opt. Express* **2014**, *22*, 30924–30933. [CrossRef] [PubMed]

14.   Lin, J.; Xu, Y.; Fang, Z.; Wang, M.; Song, J.; Wang, N.; Qiao, L.; Fang, W.; Cheng, Y. Fabrication of high-Q lithium niobate microresonators using femtosecond laser micromachining. *Sci. Rep.* **2015**, *5*, 8072. [CrossRef] [PubMed]

15.   Lin, J.; Xu, Y.; Ni, J.; Wang, M.; Fang, Z.; Qiao, L.; Fang, W.; Cheng, Y. Phase-matched second-harmonic generation in an on-chip LiNbO$_3$ microresonator. *Phys. Rev. Appl.* **2016**, *6*, 014002. [CrossRef]

16.   Lin, J.; Xu, Y.; Fang, Z.; Wang, M.; Wang, N.; Qiao, L.; Fang, W.; Cheng, Y. Second harmonic generation in a high-Q lithium niobate microresonator fabricated by femtosecond laser micromachining. *Sci. China Phys. Mech. Astron.* **2015**, *58*, 114209. [CrossRef]

17.   Hao, Z.; Wang, J.; Ma, S.; Mao, W.; Bo, F.; Gao, F.; Zhang, G.; Xu, J. Sum-frequency generation in on-chip lithium niobate microdisk resonators. *Photonics Res.* **2017**, *5*, 623–628. [CrossRef]

18.   Wang, L.; Wang, C.; Wang, J.; Bo, F.; Zhang, M.; Gong, Q.; Lončar, M.; Xiao, Y.-F. High-Q chaotic lithium niobate microdisk cavity. *Opt. Lett.* **2018**, *43*, 2917–2920. [CrossRef] [PubMed]

19.   Luo, R.; Jiang, H.; Rogers, S.; Liang, H.; He, Y.; Lin, Q. On-chip second-harmonic generation and broadband parametric down-conversion in a lithium niobate microresonator. *Opt. Express* **2017**, *25*, 24531–24539. [CrossRef] [PubMed]

20.   Jiang, W.C.; Lin, Q. Chip-scale cavity optomechanics in lithium niobate. *Sci. Rep.* **2016**, *6*, 36920. [CrossRef] [PubMed]

21.   Fang, Z.; Xu, Y.; Wang, M.; Qiao, L.; Lin, J.; Fang, W.; Cheng, Y. Monolithic integration of a lithium niobate microresonator with a free-standing waveguide using femtosecond laser assisted ion beam writing. *Sci. Rep.* **2017**, *7*, 45610. [CrossRef] [PubMed]

22.   Liang, H.; Luo, R.; He, Y.; Jiang, H.; Lin, Q. High-quality lithium niobate photonic crystal nanocavities. *Optica* **2017**, *4*, 1251–1258. [CrossRef]

23.   Wang, C.; Zhang, M.; Stern, B.; Lipson, M.; Lončar, M. Nanophotonic lithium niobate electro-optic modulators. *Opt. Express* **2018**, *26*, 1547–1555. [CrossRef] [PubMed]

24.   Wang, C.; Xiong, X.; Andrade, N.; Venkataraman, V.; Ren, X.-F.; Guo, G.-C.; Lončar, M. Second harmonic generation in nano-structured thin-film lithium niobate waveguides. *Opt. Express* **2017**, *25*, 6963–6973. [CrossRef] [PubMed]

25.   Luo, R.; He, Y.; Liang, H.; Li, M.; Lin, Q. Highly tunable efficient second-harmonic generation in a lithium niobate nanophotonic waveguide. *Optica* **2018**, *5*, 1006–1011. [CrossRef]

26. Cai, L.; Han, H.; Zhang, S.; Hu, H.; Wang, K. Photonic crystal slab fabricated on the platform of lithium niobate-on-insulator. *Opt. Lett.* **2014**, *39*, 2094–2096. [CrossRef] [PubMed]
27. Diziain, S.; Geiss, R.; Steinert, M.; Schmidt, C.; Chang, W.-K.; Fasold, S.; Füßel, D.; Chen, Y.-H.; Pertsch, T. Self-suspended micro-resonators patterned in Z-cut lithium niobate membranes. *Opt. Mater. Express* **2015**, *5*, 2081–2089. [CrossRef]
28. Krasnokutska, I.; Tambasco, J.-L.J.; Li, X.; Peruzzo, A. Ultra-low loss photonic circuits in lithium niobate on insulator. *Opt. Express* **2018**, *26*, 897–904. [CrossRef] [PubMed]
29. Wolf, R.; Breunig, I.; Zappe, H.; Buse, K. Scattering-loss reduction of ridge waveguides by sidewall polishing. *Opt. Express* **2018**, *26*, 19815–19820. [CrossRef] [PubMed]
30. Wu, R.; Zhang, J.; Yao, N.; Fang, W.; Qiao, L.; Chai, Z.; Lin, J.; Cheng, Y. Lithium niobate micro-disk resonators of quality factors above 10^7. *Opt. Lett.* **2018**, *43*, 4116–4119. [CrossRef] [PubMed]
31. Poberaj, G.; Hu, H.; Sohler, W.; Guenter, P. Lithium niobate on insulator (LNOI) for micro-photonic devices. *Laser Photonics Rev.* **2012**, *6*, 488–503. [CrossRef]
32. Sugioka, K.; Cheng, Y. Ultrafast lasers—Reliable tools for advanced materials processing. *Light Sci. Appl.* **2014**, *3*, e149. [CrossRef]
33. Boes, A.; Corcoran, B.; Chang, L.; Bowers, J.; Mitchell, A. Status and potential of lithium niobate on insulator (LNOI) for photonic integrated circuits. *Laser Photonics Rev.* **2018**, *12*, 1700256. [CrossRef]
34. Wolf, R.; Breunig, I.; Zappe, H.; Buse, K. Cascaded second-order optical nonlinearities in on-chip micro rings. *Opt. Express* **2017**, *25*, 29927–29933. [CrossRef] [PubMed]
35. Wolf, R.; Breunig, I.; Zappe, H.; Buse, K. Q-factor enhancement of integrated lithium-niobate-on-insulator ridge waveguide whispering-gallery-mode resonators by surface polishing. In Proceedings of the Laser Resonators, Microresonators, and Beam Control XIX, San Francisco, CA, USA, 20 February 2017; p. 1009002. [CrossRef]
36. Guarino1, A.; Poberaj, G.; Rezzonico, D.; Degl'Innocenti, R.; Günter, P. Electro–optically tunable microring resonators in lithium niobate. *Nat. Photonics* **2007**, *1*, 407–410. [CrossRef]

nanomaterials

Article

Pulsed Laser Fabrication of TiO$_2$ Buffer Layers for Dye Sensitized Solar Cells

Jeanina Lungu [1], Gabriel Socol [2], George E. Stan [3], Nicolaie Ştefan [2], Cătălin Luculescu [2], Adrian Georgescu [1], Gianina Popescu-Pelin [2], Gabriel Prodan [1], Mihai A. Gîrţu [1,*] and Ion N. Mihăilescu [2,*]

[1] Department of Physics, Ovidius University of Constanţa, Constanţa 900527, Romania; jmatei@univ-ovidius.ro (J.L.); contact.adriangeorgescu@gmail.com (A.G.); gprodan@univ-ovidius.ro (G.P.)

[2] National Institute for Lasers, Plasma and Radiation Physics, P.O. Box MG-36, Măgurele 077125, Romania; gabriel.socol@inflpr.ro (G.S.); stefan.nicolaie@inflpr.ro (N.S.); catalin.luculescu@inflpr.ro (C.L.); gianina.popescu@inflpr.ro (G.P.-P.)

[3] National Institute of Materials Physics, P.O. Box MG-7, Măgurele 077125, Romania; george_stan@infim.ro

* Correspondence: mihai.girtu@univ-ovidius.ro (M.A.G.); ion.mihailescu@inflpr.ro (I.N.M.)

Received: 25 April 2019; Accepted: 6 May 2019; Published: 15 May 2019

Abstract: We report on the fabrication of dye-sensitized solar cells with a TiO$_2$ buffer layer between the transparent conductive oxide substrate and the mesoporous TiO$_2$ film, in order to improve the photovoltaic conversion efficiency of the device. The buffer layer was fabricated by pulsed laser deposition whereas the mesoporous film by the doctor blade method, using TiO$_2$ paste obtained by the sol–gel technique. The buffer layer was deposited in either oxygen (10 Pa and 50 Pa) or argon (10 Pa and 50 Pa) onto transparent conducting oxide glass kept at room temperature. The cross-section scanning electron microscopy image showed differences in layer morphology and thickness, depending on the deposition conditions. Transmission electron microscopy studies of the TiO$_2$ buffer layers indicated that films consisted of grains with typical diameters of 10 nm to 30 nm. We found that the photovoltaic conversion efficiencies, determined under standard air mass 1.5 global (AM 1.5G) conditions, of the solar cells with a buffer layer are more than two times larger than those of the standard cells. The best performance was reached for buffer layers deposited at 10 Pa O$_2$. We discuss the processes that take place in the device and emphasize the role of the brush-like buffer layer in the performance increase.

Keywords: dye-sensitized solar cells; photovoltaic conversion efficiency; TiO$_2$ thin films; pulsed laser deposition

1. Introduction

Dye-sensitized solar cell technology continues to be a key technological domain as it allows for the production of low-cost energy from renewable sources [1], particularly under ambient lighting [2]. Dye-sensitized solar cells (DSSC) are photovoltaic devices consisting of a photoelectrode with a mesoporous layer of a nanocrystalline wide band gap semiconductor (such as anatase TiO$_2$) on transparent conducting oxide, sensitized with a dye, and a counter electrode, for example platinized conductive glass, with a liquid or solid state electrolyte in-between [3,4]. The working principle of the device is based on light absorption in the dye, followed by transfer of the resulting photoelectrons from the excited level of the dye into the conduction band of TiO$_2$. The electron diffuses via the semiconductor to the conducting glass substrate, passes through the external circuit and is carried by the redox electrolyte from the counter electrode back to the dye, to regenerate it [5].

For DSSCs with iodide/triiodide electrolyte, conversion efficiencies, in standard air mass 1.5 global (AM 1.5G) conditions, of more than 11% have been obtained using Ru(II)-polypyridyl complexes [6,7], which are widely used dyes in the photovoltaic devices [8,9]. More recently, by using porphyrin dyes, the efficiency reached 11.5% [10] and by replacing the iodine electrolyte with cobalt based complexes the efficiency went up to 12% [11]. Even higher performance was reported when using organic silyl-anchor dyes [12]. The use of perovskite light absorbers and organic hole conductors in a solid state cell resulted in efficiencies larger than 15% [13], which was subsequently further increased by design changes away from the DSSC structure to more than 22% [14].

Current studies dedicated to the mechanism of charge transport in DSSC have indicated that progress can be achieved through understanding and controlling the secondary processes inside cell. It has been found [15–21] that the cell performance is enhanced when employing an intermediate nanocrystalline layer of TiO_2 between the transparent conducting glass substrate (FTO—fluorine-doped tin oxide) and the mesoporous TiO_2 semiconductor. The buffer layer has the role to ensure a good mechanical contact, as well as to protect the electrodes against the dye solution action and the oxidation at high temperature and to reduce the recombination of electrons at the electrode/electrolyte interface.

A compact TiO_2 layer on the conductive glass substrate can be prepared by different methods. Examples are electrochemical deposition [22,23], spray pyrolysis [24–26], screen printing [26–28], sol–gel [15,29,30], sputtering [31–34], chemical vapor deposition [35], atomic layer deposition [36–38], dip coating [39], spin coating [21,40], etc. It has been argued that the buffer layer prevents the back transfer of electrons from the conductive substrate to the electrolyte, leading to an increase in the DSSC conversion efficiency. For that reason the thin compact buffer film of TiO_2 was also called a blocking layer. The range of the efficiency enhancement is wide, reports claiming 20% in the case of the sol–gel method [29], from 15–20% [32] up to 80% [34] for sputtering, almost 30% when mixing exfoliated titania nanosheets with anatase TiO_2 nanoparticles [41].

In contrast with the numerous reports on TiO_2 blocking layers obtained by the methods just mentioned, pulsed laser deposition (PLD) was less used for the preparation of buffer layers. An early report claimed about 4% efficiency increase when using pure TiO_2 and more than 21% for Nb-doped TiO_2 [42], whereas a later one indicated 42% efficiency enhancement [43].

In our study, we present the photovoltaic performance of DSSC devices fabricated with buffer layers obtained by PLD, to take advantage of the good adherence and the control of stoichiometry, crystallinity and purity of ablated materials. These benefits have attracted considerable interest during the last years for synthesizing high quality oxide thin films by PLD [44,45]. Moreover, the number and/or intensity of the laser pulses used for ablation, allow for the accurate control of the deposition rate [46], making it a unique method for obtaining oxide semiconductor nanostructures for DSSCs. The PLD TiO_2 compact film was deposited onto FTO to generate a barrier between the conducting oxide and the mesoporous TiO_2 layer prepared by the sol–gel method. We report on the characterization of the TiO_2 buffer interlayers fabricated by PLD in either oxygen or argon on FTO glass substrates kept at room temperature. We also studied the influence of the buffer layer on the photoelectron conversion process and the performance of DSSCs.

2. Materials and Methods

2.1. Solar Cell Fabrication

Both electrodes were obtained starting from transparent conductive glass substrates, which consisted of soda lime glass sheets of 2.2 mm thickness, covered with a conductive layer of fluorine-doped tin oxide (SnO_2:F; FTO) with a 7 Ω/square resistivity (available from Solaronix). Before the preparation of the electrodes, the conductive glass was ultrasonically cleaned for 15 min in acetone, ethanol and deionized water, respectively, to remove any impurities, and then blown dry with high purity nitrogen. The first step in the preparation of the photoelectrode was the ablation of the pure TiO_2 target on clean FTO glasses, using an excimer laser source KrF* (λ = 248 nm, τ_{FWHM} = 25 ns). The thin films deposition was performed inside a stainless steel irradiation chamber at room temperature. The target was produced from homogeneous anatase TiO_2 powder (Sigma-Aldrich Corp.,

St. Louis, MO, USA, 637,254, 99.7% purity) with nanoparticle sizes of less than 25 nm mixed in agate mortar grinder. The ground TiO_2 powder was initially pressed at 5 MPa and after that sintered for 6 h at 1100 °C in air, with a heating/cooling ratio of 20 °C/min, to obtain compact pellets. The laser beam incidence angle onto the target surface was about 45° and the target-substrate separation distance was set at 4 cm. For the deposition of one film 3×10^3 subsequent laser pulses were applied, succeeding to each other with a repetition rate of 2 Hz. The targets were irradiated with a laser fluence of 2 J/cm^2.

The buffer layers were obtained at 10 Pa and 50 Pa, by circulating high purity (99.999%) oxygen or argon inside the irradiation chamber, with the aid of a calibrated inlet. The dynamic pressure was monitored with an MKS 100 controller. The samples were labeled TO and TAR for layers deposited in oxygen or argon, respectively, and with two extra digits indicating the pressure, in Pa.

The electrodes obtained with a buffer layer under oxygen and argon atmosphere were used to further fabricate DSSC devices. The active layer was fabricated using a TiO_2 paste prepared by the Pechini type sol–gel method [47], starting from a polyester-based titanium sol. The sol contained a mixture of precursor with molar ratio of 1:4:16 [Ti(iOPr)$_4$:citric acid:ethylene glycol]. The paste was obtained by grinding in a mortar the nanocrystalline anatase TiO_2 powder (Sigma-Aldrich Corp., St. Louis, MO, USA, 637,254). The sol–gel solution had 7:1 molar ratio between TiO_2 and titanium (IV) isopropoxide [Ti(iOPr)$_4$] (Sigma-Aldrich Corp., St. Louis, MO, USA) [48,49]. The paste was spread on the TiO_2 buffer layer by the 'doctor-blade' technique. TiO_2 films were annealed at 450 °C for 1 h in air and left until cooling to room temperature [18].

The last step in the preparation of the photoelectrodes was the sensitization of the mesoporous film of nanocrystalline TiO_2 grains with the N719 (Ruthenium 535-bisTBA) dye, cis-diisothiocyanato-bis(2,20-bipyridyl-4,40-carboxylato) ruthenium(II) bis (tetrabutylammonium) [50] (from Solaronix S.A., Aubonne, Switzerland). The photoelectrodes were immersed in the dye solution (0.2 mM in absolute ethanol) at a temperature of 80 °C for 2 h, then rinsed with absolute ethanol and dried in the oven at 80 °C for 10 min.

The counter electrodes were obtained by spreading a few drops of Platisol T (Solaronix) onto the FTO and drying at 450 °C for 10 min. Both types of electrodes were stored in desiccators before use. The DSSCs were assembled by pressing the photoelectrode against the counterelectrode with small bulldog clips [51,52]. Finally, the electrolyte (Iodolyte Z-50, from Solaronix S.A., Aubonne, Switzerland) was injected between the electrodes filling up the space by capillary action.

2.2. Measurements

X-ray diffraction (XRD) was used to assess the structural properties of the samples and identify the crystalline phases. The XRD measurements were performed in Bragg-Brentano geometry with a D8 Advance diffractometer (Bruker Corp., Billerica, MA, USA), equipped with CuKα (λ = 1.5418 Å) radiation and a high efficiency one-dimensional LynxEye™ detector operated in integration mode. The patterns were recorded in the 2θ range 20°–60°, using a step size of 0.04° and a time per step of 5 s.

The surface morphology of samples was investigated by scanning electron microscopy (SEM) using a Inspect S electron microscope (FEI Co., Hillsboro, OR, USA). The SEM measurements were performed in high vacuum, at 20 kV acceleration voltage, using the secondary electrons acquisition mode. Before the SEM examination, a thin Au film was applied to coat the samples to prevent the electrical charging. The film uniformity and thickness were estimated based on cross-section SEM micrographs. The chemical analyses of the TiO_2 films were carried out by means of energy dispersive spectroscopy (EDS). Additionally, transmission electron microscopy (TEM) examinations were conducted using a *CM 120 ST* microscope (Philips N.V., Amsterdam, The Netherlands), which operated at 120 kV and had a point-to-point resolution of 0.24 nm. The samples for the TEM investigation were dispersed in ethylic alcohol and collected on 300 mesh coated grids [53,54].

The optical transmission properties were analyzed with a Cintra 10e UV-Vis spectrophotometer (GBC Scientific Equipment Pty Ltd., Braeside VIC, Australia), in the range 300–1200 nm. The electro-optic parameters of the devices, particularly the fill factor (*FF*), the photovoltaic conversion efficiency (η), the short circuit current (I_{SC}) and the open circuit voltage (V_{OC}), were measured using a home-made small area solar simulator [55], which provided AM 1.5G standard irradiation conditions. The solar simulator illuminated the surface of the DSSCs through a circular slit of 10 mm diameter, such that the area exposed to light was of about 0.785 cm^2. The measurement of current and voltage was performed using two digital *MS8050* multimeters (Mastech Group International Ltd., Hong Kong, China) and a precision decade resistance box. The measurements were carried out by changing the load resistance, at intervals of about 45 s, which allowed enough time for stable reading.

3. Results

At visual inspection, PLD TiO$_2$ films were uniform, and adherent (even at edges). TO10 and TAR10 films were compact, completely transparent, with rainbow reflection in daylight. TO50 films were faintly translucent, exhibiting a surface with porous aspect, whilst TAR50 films were slightly smoky, but compact.

Top-view SEM investigations of the TiO$_2$ layer deposited on FTO (Figure 1a) revealed nanoparticles with homogeneous shape and size dimension, in agreement with previous studies [56–58]. The buffer-layers display a brush-like compact structure consisting of TiO$_2$ nanobars (Figure 1). The cross-section SEM images allowed the estimation of the layer thickness, as shown in Table 1. The samples obtained by low pressure PLD are systematically thinner than those deposited under higher pressure. One possible explanation may be that at higher pressure, due to multiple collisions with atoms/molecules of the gas, the ablated species lose more of their kinetic energy, resulting in a more confined plasma plume and the growth of larger nanobars.

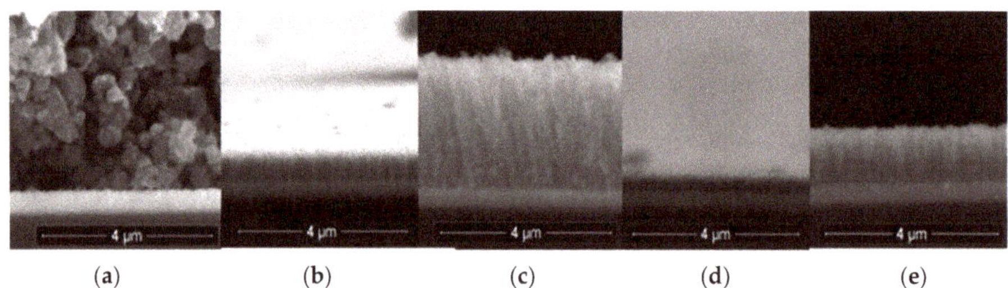

(a)   (b)   (c)   (d)   (e)

Figure 1. SEM micrographs of samples with the mesoporous TiO$_2$ deposited directly on fluorine-doped tin oxide (FTO) (**a**) and structures with TiO$_2$ buffer layers deposited by pulsed laser deposition (PLD) on FTO in oxygen at 10 Pa (TO10) (**b**), and 50 Pa (TO50) (**c**), in argon at 10 Pa (TAR10) (**d**), and at 50 Pa (TAR50) (**e**).

Table 1. Thickness of TiO$_2$ mesoporous coatings and of TiO$_2$ buffer layers deposited by PLD, on a SnO$_2$:F conductive film of 0.41 µm.

Film Type	TiO$_2$ Mesoporous Film	TO10 Buffer Layer	TO50 Buffer Layer	TAR10 Buffer Layer	TAR50 Buffer Layer
Thickness (µm)	~29.89	~0.65	~2.95	~0.33	~1.24

The top-view and cross-view SEM images of the deposited TiO$_2$ layer pointed to a mesoporous morphology (Figure 2). The EDS spectrum demonstrated that no additional elements were present in the mesoporous layer, except for carbon, which often contaminates the surface of samples kept in air. Carbon was not taken into account in the quantification of the atomic number, absorption and fluorescence (ZAF) corrections.

The XRD patterns of the mesoporous TiO$_2$ film, and the buffer layers are given in Figure 3. Except for the TAR50 sample, which seemed amorphous, all other layers suggested a monophasic structure, exhibiting the maxima only of TiO$_2$-anatase (ICDD: 00-021-1272). This phase is characteristic to PLD films annealed at 450 °C [23], along with the peaks of the electrode substrate layer (SnO$_2$, ICDD:01-077-0452), which covers the glass substrates. No texturing of the anatase films was noticed. However, in the case of the FTO layer, a clear texturing in the (200) crystalline direction was evident. This FTO peak was superimposing, in the case of the mesoporous film, or even obscuring the (004) line of anatase, in the buffer layer samples.

Figure 2. SEM micrographs, top-view (**a**) and cross-section showing a thickness of 29.89 μm (**b**), and energy dispersive spectroscopy (EDS) spectrum (**c**) of the TiO$_2$ mesoporous layer.

In the case of the TAR50 sample no TiO$_2$ peaks were evidenced, thus suggesting the amorphous status of these films. TAR50 samples were amorphous likely because at high pressure the velocity of the ablated species significantly decreased due to the collisions with "huge" Ar atoms. Accordingly, the energy necessary for the nucleation of crystals was significantly reduced in the case of these films.

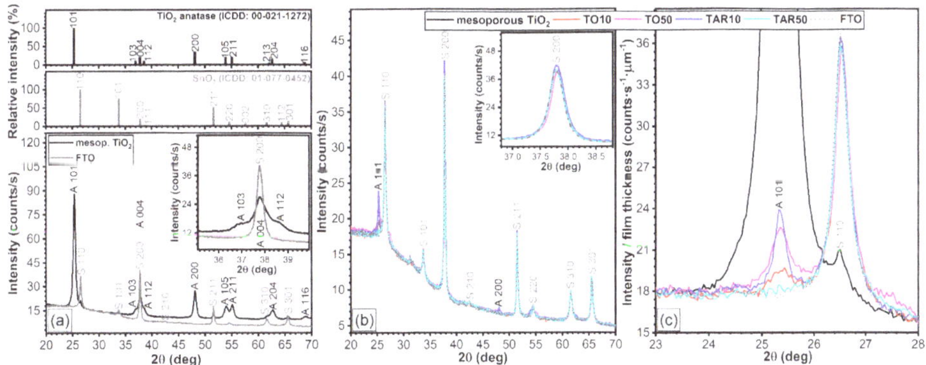

Figure 3. XRD patterns of the TiO$_2$ mesoporous film and of the TiO$_2$ buffer layers deposited by PLD on FTO glass substrates: TiO$_2$ mesoporous film vs. FTO substrate (**a**); PLD layers vs. FTO substrate (**b**) and detail of the XRD pattern characteristic to the anatase (101) peak (**c**).

A simple attenuation calculus for the anatase phase, at the Cu Kα energy, based upon a hypothetical density of ~4 g/cm^3, indicated that the X-ray beam scattered at 2θ ≈ 25.3° would be totally attenuated by a dense film with a thickness in excess of 5 μm. The average crystal coherence length, as estimated via Scherrer equation, is collected in Table 2. The crystal coherence length of the anatase coatings was estimated from the full width at half maximum (FWHM) of the (101) diffraction line. The instrumental broadening of the diffraction line was corrected using a CeO$_2$ highly crystalline laboratory control. We note that Scherrer equation assumes a negligible contribution of the lattice strain to the peak broadening.

Table 2. Buffer layers: Structural and morphological features.

TiO$_2$ Layer Type	XRD (101) Crystalline Coherent Length (nm)	TEM Mean Grain Size (nm)
TO10	~21.5	12.6
TO50	~28.1	12.0
TAR10	~40.8	16.9
TAR50	n/a	13.3
Mesoporous film	~19.0	16.4

TEM images, along with grain size histograms (lognormal fitted), for mesoporous TiO$_2$ films and PLD TiO$_2$ buffer layers are shown in Figure 4. The TEM specimens were prepared by detaching fragments from the film onto the grid [31]. The grains were identified manually with the Olympus Soft Imaging Solutions GmbH, Münster, Germany iTEM TEM imaging platform. The grain sizes were distributed between 10 nm and 30 nm, with the average grain size, reported in Table 2, in the range of 12–17 nm. Although these values are smaller than those inferred by XRD studies, the different results obtained by the two methods are not incompatible. In the case of the XRD estimations the use of a single well-defined peak is a limitation, whereas in the case of the TEM assessments, the use of a particular sample area and of a finite number of grains influence the accuracy of the grain size distribution function. Within the limits of validity of the approximations used and the corresponding error bars of the two independent estimations, we suggest that, overall, the results were consistent.

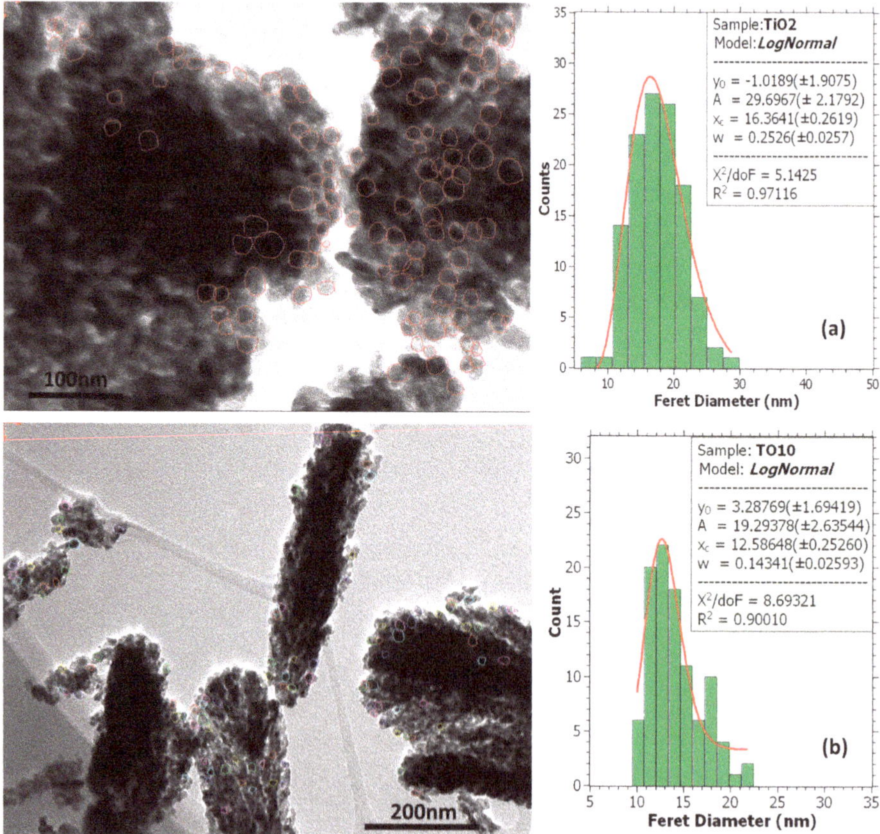

Figure 4. *Cont.*

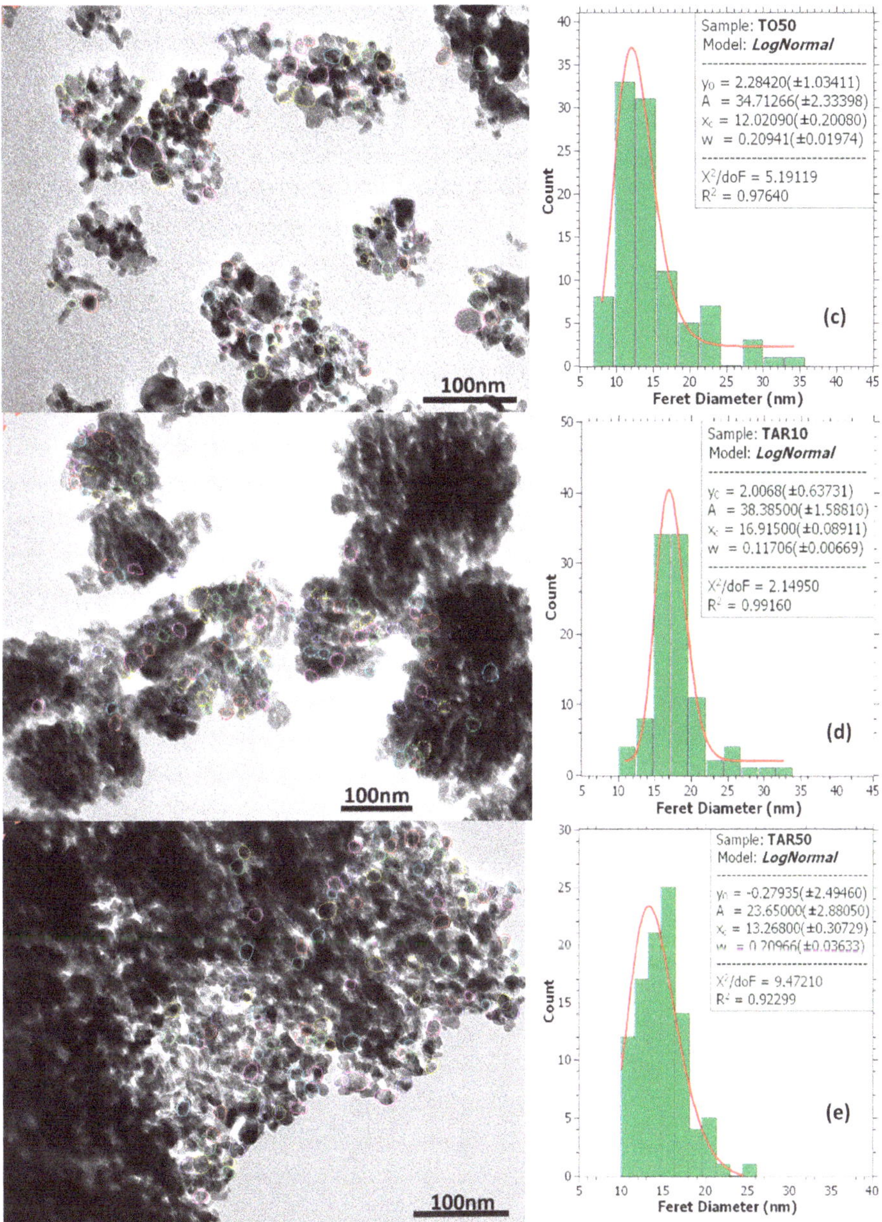

Figure 4. TEM image, selected area electron diffraction (SAED) pattern and distribution of grain size for the mesoporous TiO$_2$ film (**a**), and the buffer layers: TO10 (**b**), TO50 (**c**), TAR10 (**d**) and TAR50 (**e**).

Figure 5 presents the selected area electron diffraction (SAED) profiles extracted from the SAED patterns (see Supplementary Materials). For the mesoporous TiO$_2$ film, Figure 5 reveals strong signature peaks of the anatase phase, particularly for the (101) plane. Other noticeable peaks were indicative of the (004) and (200) planes. In contrast, the TO10 and TAR10 samples showed much lower intensity peaks, which could still be associated with the anatase phase of TiO$_2$, although the presence of the rutile phase cannot be excluded. Intermediate intensity peaks were present in the spectrum of TO50, pointing to the existence of wide crystallinity regions, with only small amorphous phase contributions. A clear mixture of anatase and rutile phases was noticeable in the TAR50 spectrum, particularly for the (110) plane of the rutile structure. For the TAR50 samples, the SAED and RDF data suggested the presence of crystalline regions along with the amorphous phase, in contrast to the XRD patterns. The differences may be caused by the irregularities of the samples, because TEM investigates a particular area of the sample whereas XRD provides a global perspective.

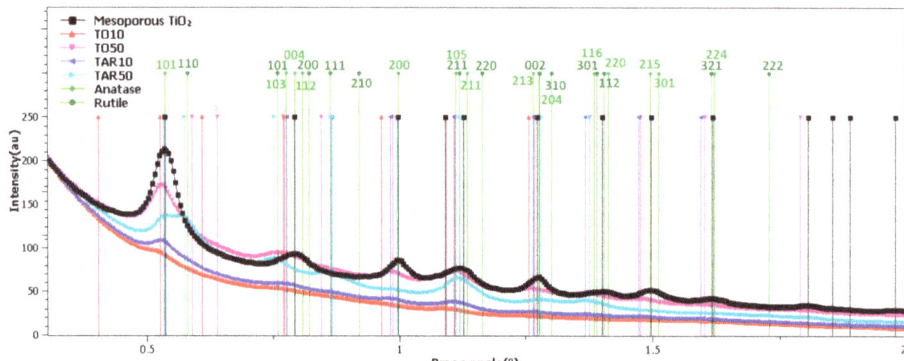

Figure 5. SAED profile obtained using CRISP2 software [59] with the ELD module [60] for the mesoporous TiO$_2$ film and the buffer layers: TO10, TO50, TAR10 and TAR50. The indexing of anatase and rutile phases, which was performed according to Refs. [61,62], is represented as: Anatase = diamonds and rutile = circles.

Figure 6 displays a typical high-resolution transmission electron microscopy (HRTEM) image of a TO10 sample with an inset showing the fast Fourier transform (FFT) power spectrum of a specific sample area. By indexing the FFT spectrum one obtains interplanar distances of 0.359 nm assigned to the (101) lattice plane reflections of the tetragonal anatase TiO$_2$ phase. The anatase phase nanocrystallites could be also identified from the 0.359 nm lattice fringes visible in the HRTEM image.

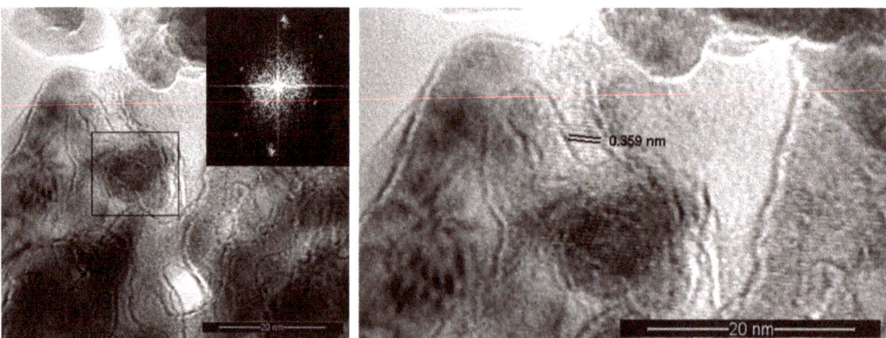

Figure 6. High resolution TEM image with inset of the corresponding fast Fourier transformer presentation of the selected area of the TiO$_2$ buffer layer deposited at 10 Pa oxygen (TO10) (**left**) and a detail showing the interplanar distance corresponding to the anatase (101) plane (**right**).

The UV/Vis transmission spectra of the TiO$_2$ buffer layers obtained by PLD on FTO glass substrate are presented in Figure 7. It can be seen that the average transmittance value, in the visible range, reached around 75–80%. The transmission of TO50 samples was slightly lower, likely due to their larger thickness.

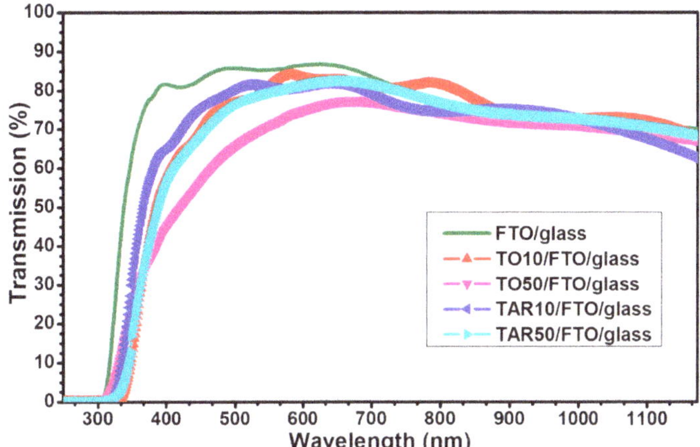

Figure 7. Transmission spectra of mesoporous TiO_2 film and of TiO_2 buffer layers deposited by PLD on FTO glass substrates.

Table 3 displays the typical values of the key parameters of the DSSCs fabricated using photo-electrodes with various PLD buffer layers, determined by electro-optical measurements. The *I-V* curves for a few selected cases are illustrated in Figure 8. For comparison, we provided values for the case of a DSSC without a buffer layer.

Table 3. Electro-optical parameters: Short circuit current (I_{sc}), open circuit voltage (V_{oc}), short circuit current density (J_{sc}), maximum power (P_{max}), fill factor (*FF*) and photovoltaic conversion efficiency (η) of typical dye-sensitized solar cells (DSSCs) measured under standard illumination conditions (Figure 9).

Sample	I_{sc} (mA)	V_{oc} (mV)	J_{sc} (mA/cm^2)	P_{max} (μW)	*FF*	η (%)
TO10	6.39	608	8.136	2228	0.57	2.84
TO50	5.92	604	7.537	2015	0.56	2.57
TAR10	6.21	594	7.907	2041	0.55	2.60
TAR50	5.90	609	7.512	1978	0.55	2.52
No buffer	2.48	590	3.166	968	0.66	1.23

One remark is that the photovoltaic conversion efficiency for solar cells fabricated with the buffer layer was more than twice larger than the efficiency of the devices with the mesoporous TiO_2 film applied directly to the FTO. As the open-circuit voltage was about the same, crucial in determining the significantly higher efficiency was the much larger short-circuit current density. By introducing a buffer layer we obtained a substantial increase of the short circuit current density, J_{sc}, from 3.166 mA/cm^2 for cells without buffer layer to 8.136 for TO10.

The fill factor was within the typical range of solar cells 0.5 to 0.8, although closer to the lower rather than the higher limit. As the highest value was reached for the device without a buffer layer, an analysis of the device losses is necessary.

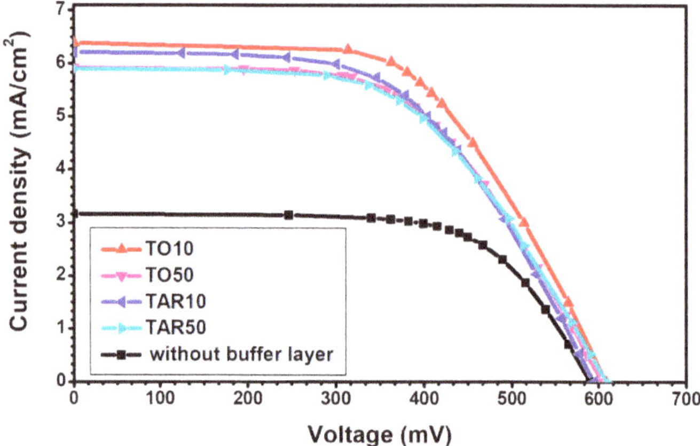

Figure 8. Current-voltage (*I-V*) curves for typical DSSCs fabricated using photo-electrodes with various buffer layers: TO10 (dashed line), TO50 (dotted line), TAR10 (dashed dotted line) or TAR50 (dashed dotted dashed line). The measurements were performed under standard air mass 1.5 global (AM 1.5G) illumination conditions.

4. Discussion

The efficiency of solar cells was affected by the losses due to parallel (shunt) resistance (R_{sh}) and series resistance (R_s), as shown in Figure 9. The equivalent circuit of a solar cell allows us to infer through Equation (1) a relation between the current and the voltage of the cell (I_{CELL}, V_{CELL}), as a function of I_{ph}, the photocurrent density, I_0, the dark current (reverse saturation current of the diode), R_s and R_{sh}, the series and shunt resistances of the cell [63]. In Equation (1) T is the absolute cell temperature, m the diode ideality factor, k_B, the Boltzmann constant and q_e the electron charge.

$$I_{CELL} = I_{ph} - I_0\left(e^{\frac{-q_e(V_{CELL}+I_{CELL}R_S)}{mk_BT}} - 1\right) + \frac{V_{CELL} + I_{CELL}R_s}{R_{sh}} \tag{1}$$

For an ideal cell, R_{sh} is infinite, whereas R_s is zero. In the *I-V* plot of the ideal cell, near the short-circuit point, the curve was roughly horizontal, indicating that the shunt resistance was high. Near the other important point, the open-circuit limit, the *I-V* curve was close to vertical, meaning that the series resistance was low.

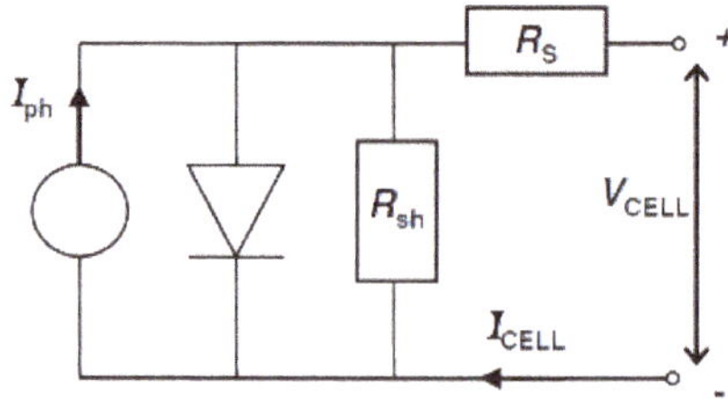

Figure 9. Equivalent circuit model of a solar cell.

When the shunt resistance is very high but the series resistance cannot be neglected, the third term in Equation (1) becomes negligible. In this case, close to the short-circuit point the curve is roughly horizontal, but near the open-circuit point the *I-V* curve is not vertical, suggesting significant series resistive losses in the cell. Therefore, the shape of the *I-V* curves in Figure 8 points to significant series losses.

Examining Tables 1 and 3 we note that, for the buffer layer deposited in oxygen, J_{sc} decreased from 8.136 mA/cm^2 to 7.907 mA/cm^2, while for the buffer layer deposited in argon J_{sc} decreased from 7.537 mA/cm^2 to 7.512 mA/cm^2. Increasing the buffer layer thickness led to a decrease of J_{sc}, as the wider films have a lower transmission (seen in Figure 7), such that less photons reach the active mesoporous film, and fewer photoelectrons are generated. Moreover, the diffusion of the photoelectrons towards the FTO was hindered by thicker buffer layers.

Finally, the highest value of efficiency (2.83%) was reached for TO10, suggesting that the lower the pressure for the deposition of the buffer layer, the higher the performance of the final device is.

To understand these results one needs to examine the basic processes that occur in the DSSCs [64]. Processes such as charge injection into TiO$_2$, charge diffusion to FTO and dye regeneration are desirable. In contrast, the nonradiative decay, the back transfer to dye and the charge interception by electrolyte are detrimental processes [64].

Our results suggest that the higher J_{sc} for the devices with the buffer layer is likely due to an increased charge transfer from the TiO$_2$ mesoporous layer to the FTO, through the buffer layer (possibly caused by the superior conductivity of the brush-like nanobar structures shown in Figure 1). Another reason may be the decrease in charge loss through charge interception by the electrolyte (which does not get in contact with the fluorine-doped transparent oxide).

The lower fill factor indicates that the introduction of a buffer layer increases the series resistance, R_s. The large R_s was reflected in the deviation from verticality of the current-voltage curve near the open-circuit point (Figure 8). We suggest that a larger buffer layer thickness, meaning an increased length over which the charges should diffuse, hinders the electron flow and facilitates the charge interception.

We conclude that although the photovoltaic conversion efficiencies obtained in our study were relatively low compared to the existing records, the fact that the buffer layer more than doubles the performance should be emphasized, as it shows promise. It strengthens the opinion that PLD has much to offer in photovoltaics [46,65,66], making it a relevant method for improving the fabrication of TiO$_2$-based DSSCs with superior conversion efficiency.

5. Conclusions

Our study aimed to increase DSSC performance by interposing an intermediary TiO$_2$ buffer layer deposited by PLD. The hint was that an improved contact between the FTO and the mesoporous TiO$_2$ film would minimize the resistive losses and increase the short circuit current density by preventing the electrolyte from getting in contact with the FTO.

The buffer layer was deposited in either ambient oxygen or argon at different pressures. The PLD films were comparatively analyzed by transmittance measurements and SEM, TEM and XRD investigations.

Top-view and cross-section SEM micrographs of the TO and TAR films revealed round particulates with homogeneous shape and size dimension. The low-pressure samples displayed thinner PLD layers, whereas the high pressure ones had longer TiO$_2$ nanobars. The XRD analyses of TAR10, TO10 and TO50 samples evidenced a monophasic structure, exhibiting maxima only of TiO$_2$-anatase. In the case of TAR50 samples no TiO$_2$ peaks were observed, thus suggesting an amorphous structure.

TEM images with SAED patterns of the films have confirmed the anatase phase of TiO$_2$. Moreover, the anatase phase has also been identified from the 0.359 nm lattice fringes, visible in the HRTEM images. TEM studies of TAR50 samples clearly indicated the presence of the rutile phase but also suggested a mixture of crystalline and amorphous regions.

Electro-optical measurements carried out, under standard AM 1.5G conditions, have shown that the insertion of a buffer layer at the interface of FTO/TiO$_2$ led to photovoltaic conversion efficiencies more than two times larger than those of the standard cells. The best performance was recorded for buffer layers deposited in 10 Pa O$_2$, which are characterized by an open circuit voltage, V_{oc} as high as 608 mV and a short circuit current, I_{sc} of 6.39 mA.

The processes that take place inside the device were discussed and the role of the brush-like buffer layer in the performance increase was emphasized. The higher I_{sc} for the devices with the buffer layer is likely associated to an increased charge transfer from the TiO_2 mesoporous layer to the FTO, through the buffer layer as well as to a decrease in charge loss due to charge interception by the electrolyte. The lower *FF* indicates that the introduction of a buffer layer increases the series resistance, R_s, due to an increased length over which the charges have to diffuse, hindering the electron flow and facilitating the charge interception.

The goal of the present study was to find the PLD parameters that optimize the performance of the DSSCs. The fact that the buffer layer more than doubled the performance is to be emphasized, as it shows promise. Despite the relatively low efficiency obtained, which indicates that the fabrication technology was not yet optimized, the increase reported here was significant and exceeds the enhancements of 15–80% stated for other techniques as well as that of up to 42% for the same method. Our results strengthen the opinion that PLD has much to offer in photovoltaics, making it an interesting method for obtaining oxide semiconductor nanostructures for DSSCs.

Finally, it should be noted that the long-term device performance is critical for applications. It would clearly verify the role of the intermediate TiO_2 buffer layer on the good mechanical contact, protection of the electrodes against the dye solution action, oxidation at operating temperatures, as well as the reduced recombination of electrons at the electrode/electrolyte interface. Such studies, which require a long time, are underway and will be subject of a subsequent report.

Supplementary Materials: The following are available online at http://www.mdpi.com/2079-4991/9/5/746/s1, Figure S1: Selected area electron diffraction (SAED) patterns for the mesoporous TiO_2 film (a), and the buffer layers: TO10 (b), TO50 (c), TAR10 (d), and TAR50 (e).

Author Contributions: Conceptualization: G.S., M.A.G. and I.N.M.; methodology: J.L., G.S., N.S., G.P.-P. and G.P.; validation: A.G., G.P., M.A.G. and I.N.M.; formal analysis: J.L., G.S., G.E.S. and G.P.; investigation: J.L., G.E.S., N.Ş., C.L. and A.G.; data curation: G.E.S. and C.L.; writing—original draft preparation: J.L.; writing—review and editing: G.P.-P., M.A.G. and I.N.M.; visualization: G.E.S. and G.P.-P.; supervision, G.S., M.A.G. and I.N.M.

Funding: This research was funded by the Romanian Ministry of Research and Innovation through the Core Programme—Contract LAPLAS VI 16N/08.02.2019.

Acknowledgments: The authors acknowledge useful discussions with Victor Ciupina and Corneliu I. Oprea.

Conflicts of Interest: The authors declare no conflict of interest.

References

1. Gong, J.; Sumathy, K.; Qiao, Q.; Zhou, Z. Review on dye-sensitized solar cells (DSSCs): Advanced techniques and research trends. *Ren. Sust. En. Rev.* **2017**, *68*, 234–246. [CrossRef]
2. Freitag, M.; Teuscher, J.; Saygili, Y.; Zhang, X.; Giordano, F.; Liska, P.; Hua, J.; Zakeeruddin, S.M.; Moser, J.-E.; Grätzel, M.; Hagfeldt, A. Dye-sensitized solar cells for efficient power generation under ambient lighting. *Nat. Photonics* **2017**, *11*, 372–379. [CrossRef]
3. O'Regan, B.; Grätzel, M. A low-cost, high-efficiency solar cell based on dye-sensitized colloidal TiO_2 films. *Nature* **1991**, *353*, 737–740. [CrossRef]
4. Gratzel, M. Photoelectrochemical Cells. *Nature* **2001**, *414*, 338–344. [CrossRef] [PubMed]
5. Hagfeldt, A.; Boschloo, G.; Sun, L.; Kloo, L.; Pettersson, H. Dye-Sensitized Solar Cells. *Chem. Rev.* **2010**, *110*, 6595–6663. [CrossRef] [PubMed]
6. Nazeeruddin, M.K.; Pechy, P.; Renouard, T.; Zakeeruddin, S.M.; Humphry-Baker, R.; Comte, P.; Liska, P.; Cevey, L.; Costa, E.; Shklover, V.; et al. Engineering of efficient panchromatic sensitizers for nanocrystalline TiO(2)-based solar cells. *J. Am. Chem. Soc.* **2001**, *123*, 1613–1624. [CrossRef]
7. Nazeeruddin, M.K.; De Angelis, F.; Fantacci, S.; Selloni, A.; Viscardi, G.; Liska, P.; Ito, S.; Bessho, T.; Grätzel, M. Combined Experimental and DFT-TDDFT Computational Study of Photoelectrochemical Cell Ruthenium Sensitizers. *J. Am. Chem. Soc.* **2005**, *127*, 16835–16847. [CrossRef]
8. Gratzel, M. Solar Energy Conversion by Dye-Sensitized Photovoltaic Cells. *Inorg. Chem.* **2005**, *44*, 6841–6851. [CrossRef]

9. Robertson, N. Catching the rainbow: Light harvesting in dye-sensitized solar cells. *Angew. Chem. Int. Ed.* **2008**, *47*, 1012–1014. [CrossRef]

10. Xie, Y.; Tang, Y.; Wu, W.; Wang, Y.; Liu, J.; Li, X.; Tian, H.; Zhu, W.-H. Porphyrin Cosensitization for a Photovoltaic Efficiency of 11.5%: A Record for Non-Ruthenium Solar Cells Based on Iodine Electrolyte. *J. Am. Chem. Soc.* **2015**, *137*, 14055–14058. [CrossRef]

11. Yella, A.; Lee, H.W.; Tsao, H.N.; Yi, C.; Chandiran, A.K.; Nazeeruddin, M.K.; Diau, E.W.G.; Yeh, C.Y.; Zakeeruddin, S.M.; Gratzel, M. Porphyrin-Sensitized Solar Cells with Cobalt (II/III)—Based Redox Electrolyte Exceed 12 Percent Efficiency. *Science* **2011**, *334*, 629–634. [CrossRef] [PubMed]

12. Kenji Kakiage, K.; Aoyama, Y.; Yano, T.; Oya, K.; Kyomen, T.; Hanaya, M. Fabrication of a high-performance dye-sensitized solar cell with 12.8% conversion efficiency using organic silyl-anchor dyes. *Chem. Commun.* **2015**, *51*, 6315–6317. [CrossRef] [PubMed]

13. Liu, M.; Johnston, M.B.; Snaith, H.J. Efficient planar heterojunction perovskite solar cells by vapour deposition. *Nature* **2013**, *501*, 395–398. [CrossRef]

14. Yang, W.S.; Park, B.-W.; Jung, E.H.; Jeon, N.J.; Kim, Y.C.; Lee, D.U.; Shin, S.S.; Seo, J.; Kim, E.K.; Noh, J.H.; et al. Iodide management informamidinium-lead-halide–basedperovskite layers for efficientsolar cells. *Science* **2017**, *356*, 1376–1379. [CrossRef]

15. Yu, H.; Zhang, S.Q.; Zhao, H.J.; Will, G.; Liu, P.B. An Efficient and Low-Cost TiO_2 Compact Layer for Performance Improvement of Dye-Sensitized Solar Cells. *Electrochim. Acta* **2009**, *54*, 1319–1324. [CrossRef]

16. Patrocinio, A.O.T.; Paterno, L.G.; Iha, N.Y.M. Layer-by-layer TiO_2 films as efficient blocking layers in dye-sensitized solar cells. *J. Photochem. Photobiol. A* **2009**, *205*, 23–27. [CrossRef]

17. Ito, S.; Murakami, T.N.; Comte, P.; Liska, P.; Gratzel, C.; Nazeeruddin, M.K.; Grätzel, M. Fabrication of Thin Film Dye Sensitized Solar Cells with Solar to Electric Power Conversion Efficiency over 10%. *Thin Solid Films* **2008**, *516*, 4613–4619. [CrossRef]

18. Wang, P.; Zakeeruddin, S.M.; Comte, P.; Charvet, R.; Humphry-Baker, R.; Grätzel, M. Enhance the Performance of Dye-Sensitized Solar Cells by Co-grafting Amphiphilic Sensitizer and Hexadecylmalonic Acid on TiO_2 Nanocrystals. *J. Phys. Chem. B* **2003**, *107*, 4336–143341. [CrossRef]

19. Kim, J.H.; Lee, K.J.; Roh, J.H.; Song, S.W.; Park, J.H.; Yer, I.H.; Moon, B.M. Ga-doped ZnO transparent electrodes with TiO_2 blocking layer/nanoparticles for dye-sensitized solar cells. *Nano. Res. Lett.* **2012**, *7*, 1–12. [CrossRef]

20. Yoo, B.; Kim, K.J.; Bang, S.Y.; Ko, M.J.; Kim, K.; Park, N.G. Chemically deposited blocking layers on FTO substrates: Effect of precursor concentration onphotovoltaic performance of dye-sensitized solar cells. *J. Electroanal. Chem.* **2010**, *638*, 161–166. [CrossRef]

21. Lungu, J.; Ştefan, N.; Prodan, G.; Georgescu, A.; Mandeş, A.; Ciupina, V.; Mihăilescu, I.N.; Gîrţu, M.A. Characterization of spin-coated TiO_2 buffer layers for dye-sensitized solar cells. *Digest J. Nanomat. Biostruct.* **2015**, *10*, 967–976.

22. Kavan, L.; O'Regan, B.; Kay, A.; Grätzel, M. Preparation of TiO_2 (anatase) films on electrodes by anodic oxidative hydrolysis of $TiCl_3$. *J. Electroanal. Chem.* **1993**, *346*, 291–307. [CrossRef]

23. Kavan, L.; Tetreault, N.; Moehl, T.; Grätzel, M. Electrochemical Characterization of TiO_2 Blocking Layers for Dye-Sensitized Solar Cells. *J. Phys. Chem. C* **2014**, *118*, 16408–16418. [CrossRef]

24. Kavan, L.; Grätzel, M. Highly efficient semiconductingTiO_2 photoelectrodes prepared by aerosol pyrolysis. *Electrochim. Acta* **1995**, *40*, 643–652. [CrossRef]

25. Cameron, P.J.; Peter, L.M. Characterization of Titanium Dioxide Blocking Layers in Dye-Sensitized Nanocrystalline Solar Cells. *J. Phys. Chem. B* **2003**, *107*, 14394–14400. [CrossRef]

26. Peng, B.; Jungmanna, G.; Jäger, C.; Haarer, D.; Schmidt, H.-W.; Thelakkat, M. Systematic investigation of the role of compact TiO_2 layer in solid state dye-sensitized TiO_2 solar cells. *Coord. Chem. Rev.* **2004**, *248*, 1479–1489. [CrossRef]

27. Ito, S.; Zakeeruddin, S.; Humphry-Baker, R.; Liska, P.; Charvet, R.; Comte, P.; Nazeeruddin, M.; Péchy, P.; Takata, M.; Miura, H.; et al. High-Efficiency Organic-Dye-Sensitized Solar Cells Controlled by Nanocrystalline-TiO_2 Electrode Thickness. *Adv. Mater.* **2006**, *18*, 1202. [CrossRef]

28. Burke, A.; Ito, S.; Snaith, H.; Bach, U.; Kwiatkowski, J.; Grätzel, M. The Function of a TiO_2 Compact Layer in Dye-Sensitized Solar Cells Incorporating "Planar" Organic Dyes. *Nano Lett.* **2008**, *8*, 977–981. [CrossRef] [PubMed]

29. Hart, J.N.; Menzies, D.; Cheng, Y.B.; Simon, G.P.; Spiccia, L. TiO_2 sol-gel blocking layers for dye-sensitized solar cells. *C. R. Chimie* **2006**, *9*, 622–626. [CrossRef]

30. Kavan, L.; Zukalova, M.; Vik, O.; Havlicek, D. Sol–Gel Titanium Dioxide Blocking Layers for Dye- Sensitized Solar Cells: Electrochemical Characterization. *ChemPhysChem* **2014**, *15*, 1056–1061. [CrossRef]

31. Ito, S.; Liska, P.; Comte, P.; Charvet, R.; Pechy, P.; Bach, U.; Schmidt-Mende, L.; Zakeeruddin, S.M.; Kay, A.; Nazeeruddin, M.K.; et al. Control of dark current in photoelectrochemical (TiO_2/I^-–I_3^-) and dye-sensitized solar cells. *Chem. Commun.* **2005**, 4351–4353. [CrossRef]

32. Xia, J.; Masaki, N.; Jiang, K.; Yanagida, S. Deposition of a Thin Film of TiOx from a Titanium Metal Target as Novel Blocking Layers at Conducting Glass/TiO_2 Interfaces in Ionic Liquid Mesoscopic TiO_2 Dye-Sensitized Solar Cells. *J. Phys. Chem. B* **2006**, *110*, 25222–25228. [CrossRef]

33. Hitosugi, T.; Yamada, N.; Nakao, S.; Hirose, Y.; Hasegawa, T. Properties of TiO_2-based transparent conducting oxides. *Phys. Status Solidi A* **2010**, *207*, 1529–1537. [CrossRef]

34. Braga, A.; Baratto, C.; Colombi, P.; Bontempi, E.; Salvinelli, G.; Drera, G.; Sangaletti, L. An ultrathin TiO_2 blocking layer on Cd stannate as highly efficient front contact for dye-sensitized solar cells. *Phys. Chem. Chem. Phys.* **2013**, *15*, 16812–16818. [CrossRef] [PubMed]

35. Thelakkat, M.; Schmitz, C.; Schmidt, H.-W. Fully vapor-deposited thin-layer titanium dioxide solar cells. *Adv. Mater.* **2002**, *14*, 577–581. [CrossRef]

36. Hamann, T.W.; Martinson, A.B.F.; Elam, J.W.; Pellin, M.J.; Hupp, J.T. Atomic Layer Deposition of TiO_2 on Aerogel Templates: New Photoanodes for Dye-Sensitized Solar Cells. *J. Phys. Chem. C* **2008**, *112*, 10303–10307. [CrossRef]

37. Van Delft, J.A.; Garcia-Alonso, D.; Kessels, W.M.M. Atomic layer deposition for photovoltaics: Applications and prospects for solar cell manufacturing. *Semicond. Sci. Technol.* **2012**, *27*, 074002. [CrossRef]

38. Chandiran, A.K.; Yella, A.; Stefik, M.; Heiniger, L.P.; Comte, P.; Nazeeruddin, M.K.; Grätzel, M. Low-Temperature Crystalline Titanium Dioxide by Atomic Layer Deposition for Dye-Sensitized Solar Cells. *ACS Appl. Mater. Interfaces* **2013**, *5*, 3487–3493. [CrossRef] [PubMed]

39. Gan, W.Y.; Lam, S.W.; Chiang, K.; Amal, R.; Zhao, H.; Brungs, M.P. Novel TiO_2 thin film with non-UV activated superwetting and antifogging behaviours. *J. Mater. Chem.* **2007**, *17*, 952–954. [CrossRef]

40. Charbonneau, C.; Cameron, P.J.; Pockett, A.; Lewis, A.; Troughton, J.R.; Jewell, E.; Worsley, D.A.; Watson, T.M. Solution processing of TiO_2 compact layers for 3rd generation photovoltaics. *Ceram. Int.* **2016**, *42*, 11989–11997. [CrossRef]

41. Bai, Y.; Xing, Z.; Yu, H.; Li, Z.; Amal, R.; Wang, L. Porous Titania Nanosheet/Nanoparticle Hybrids as Photoanodes for Dye-Sensitized Solar Cells. *ACS Appl. Mater. Interfaces* **2013**, *5*, 12058–12065. [CrossRef]

42. Lee, S.; Noh, J.H.; Han, H.S.; Yim, D.Y.; Kim, D.H.; Lee, J.-K.; Kim, J.Y.; Jung, H.S.; Hong, K.S. Nb-Doped TiO_2: A New Compact Layer Material for TiO_2 Dye-Sensitized Solar Cells. *J. Phys. Chem. C* **2009**, *113*, 6878–6882. [CrossRef]

43. Zhou, W.-Q.; Lu, Y.-M.; Chen, C.-Z.; Liu, Z.-Y.; Cai, C.-B. Effect of Li-doped TiO_2 Compact Layers for Dye Sensitized Solar Cells. *J. Inorg. Mater.* **2011**, *26*, 819–822. [CrossRef]

44. Bäuerle, D. *Laser Processing and Chemistry*; Springer: Berlin, Germany, 1996.

45. Gyorgy, E.; Socol, G.; Axente, E.; Mihailescu, I.N.; Ducu, C.; Ciuca, S. Anatase phase TiO_2 thin films obtained by pulsed laser deposition for gas sensing applications. *Appl. Surf. Sci.* **2005**, *247*, 429–433. [CrossRef]

46. Mihailescu, I.N.; Gyorgy, E. Pulsed laser deposition: An overview. In *Proceedings of the Invited Contribution to the Fourth International Commission for Optics (ICO) Book Trends in Optics and Photonics*; Asakura, T., Ed.; Springer: Berlin, Germany, 1999; 201p.

47. Pechini, M. Method of preparing lead and alkaline earth titanates and niobates and coating method using the same to form a capacitor. US Patent 3 330 697, 1967.

48. Hocevar, M.; Opara, U.; Krasovec, U.; Berginc, M.; Drazic, G.; Hauptman, N.; Topic, M. Development of TiO_2 pastes modified with Pechini sol–gel method for high efficiency dye-sensitized solar cell. *J. Sol. Gel Sci. Technol.* **2008**, *48*, 156–162. [CrossRef]

49. Opara, U.; Krasovec, U.; Berginc, M.; Hocevar, M.; Topic, M. Unique TiO_2 paste for high efficiency dye-sensitized solar cells. *Sol. Energy Mater. Sol. Cells* **2009**, *93*, 379–381. [CrossRef]

50. Nezeeruddin, M.K.; Kay, A.; Rodicio, I.; Humphry-Baker, R.; Muller, E.; Liska, P.; Vlachopoulos, N.; Grätzel, M. Conversion of Light to Electricity by cis-XzBis(2,2′-bipyridyl-4,4′-dicarboxylate)ruthenium(11) Charge-Transfer Sensitizers (X = C1-, Br-, I-, CN-, and SCN-) on Nanocrystalline TiO_2 Electrodes. *J. Am. Chem. Soc.* **1993**, *115*, 6382–6390. [CrossRef]

51. Millington, K.R.; Fincher, K.W.; King, A.L. Mordant dyes as sensitisers in dye-sensitised solar cells. *Sol. Energy Mater. Sol. Cells* **2007**, *91*, 1618–1630. [CrossRef]

52. Smestad, G.P.; Grätzel, M. Demonstrating Electron Transfer and Nanotechnology: A Natural Dye–Sensitized Nanocrystalline Energy Converter. *J. Chem. Educ.* **1998**, *75*, 752–756. [CrossRef]

53. Teodorescu, V.S.; Blanchin, M.G. Fast and Simple Specimen Preparation for TEM Studies of Oxide Films Deposited on Silicon Wafers. *Microsc. Microanal.* **2009**, *15*, 15–19. [CrossRef]

54. Dimitriu, E.; Iuga, A.; Ciupina, V.; Prodan, G.; Ramer, R. PZT-type materials with improved radial piezoelectric properties. *J. Eur. Ceram. Soc.* **2005**, *25*, 2401–2404. [CrossRef]

55. Georgescu, A.; Damache, G.; Gîrţu, M.A. Class A small area solar simulator for dye-sensitized solar cell testing. *J. Optoelectron. Adv. Mater.* **2008**, *10*, 3003–3007.

56. Mihailescu, I.N.; Gyorgy, E.; Teodorescu, V.S.; Steinbrecker, G.; Neamtu, J.; Perrone, A.; Luches, A. Characteristic features of the laser radiation-target interactions during reactive pulsed laser ablation of Si targets in ammonia. *J.Appl. Phys.* **1999**, *86*, 7123–7128. [CrossRef]

57. Mihailescu, I.N.; Teodorescu, V.S.; Gyorgy, E.; Luches, A.; Perrone, A.; Martino, M. About the nature of particulates covering the surface of thin films obtained by reactive pulsed laser deposition. *J. Phys. D Appl. Phys.* **1998**, *31*, 2236–2240. [CrossRef]

58. Mihailescu, I.N.; Teodorescu, V.S.; Gyorgy, E.; Ristoscu, C.; Cristescu, R. Particulates in pulsed laser deposition: Formation mechanisms and possible approaches to their elimination. In Proceedings of the SPIE International Conference of ALT'01, Constanta, Romania, 11–14 September 2001; Volume 4762, pp. 64–74.

59. Hovmöller, S. CRISP: Crystallographic image processing on a personal computer. *Ultramicroscopy* **1992**, *41*, 121–135.

60. Zou, X.D.; Sukharev, Y.; Hovmöller, S. ELD — a computer program system for extracting intensities from electron diffraction patterns. *Ultramicroscopy* **1993**, *49*, 147–158. [CrossRef]

61. Wyckoff, R.W.G. *Crystal Structures*; Interscience Publishers: New York, NY, USA, 1963; Volume 1, pp. 253–254.

62. Meagher, E.P.; George, A.L. Polyhedral thermal expansion in the TiO_2 polymorphs: Refinement of the crystal structures of rutile and brookite at high temperature. *Can. Miner.* **1979**, *17*, 77–85.

63. Halme, J.; Vahermaa, P.; Miettunen, K.; Lund, P. Device physics of dye solar cells. *Adv. Mater.* **2010**, *22*, 210–234. [CrossRef]

64. Hamann, T.W.; Jensen, R.A.; Martinson, A.B.F.; van Ryswykac, H.; Hupp, J.T. Advancing beyond current generation dye-sensitized solar cells. *Energy Environ. Sci.* **2008**, *1*, 66–78. [CrossRef]

65. Chrisey, D.G.; Hubler, G.K. *Pulsed Laser Deposition of Thin Films*; Wiley: New York, NY, USA, 1994.

66. Von Allmen, M.; Blatter, A. *Laser–Beam Interactions with Materials*, 2nd ed.; Springer: Berlin, Germany, 1995.

 *nanomaterials*

Article

Ge-Sb-Te Chalcogenide Thin Films Deposited by Nanosecond, Picosecond, and Femtosecond Laser Ablation

Georgiana Bulai [1], Oana Pompilian [2,3], Silviu Gurlui [4], Petr Nemec [5], Virginie Nazabal [5,6], Nicanor Cimpoesu [7], Bertrand Chazallon [2] and Cristian Focsa [2,*]

[1] Integrated Centre for Environmental Science Studies in the North-East Development Region-CERNESIM, "Al. I. Cuza" University of Iasi, 700506 Iasi, Romania; georgiana.bulai@uaic.ro

[2] Université de Lille, CNRS, UMR 8523-PhLAM-Physique des Lasers, Atomes et Molécules, CERLA-Centre d'Etudes et de Recherches Lasers et Applications, Lille F-59000, France; oana.pompilian@inflpr.ro (O.P.); bertrand.chazallon@univ-lille.fr (B.C.)

[3] National Institute for Lasers, Plasma and Radiation Physics, RO-077125 Magurele-Bucharest, Romania

[4] Faculty of Physics, "Al. I. Cuza" University of Iasi, 700506 Iasi, Romania; sgurlui@uaic.ro

[5] Faculty of Chemical Technology, University of Pardubice, 53210 Pardubice, Czech Republic; Petr.Nemec@upce.cz (P.N.); virginie.nazabal@univ-rennes1.fr (V.N.)

[6] Université de Rennes 1, CNRS, ISCR (Institut des Sciences Chimiques de Rennes)–UMR 6226, F-35000 Rennes, France

[7] Faculty of Materials Science and Engineering, "Gheorghe Asachi" Technical University of Iasi, 700050 Iasi, Romania; nicanor.cimpoesu@tuiasi.ro

* Correspondence: cristian.focsa@univ-lille.fr; Tel.: +33-320-33-64-84

Received: 23 March 2019; Accepted: 23 April 2019; Published: 1 May 2019

Abstract: Ge-Sb-Te thin films were obtained by ns-, ps-, and fs-pulsed laser deposition (PLD) in various experimental conditions. The thickness of the samples was influenced by the Nd-YAG laser wavelength, fluence, target-to-substrate distance, and deposition time. The topography and chemical analysis results showed that the films deposited by ns-PLD revealed droplets on the surface together with a decreased Te concentration and Sb over-stoichiometry. Thin films with improved surface roughness and chemical compositions close to nominal values were deposited by ps- and fs-PLD. The X-ray diffraction and Raman spectroscopy results showed that the samples obtained with ns pulses were partially crystallized while the lower fluences used in ps- and fs-PLD led to amorphous depositions. The optical parameters of the ns-PLD samples were correlated to their structural properties.

Keywords: pulsed laser deposition; chalcogenide thin films; Raman spectroscopy; spectroscopic ellipsometry

---

1. Introduction

Important advances in nonvolatile solid state memory devices were driven by the discovery of Ge-Sb-Te (GST) alloys along the GeTe-Sb_2Te_3 tie-line in the mid-1980s [1]. Phase change (PC) memories are based on changes in optical properties and electrical conductivity of chalcogenide materials upon a rapid amorphous-to-crystalline phase transition and vice versa. These two states must present a high enough contrast in electrical resistivity or other (optical) parameters in order to be identified. The rapid changes from an amorphous (high electrical resistivity) to crystalline structure (low electrical resistivity) are induced by the Joule effect using an electric current pulse [2]. Depending on the intensity and duration of the pulses, the PC memory cells can be written or erased. The rapid laser-induced crystallization with large property changes represented the grounds for many research studies [3–9]. For applications in data storage devices, other properties such as a good thermal stability of the amorphous phase and the possibility of applying a large number of write-erase cycles need to be considered. The investigations of Yamada et al. [1] on 100-nm-thick GST films deposited by electron beam evaporation revealed that their crystallization temperatures were larger than room temperature but accessible for phase transitions by electric pulses. The laser-induced crystallization time of these samples was below 70 ns, which ensured a rapid recording. The degree of the optical change n,k (crystalline)-n,k (amorphous) on the GeTe-Sb_2Te_3 pseudo-binary line increases with an increasing Ge content [10], but GST chalcogenides with higher Sb concentrations present faster phase changes. Thus, the study of Ge-Sb-Te-based compounds in various compositions is essential when developing phase change devices with remarkable characteristics [11].

Several methods for chalcogenide thin film deposition have been employed to date such as spin coating [12], magnetron sputtering [13,14], thermal evaporation [15], atomic layer deposition [16], and metal organic vapor phase epitaxy [17]. Among these, the Pulsed Laser Deposition (PLD) technique is suitable for the thin film growth of complex materials with a good adhesion to the substrate and a high homogeneity. Chalcogenide thin films with a low surface roughness were reported in Reference [14]. The films deposited by PLD presented lower bandgap values than the samples obtained by sputtering [14]. Musgraves et al. [15] compared the structural, optical, and electrical properties of Ge-Sb-S thin films deposited by two methods: thermal evaporation (TE) and laser ablation. The chemical composition analysis revealed a slight variation of the Sb/S ratio from the stoichiometric value in the TE samples, while the PLD thin films replicated the atomic percentages of the main elements from the target. The refractive indices of the as-deposited (amorphous) PLD thin films presented higher values that the ones observed for the TE samples and even than the ones of the bulk material. PLD epitaxial $Ge_2Sb_2Te_5$ thin films were obtained by Hilmi et al. [18]. However, their results also showed a decreased deposition rate as the substrate temperature was augmented, indicating a strong desorption during the deposition process. Similar observations were reported in Reference [19]. The studies done by Song et al. [20] and Boschker et al. [21] showed that the high adatom energy (proportional to the kinetic energy of the ejected particles that arrived at the substrate surface) during the pulsed laser deposition process influenced the stoichiometry and roughness of the film through preferential resputtering. However, photoexcited desorption [22] and in situ plasma plume diagnostics can offer information on the velocity of the ejected species [23–26].

This paper presents the main experimental results of an extensive systematic study on thin films of chalcogenide materials based on the ternary Ge-Sb-Te diagram. The films were deposited by laser ablation in various experimental conditions, varying laser parameters (pulse duration, repetition rate, wavelength, and fluence), target-to-substrate distance, and deposition time. The investigated materials were the endpoints of the GeTe-Sb_2Te_3 pseudo-binary line and the intermediate stable phases containing different proportions of these two structures: $GeSb_2Te_4$ (GST 124), $GeSb_4Te_7$ (GST 147), and $Ge_2Sb_2Te_5$ (GST 225).

2. Materials and Methods

The Ge-Sb-Te thin films were synthesized by Pulsed Laser Deposition using an experimental setup described in detail in previous papers [24,27–29]. Two types of lasers were used for target ablation: a Nd-YAG laser (Continuum Surelite III-10) with a 10-ns pulse width and a 10-Hz repetition rate for which we used all four harmonics (266, 355, 532, and 1064 nm) and a Ti-Sa laser (Spectra Physics) with pulse durations of 2 ps and 120 fs and with a repetition rate of up to 1 kHz. The bulk materials (GeTe, $GeSb_2Te_4$, $Ge_2Sb_2Te_5$, $GeSb_4Te_7$, and Sb_2Te_3) were prepared by the melt quenching method using high-purity elements (5N purity) and a melting temperature of 960 °C. The obtained polycrystalline targets were placed inside the stainless-steel vacuum chamber on a micrometric precision 3D-axis manipulator, while the substrate (single crystalline (100) Silicon and glass) was positioned at different distances in front of the target. The pressure during the depositions was kept in approximately the 10^{-5} Torr range using a turbomolecular pump. The other varied experimental parameters were the target-to-substrate distance (15–60 mm), fluence (0.1–10 J/cm^2), and deposition time (5–60 min). Considering the numerous deposited samples, details on the growth conditions of each film are given as the paper proceeds.

The morphological, compositional, structural, and optical properties of the synthesized films were studied using various techniques. The sample thickness was estimated using stylus profilometry (Dektak 6M). Images of the surface topography were obtained by optical microscopy (Olympus BXFM free-space confocal microscope, Olympus Europa, Hamburg, Germany) and scanning electron microscopy (Tescan Vega II LMH, Tescan, Brno, Czech Republic), using different magnifications. The chemical composition of the samples was studied by Energy-dispersive X-ray spectroscopy (EDS, Tescan Vega II LMH, Tescan, Brno, Czech Republic). Time of Flight-Secondary Ion Mass Spectroscopy (ToF-SIMS, ION-TOF 5, IONTOF, Münster, Germany) was used to analyze the distribution of the main elements on a 500 × 500 μm area on the sample surface in negative and positive polarity. In-depth profiles were obtained by sputtering 300 × 300 μm section with O_2 (for positive polarity) or Cs (for negative polarity) ion beams and analyzing a 100 × 100 μm inner (centered) surface with Bi_3^+ ion beam (25 kV, 1 pA). Raman spectroscopy measurements were performed using an InVia Reflex spectrometer (Renishaw, 250 mm focal length, Renishaw SA, Champs-sur-Marne, France) equipped with an Ar$^+$ laser source (514.5 nm wavelength, 36 mW laser power). Room temperature X-ray diffraction (Bruker AXS–Cu Kα radiation) patterns were required in the 5–65° 2θ range with 0.02° step and 5-s step times. The optical properties were investigated by variable angle spectroscopic ellipsometry (VASE, J.A. Woollam Co., Inc., Lincoln, NE, USA) in the 0.54–4.13 eV (2300–300 nm) spectral region.

3. Results and Discussion

3.1. Topography, Chemical Composition, and Structural Properties

3.1.1. Nanosecond Laser Ablation

The optical microscopy images and thickness profiles revealed that the surface of the sample deposited by nanosecond laser ablation was affected by the presence of droplets, their density being dependent on the fluence. These microscopic particles deposited on the substrate/film surface can have several origins: dislodging of existing or laser-produced protruding target surface features, subsurface superheating, splashing of the molten surface layer, or condensation from vapor species due to supersaturation [30]. In femtosecond laser ablation, there are mainly nonthermal processes involved which end with the Coulomb explosion as the main ejection mechanism, while in nanosecond laser ablation, the longer pulse width leads to strong thermal effects. In this temporal range, the thermal mechanisms are predominant and determine the thermal damage of the lattice (homogeneous melting).

Another important parameter that can have great influence on the microstructure of the deposited samples is the laser wavelength which, depending on the thermal properties of the material, can determine the ejection of different sized particles. Near-UV wavelengths offer higher photon energies and shorter penetration depths which can reduce the thermal effects when nanosecond pulse lasers are used. Thus, the deposition with the 266-nm laser radiation presents an advantage, especially when lower fluences are used. Large area depositions of GeTe were accomplished using the 266-nm radiation of the Nd-YAG laser and a lower fluence (1.2 J/cm^2) compared to the other GeTe samples. The glass substrate was placed at a distance of 6 cm from the target, and the deposition time was 30 min. With these deposition conditions, improved results related to surface microstructure were obtained. The lower fluence used for ablation determined the deposition not only of a few droplets but also of a thinner thin film. However, one should consider that the thickness value of 120 nm was obtained when analyzing the ends of the deposited area but that the thin film can present a greater thickness in the center region due to the strong directionality of the ablation plume on the normal to the target surface. Although a larger target-to-substrate distance, a shorter deposition time, and a lower fluence were used to deposit the film on glass substrate, the thickness was still reasonable (120 nm) while the uniformity was improved.

For a comparative study, Ge-Sb-Te thin films were deposited using the 266-nm laser wavelength in the same other conditions: laser fluence 3.81 J/cm^2, deposition time 60 min, and target–substrate distance of 3 cm. Table 1 summarizes the stylus profilometer thin films thickness measurements and their elemental composition, as measured by EDS. The latter reveals a Ge over-stoichiometry in the GeTe samples and an increased Sb content in the Sb_2Te_3 and Ge-Sb-Te-based thin films. The concentration errors were approximately 1–2 at%. These main trends in composition variation were also observed when different conditions were used for thin film deposition. However, smaller deviations from the nominal values were found for the GeTe and Sb_2Te_3 samples deposited at higher target-to-substrate distances. For the intermediate compositions ($GeSb_2Te_4$, $GeSb_4Te_7$, and $Ge_2Sb_2Te_5$), this content evolution with a target-to-substrate distance was not observed.

Table 1. The thickness (stylus profilometry) and elemental composition (EDS) of the thin films deposited using the 266-nm radiation with a 3.81 J/cm^2 fluence, a 60-min deposition time, and a 3-cm target–substrate distance.

Target Nominal at% Composition	Thickness (nm)	EDS Measured Thin Film Composition (at%)		
		Ge	Sb	Te
GeTe $Ge_{50}Te_{50}$	600	62.68	-	37.32
Sb_2Te_3 $Sb_{40}Te_{60}$	620	-	48.78	51.22
$GeSb_2Te_4$ $Ge_{14.28}Sb_{28.57}Te_{57.14}$	600	13.14	35.78	51.08
$GeSb_4Te_7$ $Ge_{8.33}Sb_{33.33}Te_{58.33}$	600	9.57	39.8	50.63
$Ge_2Sb_2Te_5$ $Ge_{22.22}Sb_{22.22}Te_{55.55}$	690	12.75	31.81	55.45

A higher Ge concentration was also reported in References [6,31,32] for Ge-Sb-Te thin films deposited by PLD using a KrF excimer laser (248 nm, 30 ns, 20 Hz) at a fluence of 2.6 J/cm^2. However, the recorded deviations were lower than the ones reported in this paper. The large deviations from the nominal composition in this study can be explained by the higher fluence used for target ablation. The high temperature induced at laser-target interaction could induce a more rapid evaporation of Te with respect to Ge, depending on the chemical properties of each species. The tellurium deficiency can be caused by its higher volatility compared to Ge or Sb [24].

The distribution of Ge, Sb, and Te in the thin film volume was analyzed by ToF-SIMS depth profiling. Figure 1 presents the obtained profiles for the $Ge_1Sb_4Te_7$ thin film.

A uniform depth profile distribution was observed for $Ge_1Sb_4Te_7$ thin films where three samples with different thicknesses were analyzed. However, an elevated Te content was observed at the region close to the thin film surface. This can come also from matrix effect due to oxidation at interface. As Te has a more pronounced metallic character, it can be more affected by this. The same behavior was observed in the GST 225 thin films deposited by Krusin-Elbaum et al. [33] by magnetron sputtering on Si substrate. Their work revealed that the deposited samples present a Te segregation on grain boundaries and surfaces. The composition of the main elements on the thin film surface was determined in Reference [32] by proton-induced X-ray emission (PIXE) and Rutherford back-scattering (RBS).

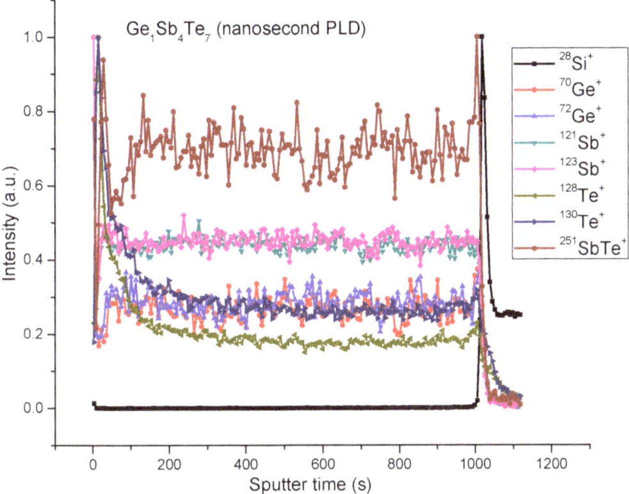

Figure 1. The ToF-SIMS depth profiles obtained in a positive polarity of the $Ge_1Sb_4Te_7$ sample.

Information on the structural properties of the deposited samples was obtained by Raman spectroscopy and X-ray diffraction. The XRD patterns of the $GeSb_4Te_7$, $GeSb_2Te_4$, and $Ge_2Sb_2Te_5$ thin films indicated the formation of a face-centered cubic (fcc) crystalline structure (Figure 2). The peaks found at approximately 29° and 42° 2θ angles were correlated to the (200) and (220) diffraction lines of the cubic phase. The GeTe and Sb_2Te_3 samples presented a different behavior. While the XRD measurements for the GeTe film suggested an amorphous deposition, the ones for Sb_2Te_3 sample presented peaks that can be associated with two types of structures: one characterized by wider peaks (thus, smaller crystallite dimensions) and another one represented by the narrower diffraction line at the same 2θ angle as the fcc structure of the GST based samples. The first mentioned phase can be due to the excess of Sb in the Sb_2Te_3 sample. The larger diffraction lines from our XRD pattern were found at the same 2θ angles as the ones observed by Prokhorov et al. [34] when analyzing thin films of Sb-Te with a higher Sb atomic content.

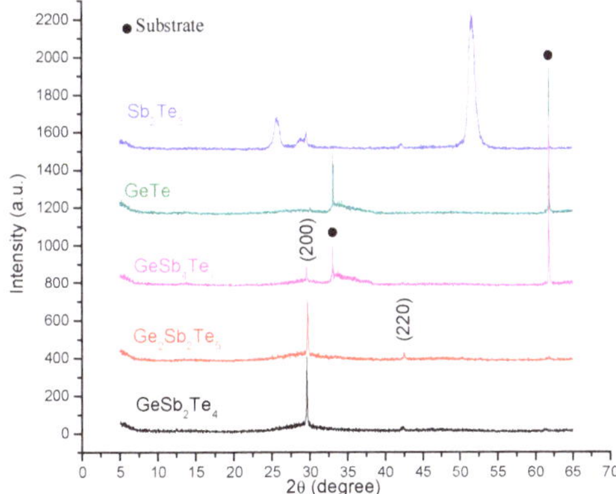

Figure 2. The XRD patterns of the chalcogenide thin films deposited using the 266-nm harmonic of the Nd-YAG laser.

The Raman spectra of the five thin films mentioned before (Table 1) are presented in Figure 3. The reported studies on Raman spectroscopy of GeTe materials revealed that the amorphous GeTe (a-GeTe) presents four peaks at 83, 125, 162, and 218 cm^{-1}, and the crystalline GeTe (c-GeTe) shows dominant vibrational modes at about 80 and 120 cm^{-1}. Moreover, Andrikopoulos et al. [35] observed several similarities between the peaks of the a- and c-GeTe samples. These were related to the much wider peak of the crystalline phase and to the narrower peak of the amorphous sample compared to the Raman response of other materials. These Raman features indicated that GeTe chalcogenide crystals present a distorted rock salt structure, while the a-GeTe seems more ordered than other glasses. In our study, the wide band in the 110–200 cm^{-1} region observed for the GeTe thin film is probably due to a combination between a crystalline structure and an amorphous phase.

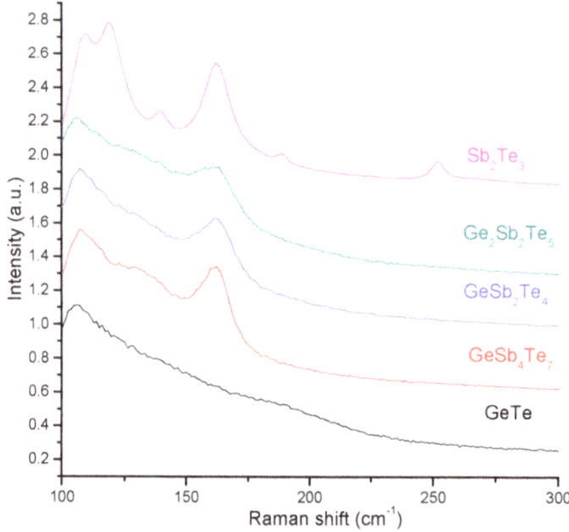

Figure 3. The Raman spectra of the five thin films deposited by ns-PLD at 266 nm (see Table 1 for deposition conditions).

Sb$_2$Te$_3$ has a rhombohedral (D_{3d}^5 symmetry) structure, with the following centre of the Brillouin zone representation:

$$\Gamma = 2\left(A_{1g} + E_g\right) + 3(E_u + A_{2u}) \tag{1}$$

where the ungerade(u)-modes are Raman active and gerade (g)-modes are IR active. Using density the functional perturbation theory, Sosso et al. [36] represented the IR and Raman spectra of crystalline Sb$_2$Te$_3$. The good agreement of their observations to the experimental vibrational spectra allowed them to assign each peak to specific phonons: E_g at 46 and 113 cm^{-1} and A_{1g} at 62 and 166 cm^{-1}. A sketch of the displacement patterns of phonons is also presented in Reference [37]. Two peaks at 110 and 165 cm^{-1} were also observed by Nemec et al. [6] in the Raman spectroscopy study on Sb$_2$Te$_3$ bulk materials used as targets in the deposition process. In our case, the spectra recorded for this type of material are described by two peaks: 110 and 163 cm^{-1}. In accordance with the data published in References [6,36,37], the first peak was attributed to the active Raman E_g mode, while the second one to the A_{1g} vibrational mode. However, for the Sb$_2$Te$_3$ film deposited using the 266 nm radiation of the Nd-YAG laser, additional vibrational modes were detected which were associated with the antimony-rich phase observed through XRD measurements.

The Raman spectra of the Ge-Sb-Te based compounds indicated the formation of a crystalline structure, presenting two peaks at 110 and 160 cm^{-1}. The same Raman response was obtained by Nemec et al. [6] when analyzing Ge-Sb-Te bulk materials. Based on the interpretation of the Raman spectra of GeTe and Sb$_2$Te$_3$ crystals, the bands of GST materials found at approximately 115–110 cm^{-1} and 165 cm^{-1} were assigned to the $\Gamma_1(A_1)$, $E_g(2)$, and $A_{1g}(2)$ modes, respectively [6]. However, since the two peaks do not present a narrow width, we should consider the presence of an amorphous phase in the deposited samples, an observation that is sustained by the ellipsometry measurements as well.

The presence of an amorphous phase can also be deduced from the XRD measurements where only the most intense peak of the fcc crystalline structure is observed.

3.1.2. Femtosecond and Picosecond Laser Ablation

Several thin films of Ge-Sb-Te based materials were also deposited by fs- and ps-PLD using a Ti-Sa laser with a wavelength of 800 nm, 1.6 mJ laser energy, and 1 kHz repetition rate. Other experimental parameters were the deposition time (5 to 30 min), target–substrate distance (1.5 to 6 cm), and laser fluence (0.1 to 0.5 J/cm^2). In the case of fs-PLD, the fluence used results in irradiance values in the approximate range of 1–4 TW/cm^2. Although these values might seem quite high, comparable irradiances have been already used in other fs-PLD studies [29,38,39], leading to good quality thin films. The electron density was not measured in the current work; we note, however, that laser ablation of solid targets in similar pulse duration and irradiance conditions led to values well below the critical density (which is in the range of 10^{21} cm^{-3}). For instance, Anoop et al. [40], using 40 fs pulses and fluences in the range 0.45–77 J/cm^2, measured an electron density of the order of 10^{17} cm^{-3} close to the target. When the electron density was measured farther from the target (which can present more practical interest for PLD experiments), values in the range 10^{10}–10^{13} cm^{-3} were observed [41–43].

Compared to the thin films deposited by ns-PLD, the optical microscopy and stylus profilometry measurements revealed that these samples present more uniform surface, without large droplets (see Figure 4). In most cases, the thickness variation with the modified experimental parameters is evident. For example, a decrease of the GeTe sample thickness from 900 nm to 140 nm was observed when the laser fluence was decreased from 0.3 J/cm^2 to 0.1 J/cm^2. Also, a twofold (from 1400 nm to 700 nm) thinner GeTe film was obtained when the target–substrate distance was increased from 4 to 6 cm.

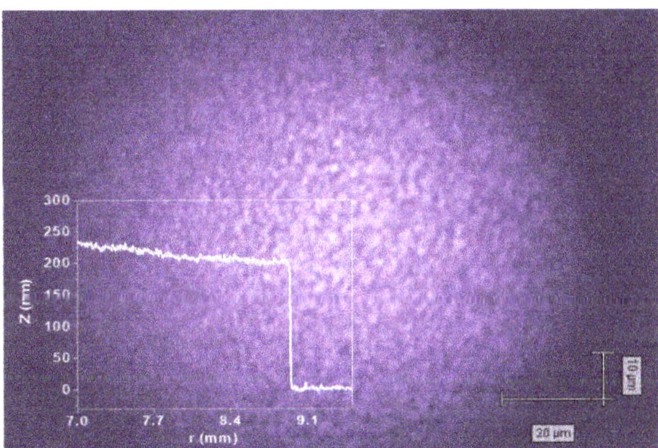

Figure 4. An optical microscopy image and surface topography of a Ge$_2$Sb$_2$Te$_5$ thin film deposited by fs-PLD.

Beside a more uniform surface, the Ge-Sb-Te based thin films presented also an improved chemical composition. Table 2 summarizes the representative Ge, Sb, and Te concentrations of three Ge$_2$Sb$_2$Te$_5$ thin films deposited using lasers with different pulse duration. Deviations from the nominal (stoichiometric) concentrations were also recorded for the films deposited by fs- and ps-PLD; however, they were smaller (usually below 4 at%) than the ones observed for the ns-deposited thin films.

Table 2. Representative concentrations for Ge, Sb, and Te of three $Ge_2Sb_2Te_5$ thin films deposited in different temporal regimes.

Pulse Duration	Deposition Conditions	Nominal Composition		
		Atomic %		
		Ge	Sb	Te
		22.22	**22.22**	**55.55**
Nanosecond	Nd-YAG laser (266 nm); Target-to-substrate distance = 3 cm; Fluence = 3.8 J/cm^2 Deposition time = 30 min	25.23	26.81	47.97
Picosecond	Ti-Sa laser; Target-to-substrate distance = 4 cm; Fluence = 0.3 J/cm^2; Deposition time = 60 min	23.45	21.69	54.86
Femtosecond	Ti-Sa laser; Target-to-substrate distance = 4 cm; Fluence = 0.3 J/cm^2; Deposition time = 30 min	22.58	22.28	55.13

The elemental composition of the deposited samples was also probed by ToF-SIMS depth profiling. An example is given in Figure 5 for the $Ge_2Sb_2Te_5$ thin film deposited by ps-PLD. As observed for the ns-PLD samples, a uniform distribution was recorded for the three elements.

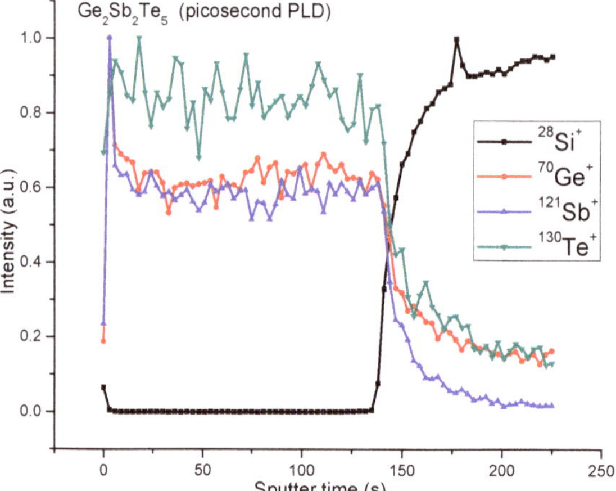

Figure 5. The ToF-SIMS depth profiles for the $Ge_2Sb_2Te_5$ thin film deposited by ps-PLD (target-to-substrate distance = 3 cm, fluence = 0.3 J/cm^2, deposition time = 60 min).

The structural properties of the samples deposited using the Ti-Sa laser were analyzed using the same two methods mentioned in the previous section. The XRD patterns revealed an amorphous phase deposition for the Sb_2Te_3 and $Ge_2Sb_2Te_5$ samples. Figure 6a presents the Raman spectra of two GeTe thin films deposited using different fluences of 0.1 J/cm^2 and 0.5 J/cm^2 (the target–substrate distance (6 cm), deposition time (30 min), and pressure (10^{-5} Torr) were kept constant). While the first film present two narrow peaks centered at 120 and 140 cm^{-1}, the sample deposited using a higher fluence (thus, an increased thickness) showed a wider band with the maximum value positioned at 120 cm^{-1} which can be associated with the amorphous structure of the GeTe material [6]. For a clear assignment of the peaks found for the film deposited at 0.1 J/cm^2, we took into consideration the Raman response of the Te phase. The bulk Te Raman spectra presents two peaks: one at 121 cm^{-1} which represents the A$_1$ mode and a second one at 140.8 cm^{-1} which represent E$_{TO}$ modes in crystalline Te-Te chain [44]. Considering that Te crystallizes at room temperature [45], a more adequate assignment of the two Raman peaks observed in our study would be based on Te segregation, which could be related as often reported to the photosensitivity of the GeTe film under laser irradiation of the Raman spectrometer when the film thickness is thinner rather than GeTe crystallization. We note, however, that parallel measurements using 785 nm excitation showed no difference in the resulting Raman spectra. Moreover, the main Raman feature of Ge is a vibrational mode at 300 cm^{-1}. However, the peak from our study found at the same wavenumber is due to the contribution of the silicon substrate and not to Ge segregation, considering its high crystallization temperature (250 °C) [46].

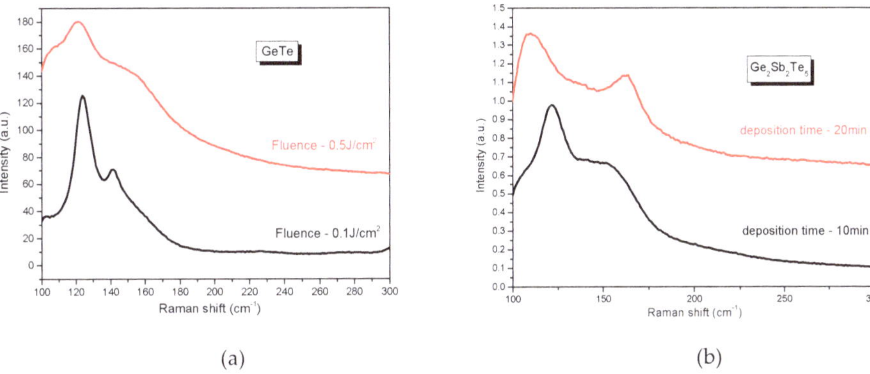

(a) (b)

Figure 6. Raman spectra of the (**a**) GeTe and (**b**) Ge$_2$Sb$_2$Te$_5$ thin films deposited in different conditions by fs-PLD.

Figure 6b shows the Raman spectroscopy results of two Ge$_2$Sb$_2$Te$_5$ thin films deposited in 10 and 20 min respectively (the target–substrate distance (6 cm), fluence (0.2 J/cm^2) and pressure (10^{-5} Torr) were kept constant). Again, a different Raman response was observed for the two samples with different thicknesses. While the first (10 min) film is characterized by an amorphous phase, the second one presents two peaks around 110 cm^{-1} and 160 cm^{-1} which could be associated with E$_g$ and A$_{1g}$ vibrational modes, respectively [6].

3.2. Optical Properties

Due to possible applications in phase-change optical storage and optical waveguides [47], an important parameter to be considered for this chalcogenide thin films is their reflectivity, which can be derived from [47,48]:

$$R(E) = \frac{(n(E)-1)^2 + k^2(E)}{(n(E)+1)^2 + k^2(E)} \tag{2}$$

where E, n, and k are the photon energy, refractive index, and extinction coefficient, respectively (we recall that the refractive index is related to the complex dielectric constant for which the imaginary part can be measured experimentally and the real part can be evaluated using the Kramers–Krönig transformation). The optical properties (refractive indices and extinction coefficients) of several samples were investigated using variable angle spectroscopic ellipsometry (VASE, J.A. Woollam) with an automated rotating analyzer.

When analyzing the Ge$_2$Sb$_2$Te$_5$ series of thin films, we observed that the optical properties were not significantly influenced by the varied experimental parameters; thus, we continued by focusing on the Ge–Sb–Te thin films deposited in the same conditions. Figure 7a–c presents the spectral dependence of the refractive index, the extinction coefficient, and the reflectivity for the samples deposited by ns laser ablation. The optical band gap values were derived from the Tauc plots $(\alpha E)^{1/2} = f(E)$ (see Figure 7d). The absorption coefficient (α) was calculated using the well-known relationship with the imaginary part of the refraction index [49]:

$$\alpha = \frac{4\pi k}{\lambda} \tag{3}$$

The obtained optical band gap values are presented in the inset of Figure 7d.

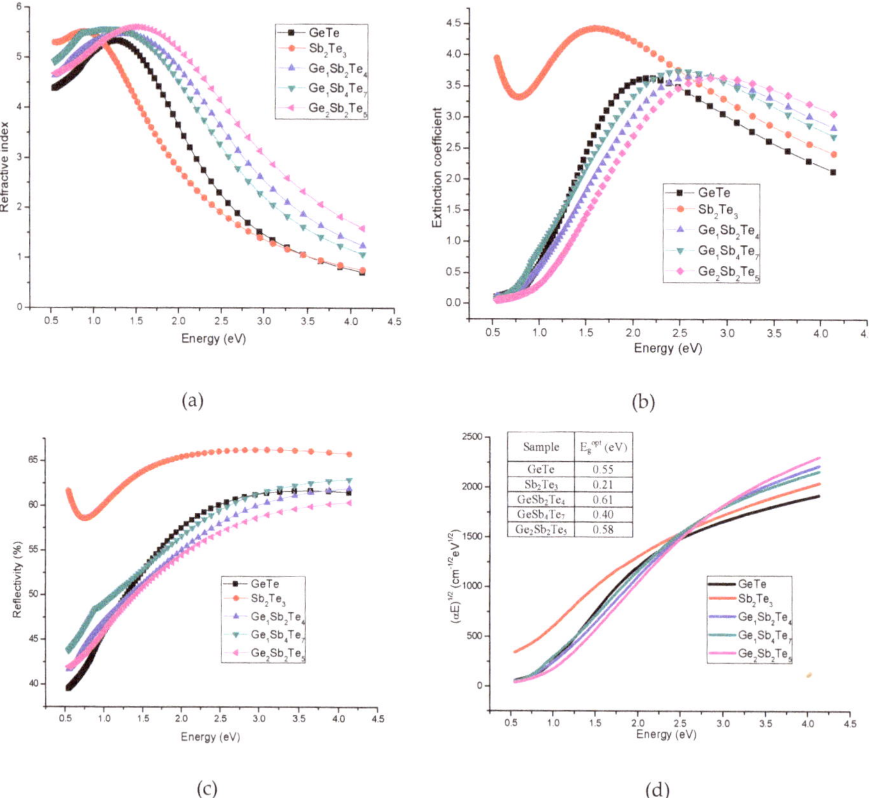

Figure 7. The ellipsometry results: the energy dependence of refractive index (**a**), extinction coefficient (**b**), reflectivity (**c**), and $(\alpha E)^{1/2}$ (**d**) for the Ge-Sb-Te based thin films deposited using the 266 nm radiation of the Nd-YAG laser.

Comparing these results with the ones reported by Nemec et al. [31,50], some similarities were observed: The refractive index presented an increase up to 1.2 eV and then a decrease for higher energies. The same type of dependence was observed for the extinction coefficient which reached its maximum value around 2.5 eV. However, the optical response of our samples presented features between those found for the amorphous and crystalline samples reported in Reference [31]. Comparing the plots of the refractive indexes, we observed higher maximum values than the amorphous chalcogenide thin films deposited by Nemec et al. [31] (and in the same time lower than the crystallized samples), but a more rapid decrease at large energy values was recorded, comparable with the optical features of the crystalline thin films. These results indicate a partial crystallization of our deposited samples. Moreover, we did not observe a clear dependence of the maximum values of the refractive index or extinction coefficient with the Sb_2Te_3 content. This can be explained by the different structural characteristics mentioned in the previous sections. Regarding the bandgap energies, our calculations (see inset Figure 7d) showed values comparable with the ones reported in Reference [31]. However, the poor microstructural quality of the deposited films led, in some cases, to slightly increased error values. Lee et al. [51] found that the optical bandgaps of their $Ge_2Sb_2Te_5$ samples (deposited by RF magnetron) with amorphous, fcc, and hexagonal structures were 0.7, 0.5, and 0.5 eV, respectively, our data being closer to the values observed for the crystalline phase. An unusual value of the optical band gap was observed for the Sb_2Te_3 sample. However, the XRD results of this thin film revealed the presence of two phases, one of them induced by the increased Sb content. Preliminary ellipsometry measurements were also performed on fs and ps deposited samples, but due to the lower thickness of these samples, only the transparent region (up to 1 eV) could be investigated for the moment. These measurements will be extended in the near future using thicker fs and ps samples. Due to the narrow range in which we obtained a signal, no calculations of the E_g^{opt} were allowed.

4. Conclusions

Ge-Sb-Te thin films were deposited using various experimental conditions and lasers with ns, ps, and fs pulse duration. Analyzing the thin films deposited by nanosecond laser ablation, we observed that the sample thickness was influenced by the laser wavelength, fluence, target-to-substrate distance, and deposition time. Most of the deposited samples in this temporal regime presented droplets on the surface which decreased as the laser fluence was diminished. The EDS results revealed Te atomic percentages lower than the nominal value for all five compositions considered, which was explained by its lower vaporization heat. An over-stoichiometric Sb concentration was observed for the Ge-Sb-Te-based samples. The ToF-SIMS images and depth profiles revealed a uniform distribution of the main elements and of their combinations on the surface and in the volume of the films. The Raman spectroscopy and XRD analysis confirmed the formation of an fcc structure together with an amorphous growth for the Ge-Sb-Te based samples. While the XRD pattern of the GeTe thin film revealed an amorphous deposition, the Sb_2Te_3 thin film presented additional diffraction lines indicating multiphase formation. Ellipsometric measurements done on the Ge-Sb-Te thin films revealed n and k values between the ones of amorphous and crystalline samples reported in Reference [31], confirming the structural analysis results. The thin films deposited with femtosecond- and picosecond-pulsed lasers presented an improved morphology with no large droplets on the surface. Also, in their case, the thickness was found to be influenced by the deposition time, laser fluence, and target–substrate distance. The Ge, Sb, and Te concentrations obtained by EDS were much closer to the nominal values than the ones of the ns-PLD, their variations being smaller than the measurement error bar. The structural analysis results revealed an amorphous deposition for the Ge-Sb-Te-based fs-PLD films. The lower fluence used in fs-PLD determined the ejection of particles with lower kinetic energies than the ones generated by nanosecond ablation. This can influence the energetic transfer at substrate surface and thus the crystallization process. Moreover, on a fundamental background, the different ejection mechanisms involved for the various laser pulse durations (see above) favor a droplet formation in the case of ns-pulses vs a nanoparticle formation for ultrashort pulses. Overall, our results confirm that (high-repetition-rate) femtosecond PLD is a useful technique to obtain uniform, amorphous, and stoichiometric thin films in a short deposition time.

Author Contributions: Conceptualization, C.F. and O.P.; investigation, C.F., S.G., O.P., P.N., V.N., N.C. and B.C.; writing—original draft preparation, O.P. and G.B.; writing—review and editing, G.B. and C.F.

Funding: This research was supported by the PHC projects Brancusi n° 29551XB and Barrande n° 28449NK; the Romanian Space Agency (ROSA) within the Space Technology and Advanced Research (STAR) Program, grant number 169/20.07.2017; the Czech Science Foundation (Project No. 18-03823S); and the Ministry of Education, Youth, and Sports of the Czech Republic (ProjectLM2015082).

Conflicts of Interest: The authors declare no conflict of interest.

References

1. Yamada, N.; Ohno, E.; Akahira, N.; Nishiuchi, K.; Nagata, K.; Takao, M. High Speed Overwritable Phase Change Optical Disk Material. *Jpn. J. Appl. Phys.* **1987**, *26*, 61–66. [CrossRef]
2. Wang, L.; Yang, C.; Wen, J.; Xiong, B. Amorphization Optimization of $Ge_2Sb_2Te_5$ Media for Electrical Probe Memory Applications. *Nanometer* **2018**, *8*, 368. [CrossRef] [PubMed]

3. Kolobov, A.V.; Fons, P.; Tominaga, J.; Frenkel, A.I.; Ankudinov, A.L.; Yannopoulos, S.N.; Andrikopoulos, K.S.; Uruga, T. Why Phase-Change Media Are Fast and Stable: A New Approach to an Old Problem. *Jpn. J. Appl. Phys.* **2005**, *44*, 3345–3349. [CrossRef]

4. Kolobov, A.V.; Fons, P.; Tominaga, J.; Uruga, T. Why DVDs work the way they do: The nanometer-scale mechanism of phase change in Ge–Sb–Te alloys. *J. Non-Cryst. Solids* **2006**, *352*, 1612–1615. [CrossRef]

5. Kolobov, A.V.; Tominaga, J. Metastability and Phase Change Phenomena. In *Chalcogenides*; Springer: Berlin/Heidelberg, Germany, 2012.

6. Nemec, P.; Nazabal, V.; Moreac, A.; Gutwirth, J.; Beneš, L.; Frumar, M. Amorphous and crystallized Ge–Sb–Te thin films deposited by pulsed laser: Local structure using Raman scattering spectroscopy. *Mater. Chem. Phys.* **2012**, *136*, 935–941. [CrossRef]

7. Raoux, S.; Ielmini, D.; Wuttig, M.; Karpov, I. Phase Change Materials. *MRS Bull.* **2012**, *37*, 118–123. [CrossRef]

8. Zhang, W.; Mazzarello, R.; Wuttig, M.; Ma, E. Designing crystallization in phase-change materials for universal memory and neuro-inspired computing. *Nat. Rev. Mater.* **2019**, *4*, 150–168. [CrossRef]

9. Olivier, M.; Němec, P.; Boudebs, G.; Boidin, R.; Focsa, C.; Nazabal, V. Photosensitivity of pulsed laser deposited Ge-Sb-Se thin films. *Opt. Mater. Express* **2015**, *5*, 781. [CrossRef]

10. Yamada, N. Origin, secret, and application of the ideal phase-change material GeSbTe. *Phys. Status Solidi* **2012**, *249*, 1837–1842. [CrossRef]

11. Wang, J.J.; Xu, Y.Z.; Mazzarello, R.; Wuttig, M.; Zhang, W. A review on disorder-driven metal-insulator transition in crystalline vacancy-rich GeSbTe phase-change materials. *Materials* **2017**, *10*, 862. [CrossRef]

12. Vlček, M.; Schroeter, S.; Čech, J.; Wágner, T.; Glaser, T. Selective etching of chalcogenides and its application for fabrication of diffractive optical elements. *J. Non-Cryst. Solids* **2003**, *326–327*, 515–518. [CrossRef]

13. Gutwirth, J.; Wagner, T.; Bezdicka, P.; Hrdlicka, M.; Vlcek, M.; Frumar, M. On angle resolved RF magnetron sputtering of GeSbTe thin films. *J. Non-Cryst. Solids* **2009**, *355*, 1935–1938. [CrossRef]

14. Nazabal, V.; Charpentier, F.; Adam, J.L.; Nemec, P.; Lhermite, H.; Brandily-Anne, M.L.; Charrier, J.; Guin, J.P.; Moréac, A. Sputtering and pulsed laser deposition for near- and mid-infrared applications: A comparative study of $Ge_{25}Sb_{10}S_{65}$ and $Ge_{25}Sb_{10}Se_{65}$ amorphous thin films. *Int. J. Appl. Ceram. Technol.* **2011**, *8*, 990–1000. [CrossRef]

15. Musgraves, J.D.; Carlie, N.; Hu, J.; Petit, L.; Agarwal, A.; Kimerling, L.C.; Richardson, K.A. Comparison of the optical, thermal and structural properties of Ge–Sb–S thin films deposited using thermal evaporation and pulsed laser deposition techniques. *Acta Mater.* **2011**, *59*, 5032–5039. [CrossRef]

16. Lee, J.; Choi, S.; Lee, C.; Kang, Y.; Kim, D. GeSbTe deposition for the PRAM application. *Appl. Surf. Sci.* **2007**, *253*, 3969–3976. [CrossRef]

17. Mussler, G.; Ratajczak, A.; von der Ahe, M.; Du, H.; Grützmacher, D. Metal organic vapor phase epitaxy of $Ge_1Sb_2Te_4$ thin films on Si(111) substrate. *Appl. Phys. A* **2019**, *125*, 1–7. [CrossRef]

18. Hilmi, I.; Rauschenbach, B.; Gerlach, J.W.; Thelander, E.; Schumacher, P.; Gerlach, J.W.; Rauschenbach, B. Epitaxial $Ge_2Sb_2Te_5$ films on Si(111) prepared by pulsed laser deposition. *Thin Solid Films* **2016**, *619*, 81–85. [CrossRef]

19. Thelander, E.; Gerlach, J.W.; Ross, U.; Lotnyk, A.; Rauschenbach, B. Low temperature epitaxy of Ge-Sb-Te films on BaF_2(111) by pulsed laser deposition. *Appl. Phys. Lett.* **2014**, *105*, 1–6. [CrossRef]

20. Song, J.H.; Susaki, T.; Hwang, H.Y. Enhanced Thermodynamic Stability of Epitaxial Oxide Thin Films. *Adv. Mater.* **2008**, *20*, 2528–2532. [CrossRef]

21. Boschker, J.E.; Folven, E.; Monsen, A.F.; Wahlström, E.; Grepstad, J.K.; Tybell, T. Consequences of high adatom energy during pulsed laser deposition of $La_{0.7}Sr_{0.3}MnO_3$. *Cryst. Growth Des.* **2012**, *12*, 562–566. [CrossRef]

22. Mihesan, C.; Gurlui, S.; Ziskind, M.; Chazallon, B.; Martinelli, G.; Zeghlache, H.; Guignard, M.; Nazabal, V.; Smektala, F.; Focsa, C. Photo-excited desorption of multi-component systems: Application to chalcogenide glasses. *Appl. Surf. Sci.* **2005**, *248*, 224–230. [CrossRef]

23. Irimiciuc, S.; Boidin, R.; Bulai, G.; Gurlui, S.; Nemec, P.; Nazabal, V.; Focsa, C. Laser ablation of $(GeSe_2)_{100-x}(Sb_2Se_3)_x$ chalcogenide glasses: Influence of the target composition on the plasma plume dynamics. *Appl. Surf. Sci.* **2016**, *418*, 594–600. [CrossRef]

24. Pompilian, O.G.; Gurlui, S.; Nemec, P.; Nazabal, V.; Ziskind, M.; Focsa, C. Plasma diagnostics in pulsed laser deposition of GaLaS chalcogenides. *Appl. Surf. Sci.* **2013**, *278*, 352–356. [CrossRef]

25. Ursu, C.; Pompilian, O.G.; Gurlui, S.; Nica, P.; Agop, M.; Dudeck, M.; Focsa, C. Al_2O_3 ceramics under high-fluence irradiation: plasma plume dynamics through space- and time-resolved optical emission spectroscopy. *Appl. Phys. A* **2010**, *101*, 153–159. [CrossRef]

26. Pompilian, O.G.; Dascalu, G.; Mihaila, I.; Gurlui, S.; Olivier, M.; Nemec, P.; Nazabal, V.; Cimpoesu, N.; Focsa, C. Pulsed laser deposition of rare-earth-doped gallium lanthanum sulphide chalcogenide glass thin films. *Appl. Phys. A* **2014**, *117*, 197–205. [CrossRef]

27. Dascalu, G.; Pompilian, G.; Chazallon, B.; Caltun, O.F.; Gurlui, S.; Focsa, C. Femtosecond pulsed laser deposition of cobalt ferrite thin films. *Appl. Surf. Sci.* **2013**, *278*, 38–42. [CrossRef]

28. Focsa, C.; Nemec, P.; Ziskind, M.; Ursu, C.; Gurlui, S.; Nazabal, V. Laser ablation of AsxSe100−x chalcogenide glasses: Plume investigations. *Appl. Surf. Sci.* **2009**, *255*, 5307–5311. [CrossRef]

29. Dascalu, G.; Pompilian, G.; Chazallon, B.; Nica, V.; Caltun, O.F.; Gurlui, S.; Focsa, C. Rare earth doped cobalt ferrite thin films deposited by PLD. *Appl. Phys. A* **2012**, *110*, 915–922. [CrossRef]

30. Chrisey, D.B.; Hubler, G.K. *Pulsed Laser Deposition of Thin Films*; John Wiley & Sons, Inc.: New York, NY, USA, 1994.

31. Nemec, P.; Přikryl, J.; Nazabal, V.; Frumar, M. Optical characteristics of pulsed laser deposited Ge–Sb–Te thin films studied by spectroscopic ellipsometry. *J. Appl. Phys.* **2011**, *109*, 073520. [CrossRef]

32. Bouška, M.; Pechev, S.; Simon, Q.; Boidin, R.; Nazabal, V.; Gutwirth, J.; Baudet, E.; Němec, P. Pulsed laser deposited GeTe-rich GeTe-Sb_2Te_3 thin films. *Sci. Rep.* **2016**, *6*, 26552. [CrossRef]

33. Krusin-Elbaum, L.; Cabral, C.; Chen, K.N.; Copel, M.; Abraham, D.W.; Reuter, K.B.; Rossnagel, S.M.; Bruley, J.; Deline, V.R. Evidence for segregation of Te in $Ge_2Sb_2Te_5$ films: Effect on the "phase-change" stress. *Appl. Phys. Lett.* **2007**, *90*, 141902. [CrossRef]

34. Prokhorov, E.; Gonzalez-Hernandez, J.; Hernandez-Landaverde, M.A.; Chao, B.; Morales-Sanchez, E. Crystallization mechanism in Sb:Te thin film.pdf. *J. Phys. Chem. Solids* **2007**, *68*, 883–886. [CrossRef]

35. Andrikopoulos, K.S.; Yannopoulos, S.N.; Voyiatzis, G.A.; Kolobov, A.V.; Ribes, M.; Tominaga, J. Raman scattering study of the a-GeTe structure and possible mechanism for the amorphous to crystal transition. *J. Phys. Condens. Matter* **2006**, *18*, 965–979. [CrossRef]

36. Sosso, G.C.; Caravati, S.; Bernasconi, M. Vibrational properties of crystalline Sb_2Te_3 from first principles. *J. Phys. Condens. Matter* **2009**, *21*, 095410. [CrossRef] [PubMed]

37. Xu, Z.; Chen, C.; Wang, Z.; Wu, K.; Chong, H.; Ye, H. Optical constants acquisition and phase change properties of $Ge_2Sb_2Te_5$ thin films based on spectroscopy. *RSC Adv.* **2018**, *8*, 21040–21046. [CrossRef]

38. Murray, M.; Jose, G.; Richards, B.; Jha, A. Femtosecond pulsed laser deposition of silicon thin films. *Nanoscale Res. Lett.* **2013**, *8*, 1–6. [CrossRef]

39. Katsuno, T.; Godet, C.; Orlianges, J.C.; Loir, A.S.; Garrelie, F.; Catherinot, A. Optical properties of high-density amorphous carbon films grown by nanosecond and femtosecond pulsed laser ablation. *Appl. Phys. A* **2005**, *81*, 471–476. [CrossRef]

40. Anoop, K.K.; Harilal, S.S.; Philip, R.; Bruzzese, R.; Amoruso, S. Laser fluence dependence on emission dynamics of ultrafast laser induced copper plasma. *J. Appl. Phys.* **2016**, *120*, 185901. [CrossRef]

41. Anoop, K.K.; Ni, X.; Wang, X.; Amoruso, S.; Bruzzese, R. Fast ion generation in femtosecond laser ablation of a metallic target at moderate laser intensity. *Laser Phys.* **2014**, *24*, 105902. [CrossRef]

42. Irimiciuc, S.A.; Gurlui, S.; Bulai, G.; Nica, P.; Agop, M.; Focsa, C. Langmuir probe investigation of transient plasmas generated by femtosecond laser ablation of several metals: Influence of the target physical properties on the plume dynamics. *Appl. Surf. Sci.* **2017**, *417*, 108–118. [CrossRef]

43. Nica, P.; Gurlui, S.; Osiac, M.; Agop, M.; Ziskind, M.; Focsa, C. Investigation of femtosecond laser-produced plasma from various metallic targets using the Langmuir probe characteristic. *Phys. Plasmas* **2017**, *24*, 103119. [CrossRef]

44. Vinod, E.M.; Singh, A.K.; Ganesan, R.; Sangunni, K.S. Effect of selenium addition on the GeTe phase change memory alloys. *J. Alloys Compd.* **2012**, *537*, 127–132. [CrossRef]

45. Vinod, E.M.; Naik, R.; Ganesan, R.; Sangunni, K.S. Signatures of $Ge_2Sb_2Te_5$ film at structural transitions. *J. Non-Cryst. Solids* **2012**, *358*, 2927–2930. [CrossRef]

46. Van Eijk, J.M. Structural Analysis of Phase-Change Materials Using X-ray Absorption Measurements. Ph.D. Thesis, RWTH Aachen University, Aachen, Germany, 17 December 2010.

47. Wei, S.; Wu, S.; Pei, F.; Li, J.; Wang, S.; Chen, L. Theoretical and Experimental Investigations of the Optical Properties of $Ge_2Sb_2Te_5$ for Multi-State Optical Data Storage. *J. Korean Phys. Soc.* **2008**, *53*, 2265–2269.

48. Hilton, A.R. Chalcogenide Glasses for Infrared. Optical Materials. *Appl. Opt.* **1966**, *5*, 1877–1882. [CrossRef] [PubMed]

49. Nechache, R.; Harnagea, C.; Li, S.; Cardenas, L.; Huang, W.; Chakrabartty, J.; Rosei, F. Bandgap tuning of multiferroic oxide solar cells. *Nat. Photonics* **2014**, *61*, 61–67. [CrossRef]

50. Nemec, P.; Moreac, A.; Nazabal, V.; Pavlišta, M.; Přikryl, J.; Frumar, M. Ge-Sb-Te thin films deposited by pulsed laser An ellipsometry and Raman scattering spectroscopy study. *J. Appl. Phys.* **2009**, *106*, 103509. [CrossRef]

51. Lee, B.-S.; Abelson, J.R.; Bishop, S.G.; Kang, D.-H.; Cheong, B.; Kim, K.-B. Investigation of the optical and electronic properties of $Ge_2Sb_2Te_5$ phase change material in its amorphous, cubic, and hexagonal phases. *J. Appl. Phys.* **2005**, *97*, 093509. [CrossRef]

Communication

The Effects of ZnTe:Cu Back Contact on the Performance of CdTe Nanocrystal Solar Cells with Inverted Structure

Bingchang Chen [1], Junhong Liu [1], Zexin Cai [1], Ao Xu [1], Xiaolin Liu [1], Zhitao Rong [1], Donghuan Qin [1,*], Wei Xu [1,*], Lintao Hou [2,*] and Quanbin Liang [1,3]

[1] School of Materials Science and Engineering, South China University of Technology, Guangzhou 510640, China; ChenBC_999@163.com (B.C.); jhliu1@163.com (J.L.); 13005112535@163.com (Z.C.); m13856403915@163.com (A.X.); liulin9708@163.com (X.L.); rzt1512388848@163.com (Z.R.); l.quanbin@mail.scut.edu.cn (Q.L.)

[2] Guangdong Provincial Key Laboratory of Optical Fiber Sensing and Communications, Guangzhou Key Laboratory of Vacuum Coating Technologies and New Energy Materials, Siyuan Laboratory, Department of Physics, Jinan University, Guangzhou 510632, China

[3] Institute of Polymer Optoelectronic Materials & Devices, State Key Laboratory of Luminescent Materials & Devices, South China University of Technology, Guangzhou 510640, China

* Correspondence: qindh@scut.edu.cn (D.Q.); xuwei@scut.edu.cn (W.X.); thlt@jnu.edu.cn (L.H.)

Received: 1 March 2019; Accepted: 12 April 2019; Published: 17 April 2019

Abstract: CdTe nanocrystal (NC) solar cells have received much attention in recent years due to their low cost and environmentally friendly fabrication process. Nowadays, the back contact is still the key issue for further improving device performance. It is well known that, in the case of CdTe thin-film solar cells prepared with the close-spaced sublimation (CSS) method, Cu-doped CdTe can drastically decrease the series resistance of CdTe solar cells and result in high device performance. However, there are still few reports on solution-processed CdTe NC solar cells with Cu-doped back contact. In this work, ZnTe:Cu or Cu:Au back contact layer (buffer layer) was deposited on the CdTe NC thin film by thermal evaporation and devices with inverted structure of ITO/ZnO/CdSe/CdTe/ZnTe:Cu (or Cu)/Au were fabricated and investigated. It was found that, comparing to an Au or Cu:Au device, the incorporation of ZnTe:Cu as a back contact layer can improve the open circuit voltage (V_{oc}) and fill factor (FF) due to an optimized band alignment, which results in enhanced power conversion efficiency (PCE). By carefully optimizing the treatment of the ZnTe:Cu film (altering the film thickness and annealing temperature), an excellent PCE of 6.38% was obtained, which showed a 21.06% improvement compared with a device without ZnTe:Cu layer (with a device structure of ITO/ZnO/CdSe/CdTe/Au).

Keywords: nanocrystal; CdTe; Cu-doped; ZnTe; solar cells; solution processed

---

1. Introduction

CdTe is a II–VI group of semiconductor materials with a moderate band gap of 1.45 eV and a high optical-adsorption coefficient (over 10^4 cm^{-1} in the optical range), which is promising for light harvesting, and regarded as an attractive material for solar-cell applications [1–3]. One of the major problems for CdTe solar cells is the difficulty in obtaining low and stable ohmic contact to CdTe (the low resistance of CdTe/back contact electrode). It is noted that CdTe has a very high electron affinity (x = 4.5 eV) and low carrier concentration (~10^{14}/cm^3). There are almost no metals with such a high work function to form ohmic contact to CdTe [4]. In order to obtain low-resistance contacts to CdTe, there are several ways to establish an interface that provides suitable electrical properties. The first uses interlayer contact materials with a high work function, such as MoOx [5,6], WOx [7], CuSCN [8], and cobalt phthalocyanine [9] before electrode deposition. Formation of a heavily doped region at the surface of a CdTe thin film is also an important way to reduce the barriers for hole-collecting. Cu is the most commonly used metal to create acceptor states with concentrations as high as $\sim 10^{19}/cm^3$ in CdTe [10–12]. An investigation shows that Cu exists as $Cu_i{}^+$ and forms shallow donor states or deep acceptor states $Cu_{Cd}{}^-$ in CdTe [13,14]. It was found that copper atoms can move quickly in a cadmium telluride thin film after heat treatment [15]. Excess-doped Cu leads to Cu gathering at the interface of the p-n junction and forming an N-type compensation caused by Cu^+, which leads to a decline in device performance [16,17]. Therefore, controlling the effective doping of copper is the key to obtaining high efficiency and stable thin-film solar cells. In order to increase stability, the chemical etching of the cadmium telluride surface is essential prior to Cu deposition in order to create a Te-rich surface [18,19]. The used etching solution includes a nitric acid solution and a bromo methanol solution. The tellurium-rich layer formed on the surface can combine with the deposited copper to form Cu_xTe to prevent the excessive diffusion of copper. Another way to control the Cu doped in CdTe is by using ZnTe:Cu as a back contact layer. As the ZnTe has a similar lattice parameter and valence band as that of CdTe, it is an ideal back-contact material for CdTe solar cells as the well-matched interface, which works as an electron blocking layer [20,21]. On the other hand, high-acceptor-level doping of ZnTe ($>10^{20}/cm^3$) can be easily obtained by varying the Cu-doped content, which allows for the formation of an ideal tunnel junction to the contact-metal layer [22].

Although back-contact techniques have been well developed for CdTe-based thin-film solar cells fabricated by vacuum technology, there are still few reports on solution-processed CdTe nanocrystal-(NC) based solar cells [23–25]. There are many merits of solution processed CdTe NC solar cells, for example, low materials consumed, low cost and simple fabricating techniques. As much of the grain boundary existed in the CdTe NC layer (CdTe grain size is about 100–200 nm), common corrosion technology cannot obtain a Te-rich surface, as corrosion solvents quickly diffuse along the grain boundary and make device decay or shunting, which has been confirmed in previous work [26]. Inserting a hole-transport layer with high work function before the deposition of an electrode metal contact has been widely accepted for improving the ohmic contact for CdTe solar cells [9,27]. Recently, inspired by the use of MoO_x in organic solar cells, MoO_x was developed as hole transfer layer for CdTe NC solar cells, and improved efficiency is found in this case [28]. Following this, organic hole-transport materials with a high work function were also developed as the back contact layer, with the advantage of low-cost solution processing. For example, Spiro-OMeTAD (2,2,7,7-tetrakis(N,N-di-4-methoxyphenylamino)-9,9-spirobitluorene) [29] and P3KT [30] are used as a back contact layer for reducing interface recombination and improving band alignment, so drastic improvement in device performance is obtained. Most recently, a new cross-linkable conjugated interface polymer TPA was successfully applied in the CdTe NC-based solar cells, and power conversion efficiency (PCE) as high as 8.34% was obtained, which is the highest value ever reported for any solution processed CdTe NCs solar cells with an inverted structure [31]. In the previous report, ZnTe:Cu contact was first applied in CdTe NC solar cells with the configuration of FTO/SnO$_2$/CdS/CdTe/ZnTe:Cu/Ti [32]. Although good ohmic contact was obtained in this case, the short circuit current density J_{sc} was low due to low junction quality. In previous reports [33–36], it was found that high junction quality is expected when a commonly used n-type partner (such as CBD-CdS [37] or TiO$_2$ [38,39] prepared by precursor decomposition etc.) is replaced by solution-processed CdSe or CdS NC due to the similar size and structure as CdTe NCs, which significantly reduced interface recombination and resulted in high device performance. Herein, we fabricated CdTe NC solar cells via solution-processed CdTe and CdSe NC as donor/acceptor materials with the configuration of ITO/ZnO/CdSe/CdTe/back contact/Au. Cu-doped ZnTe and Cu:Au were selected as back contact, while a device without any back contact was also fabricated and investigated. A PCE of 6.38% was observed from the device with ZnTe:Cu as back contact layer.

2. Experiments

CdTe, CdSe NCs, and the Zn^{2+} precursor were prepared according to the previous reported method [33]. The CdSe and CdTe NC films were deposited using layer-by-layer spin-coating and sintering under ambient conditions. A typical fabrication process is presented in the following. The ZnO thin film was prepared by spin-casting Zn^{2+} precursor on the ITO substrate and annealing at 400 °C for 10 min. Several drops of CdSe NC solution (a mixture of pyridine and 1-propanol with a volume ratio of 1:1 at 30 mg/mL) were put on top of the ITO/ZnO substrate and spin-casted at 3000 rpm for 20 s. Then the substrate was placed on a hotplate at 150 °C for 10 min and transferred to another hotplate at 350 °C for 40 s. This process was repeated two times and the CdSe NC film thickness was about 80 nm (40 nm per layer). Following this, several drops of CdTe NC solution (with 45 mg/mL in pyridine and 1-propanol with a volume ratio of 1:1) were deposited on the ITO/ZnO/CdSe thin film and spin-casted at 1100 rpm for 20 s. The ITO/ZnO/CdSe/CdTe samples were then placed at a hot place at 150 °C for 3 min. Then the substrate was dipped into saturated $CdCl_2/CH_3OH$ solution for 10 s and rinsed in 1-propanol. Then the substrate was placed immediately on a hotplate at 350 °C for 40 s (~100 nm each layer). This process was repeated for five times in order to obtain an optimized active layer thickness. The thickness of ZnO, CdSe and CdTe were ~40 nm, ~80 nm and ~500 nm respectively. Detail process can be found in the literature [31]. After washing and cleaning, ZnTe films were deposited onto the ITO/ZnO/CdSe/CdTe substrate via a thermal evaporation process at a rate of 8 Å/s through a shadow mask with an active area of 0.16 cm^2 under a vacuum pressure of 4×10^{-4} Pa. Following this, Cu and Au (60 nm) were deposited on the ZnTe through a shadow mask by thermal evaporation. The ITO/ZnO/CdSe/CdTe/ZnTe:Cu/Au and ITO/ZnO/CdSe/CdTe/Cu/Au thin films were then placed on a hot place and annealed at different temperatures. The area of all the NC solar cells is 0.16 cm^2. The atomic force microscopy (AFM) imagines were obtained using a NanoScope NS3A system (Veeko, CA, USA). The morphology and structure were further characterized by scanning electron microscopy (SEM, Nova NanoSEM430, Thermo Fisher Scientific, Eindhoven, The Netherlands) and X-ray diffraction (XRD, X'pert Pro M, Philips, Amsterdam, The Netherlands). The external quantum efficiency (EQE) was measured by a Zolix instrument (Solar Cell Scan100, Zolix Instruments Co., Ltd., Beijing, China). The *J–V* characteristics were measured with a Keithley 2400 under an illumination of 100 mW/cm^2 with an air mass 1.5 (AM 1.5) solar simulator (Oriel model 91192). The transient photovoltage measurements (TPV) were taken out by using the OmniFluo system (Zolix, Beijing, China).

3. Results and Discussion

To investigate the effect of ZnTe on the surface morphology of the CdTe NC layer, the morphology of ITO/ZnO/CdSe/CdTe/ZnTe with different ZnTe thickness was characterized by atomic force microscopy (AFM). The ZnTe film was deposited on the CdTe NC layer by thermal evaporation through a shadow mask and annealed at 200 °C for 30 min. As shown in Figure 1a, without ZnTe, the surface of the CdTe NC showed a uniform and compact structure, with a root-mean-square (RMS) roughness of 8.26 nm. It was clear that, with the ZnTe thickness increase from 10 nm to 50 nm, grain size increased linearly. The RMS values for ZnTe with thickness of 10 nm, 20 nm, and 50 nm were 8.93 nm, 7.62 nm, and 10.8 nm, respectively. There were many small particles for the 10 nm ZnTe sample (Figure 1b), while a compact and uniform surface was found for the 20 nm and 50 nm ZnTe samples. The smooth and compact surface obtained in the ZnTe sample implied good physical contact between CdTe and ZnTe, which was essential to decrease the interfacial contact resistance and improve the fill factor of the solar-cell device.

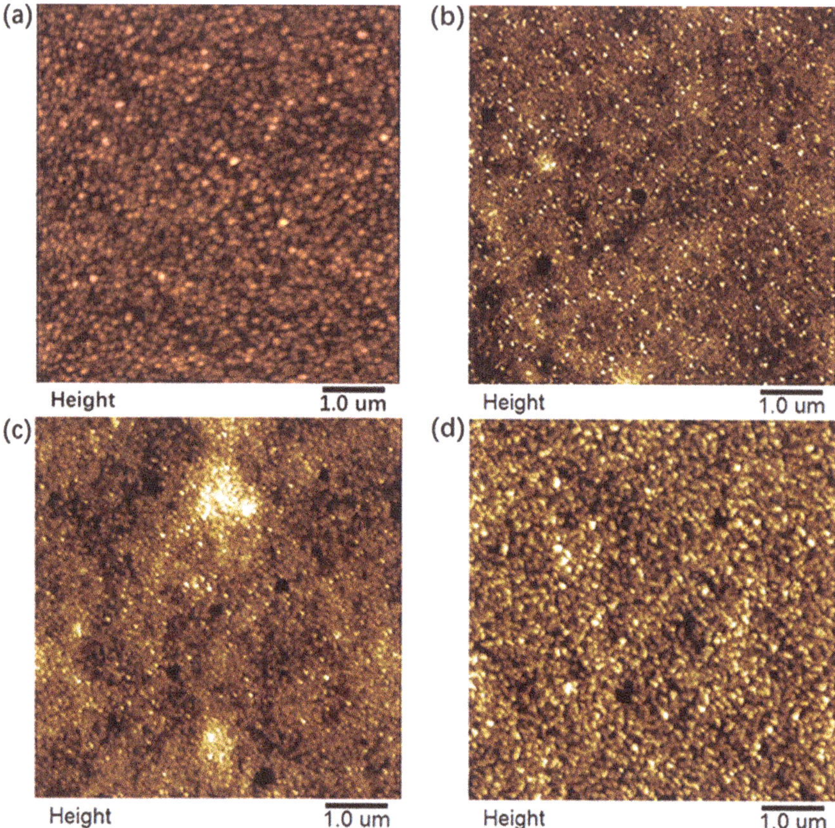

Figure 1. Atomic force microscopy (AFM) images of ITO/ZnO/CdSe/CdTe/ZnTe with different ZnTe thickness. (**a**) Without ZnTe; (**b**) 10 nm ZnTe; (**c**) 20 nm ZnTe; (**d**) 50 nm ZnTe.

The structure and composition of ZnTe were further characterized by X-ray diffraction (XRD) and energy dispersive spectrometry (EDS). The XRD sample was prepared by depositing 100 nm on the Si substrate and then annealing at different temperatures for 30 min. As shown in Figure 2a, diffraction patterns with peaks at about 25.2°, 29.3°, 41.87°, 49.6°, 51.8°, 60.7°, 66.8°,68.9°, 76.6°, and 81.9° were identified from the XRD pattern, corresponding to the (111), (200), (220), (311), (222), (400), (331), (420), (422), and (511) planes, respectively, of the ZnTe zinc blend structure. It is noted that pure ZnTe peaks were found for samples without annealing. On the contrary, there were many peaks corresponding to the pure Te element that emerged when the samples were annealed at a temperature of up to 300 °C, which may be due to the decomposition of ZnTe at high temperatures. From the energy dispersive spectrum (Figure 2b), the relative amount of Zn to Te in the sample was close to 1:1, illuminating the formation of a ZnTe alloy. The presence of Si and O is attributed to the SiO_2 substrate used for the deposition of the ZnTe film.

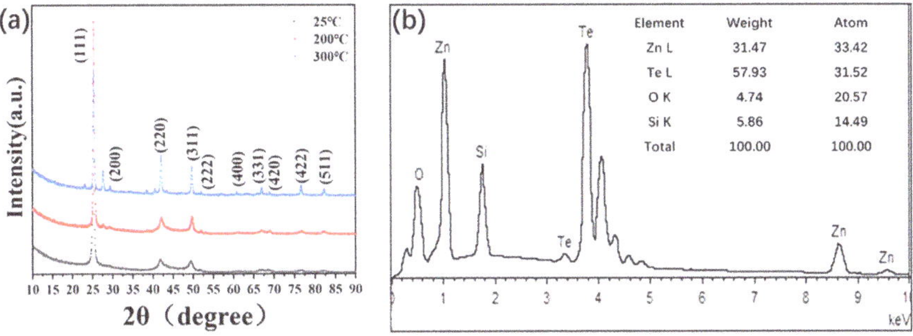

Figure 2. (**a**) X-ray diffraction (XRD) pattern of the ZnTe films at different annealing temperatures; (**b**) Energy dispersive spectrometry (EDS) of the as-prepared ZnTe sample.

The typical device architecture of CdTe NC solar cells with a ZnTe:Cu back contact layer is presented in Figure 3a. The ZnO thin film prepared by decomposition of the Zn carboxyl precursor was selected as the electron-transfer layer and prevented from casting shunt during the deposition of the active CdTe/CdSe layer. The active CdTe/CdSe layer was prepared with a layer-by-layer process. Prior to ZnTe deposition, the CdTe NC thin film was cleaned by ultrasound in methanol several times. Figure 3b shows the band alignment of ITO, ZnO, CdSe, CdTe, ZnTe:Cu, and Au. The introduction of ZnTe between CdTe and the anode optimizes energy-level alignment and decreases contact resistance, which improves charge collection and reduces interfacial carrier recombination. The surface and cross-section images of ITO/ZnO/CdSe/CdTe/ZnTe:Cu/Au were investigated using scanning electron microscopy. From Figure 3c, it is evident that CdTe grain size after chemical treatment/annealing was larger than 100 nm, and the whole film was compact and smooth, which is preferable for efficient carrier collecting due to reduced grain boundaries and interface defects. The cross-sectional SEM image of CdTe NC solar cells was included in Figure 3d. Active-layer (CdTe/CdSe) thickness was ~600 nm, prepared by depositing two layers of CdSe and five layers of CdTe NC onto ITO/ZnO in sequence.

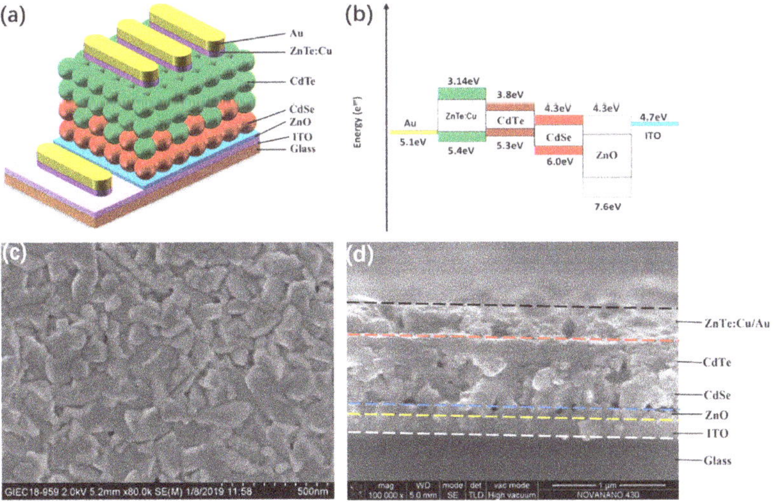

Figure 3. (**a**) Schematic of nanocrystal (NC) solar cells with a structure of ITO/ZnO/CdSe/CdTe/ZnTe:Cu/Au; (**b**) band alignment of ITO, ZnO, CdSe, CdTe, ZnTe:Cu, and Au; (**c**) SEM images of CdTe NC thin film; (**d**) cross-section SEM images of ITO/ZnO/CdSe/CdTe/ZnTe:Cu/Au.

It is well known that the most efficient CdTe thin-film solar cells are based on Cu: Au back contact [40,41]. CdTe NC solar cells with a configuration of ITO/ZnO/CdSe/CdTe/Cu/Au were fabricated by using different Cu thicknesses. In Figure S1a, we show the current density-voltage (J–V) curves obtained for CdTe NC solar cells with 1.3 nm and 2.1 nm Cu as back contact. The optimized CdTe NC solar cells with 2.1 nm Cu yielded a V_{oc} of 0.62 V, a J_{sc} of 13.09 mA/cm^2, an FF of 55.84%, leading to a PCE of 4.53%, while these values were 0.53 V, 8.91 mA/cm^2, 49.12%, 2.32%, for the 1.3 nm Cu device. The effects of annealing temperature on device performance are shown in Figure S1b and the solar cells parameters are summarized in Table S1. PCE increased as annealing temperature increased from room temperature to 200 °C, then dropped down linearly when we further increased the annealing temperature. Therefore, the PCE obtained for solar cells with a Cu/Au back contact was significantly lower than the device without Cu contact, reported before [31], which was mainly attributed to the low obtained J_{sc} in this case. We speculate that this was because the grain size of the CdTe NC active layer was only 100 nm, while this value was up to 1 μm for CdTe thin-film solar cells prepared by the CSS method. Therefore, Cu can quickly diffuse in the entire active layer after annealing due to the large grain boundary, which may result in mid-gap recombination and lead to a low-output current. By using the ZnTe buffer layer, Cu diffusion is restricted, and high device performance is expected. In the case of CdTe NC thin-film solar cells, the thickness of the ZnTe and annealing temperature should have significant effects on Cu diffusion and the device's contact resistance. ZnTe thickness for efficient CdTe thin-film solar cells prepared by the CSS method was about 100 nm. However, the thickness of CdTe NC solar cells was significantly lower than those CdTe solar cells prepared by the CSS method. Furthermore, the surface roughness of the CdTe NC thin film was small. Therefore, the thickness of ZnTe for efficient CdTe NC solar cells may be different from those reported before. To investigate the effect of thickness on device performance, devices with a ZnTe thickness between 10 nm and 100 nm were fabricated by thermally evaporating ZnTe on the ITO/ZnO/CdSe/CdTe substrate. For comparison, a controlled device with Au back contact was also fabricated. Figure 4a presents the J–V curves of the best CdTe NC solar cells with different back contact under light conditions (The J–V curves of CdTe NC with different thickness of ZnTe:Cu and annealing temperature are presented Figure S2a,b), while Table 1 summarizes the solar cells parameters. From the J–V curves, we can see that the controlled devices (ITO/ZnO/CdSe/CdTe/Au) showed a PCE of 5.27%, J_{sc} of 19.97 mA/cm^2, V_{oc} of 0.56 V, and FF of 47.15%. The low FF value implied that large contact resistance existed on the interface of CdTe/Au. In contract, the best device, with a 20 nm ZnTe/1 nm Cu/Au contact, yielded a J_{sc} of 19.73 mA/cm^2, V_{oc} of 0.65 V, FF of 49.75%, delivering a high PCE of 6.38%. The observed PCEs from NC solar cells with a ZnTe:Cu/Au back contact were more than 20% higher than those of the controlled device. The improvement in device performance was mainly attributed to the increased in FF and V_{oc}. From the J–V curves under dark, the current at the reversed bias from a device with ZnTe:Cu/Au back contact is lower than that from a device without ZnTe:Cu back contact. The low current in reverse bias voltage under dark and improved R_{sh} imply that the ZnTe:Cu can serve as an electron blocking layer to effectively prevent the leakage currents [42]. As shown in Table 1, when the thickness of ZnTe increases from 10 nm to 100nm, the V_{oc} of NC devices remain at ~0.6 V, while the current density increases from 14.36 mA/cm^2 (10 nm ZnTe) to 19.73 mA/cm^2 (20 nm ZnTe), then drops down to 14.40 mA/cm^2 (100 nm ZnTe). The changes in the fill factor have a similar behavior as that of J_{sc} or V_{oc}. On the other hand, when the thickness of ZnTe:Cu is fixed, the annealing temperature has significant effects on the NC solar cells performance. The best device is obtained at an annealing temperature of 200 °C. We speculate that the interface defects of CdTe/ZnTe:Cu and the diffusion of Cu dominate the device performance. At an annealing temperature of 200 °C and thickness of 20 nm, the interface defects is low and the diffusion of Cu in the ZnTe thin film is homogeneous, which will result in high device performance. On the contrary, too high of an annealing temperature may result in Cu accumulation or inadequate diffusion of Cu in the ZnTe, leading to low device performance.

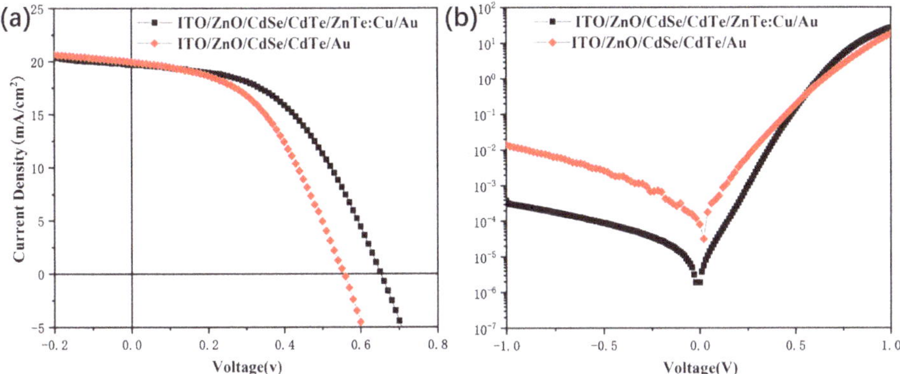

Figure 4. *J–V* curves of ITO/ZnO/CdSe/CdTe/ZnTe:Cu/Au and ITO/ZnO/CdSe/CdTe/Au (**a**) under light and (**b**) under dark conditions.

Table 1. Summary of the photovoltaic parameters of the nanocrystal (NC) solar cells prepared under different conditions.

Annealing Temperature (°C)	ZnTe Layer Thickness (nm)	Cu Layer Thickness (nm)	V_{oc} (V)	J_{sc} (mA/cm^2)	FF (%)	PCE (%)	R_s ($\Omega \cdot$cm^2)	R_{sh} ($\Omega \cdot$cm^2)
200	10	1	0.57	14.36	45.61	3.73	20.65	387.55
200	30	1	0.63	16.21	50.12	5.12	12.70	425.71
200	50	1	0.62	14.94	48.55	4.5	15.63	380.11
200	100	1	0.59	14.40	41.37	3.51	20.11	250.08
no	20	1	0.49	11.64	32.94	1.88	30.94	111.04
100	20	1	0.59	13.40	40.37	3.19	23.70	213.32
160	20	1	0.60	18.08	44.66	4.84	12.70	245.71
180	20	1	0.64	18.63	47.23	5.63	11.84	346.00
200	20	1	0.65	19.73	49.75	6.38	11.24	349.59
220	20	1	0.61	18.03	48.73	5.36	13.80	201.84
240	20	1	0.60	15.97	45.60	4.37	15.95	381.51
260	20	1	0.61	14.29	41.64	3.63	29.22	261.08
no	0	0	0.56	19.97	47.15	5.27	11.11	209.99

To investigate the performance improvement in ZnTe:Cu contact NC solar cells, external quantum efficiency (EQE) measurements for the device without a ZnTe:Cu buffer layer were taken and are presented in Figure 5a. It was found that the ZnTe:Cu contact device showed a higher EQE value between 450–650 nm, while it had a lower EQE value between 650–800 nm, when compared to a controlled device. When integrated, the J_{sc} was calculated to be 18.41 mA/cm^2 and 18.70 mA/cm^2 for device without a ZnTe:Cu contact, which agrees well with the data from the *J–V* curves. As the electron affinity for CdTe and high resistance, there are no metals that can form ohmic contact to CdTe, and Fermi-level pinning was found for all metals [12]. The low EQE response, between 600 nm to 800 nm was attributed to the interfacial recombination of CdTe/Au. When the ZnTe:Cu back contact layer was introduced, the diffusion of Cu in the interface of CdTe and the created acceptor states was at a high concentration [14] and it aligned the Fermi level to the valence band (VB) of CdTe. Then the carriers recombination will be decreased. On the other hand, the dropdown in the EQE value for the ZnTe:Cu device between 650–800 nm may have come from the accumulate of Cu in the CdTe/CdSe junction, which is also found in the CdTe/CdS solar cells prepared by CSS method [43]. It should be pointed out that the J_{sc} of the best cell with ZnTe:Cu back contact is still less than that of the control cell in this study. Furthermore, it is noted that the EQE response of devices with a ZnTe:Cu contact was significantly lower than those devices with Si-TPA as back contact, as reported before. We anticipated that, as the diffusion of Cu in ZnTe and CdTe was not homogeneous, few Cu$_x$Te could be formed as no

Te rich interface. The highly doped p^+ region via wet etching of the CdTe NC surface using dilute bromine/methanol was exclusively faint, as the acid solvent can quickly be transferred from the grain boundary, which was also confirmed by Matthew et al. Therefore, it was difficult to form a $Cu_{2-x}Te$ layer at the CdTe interface, and contact resistance was still high, further restricting improvement in the PCE. To further investigate the effects of the ZnTe:Cu buffer layer on the recombination process of NC solar cells, transient photovoltage (TPV) was used to measure the charge recombination in NC solar cells with/without a ZnTe:Cu buffer layer. During the TPV measurement, we obtained a steady-state equilibrium by placing NC solar cells under a white-light bias. By applying another weak laser pulse to NC solar cells, additional charges were generated. The charge recombination of NC solar cells was investigated by tracking the transient voltage associated with perturbations in the charge population. From Figure 5b, we can see that charge-recombination time for NC devices without a ZnTe:Cu buffer layer was 1.00 µs, while this value was 1.67 µs for the device with a ZnTe:Cu buffer layer, which implies the charge-recombination rate was lower in the ZnTe:Cu device compared to devices without a ZnTe:Cu.

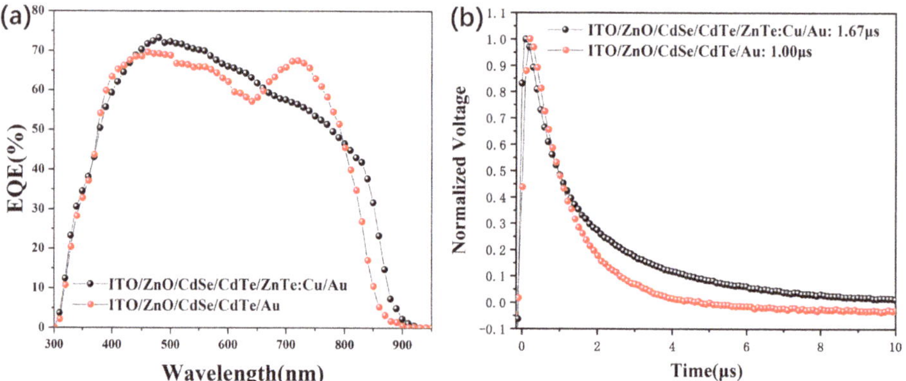

Figure 5. (**a**) External quantum efficiency (EQE) spectrum of a device without a ZnTe:Cu back contact layer; (**b**) transient photovoltage measurements of NC solar cells without a ZnTe:Cu back contact layer.

4. Conclusions

In conclusion, we described a developed ZnTe:Cu back contact layer for CdTe NC solar cells. Comparing it to a device without any back contact layer, we saw that improved solar-cell performance was attained by using ZnTe:Cu as a back contact layer. The improvement in V_{oc} and FF could be attributed to a better band alignment and low contact resistance forming on the CdTe surface, which decreased charge recombination in the interface and improved carrier collecting efficiency. A PCE of 6.38% was observed from the NC solar cells with a device structure of ITO/ZnO/CdSe/CdTe/ZnTe:Cu/Au, which was significantly higher than the controlled device with a structure of ITO/ZnO/CdSe/CdTe/Au (5.27%). Our results suggest that the back contact layer recipe involving ZnTe:Cu is applicable to solution-processed efficient CdTe NC solar cells if the ZnTe:Cu film is subjected to more optimized processing.

Supplementary Materials: The following are available online at http://www.mdpi.com/2079-4991/9/4/626/s1, Figure S1: (a) *J–V* characteristic of NC solar cells with different thickness of Cu film (all devices annealing at 200 °C); (b) *J–V* characteristic of NC solar cells with different annealing temperature (all devices with 2.1 nm Cu film), Figure S2: (a) *J–V* characteristic of NC solar cells of ITO/ZnO/CdSe/CdTe/ZnTe/Cu (1 nm)/Au structure with different thickness of ZnTe film (all devices annealing at 200 °C); (b) *J–V* characteristic of NC solar cells with different annealing temperature (all devices with 20 nm ZnTe film and 1 nm Cu film), Table S1: Summary of the photovoltaic parameters of the NC solar cells prepared under different conditions from Figure S1.

Author Contributions: B.C. and D.Q. conceived and designed the experiments; B.C., J.L., Z.C., A.X., X.L. and Z.R. performed the experiments; B.C. and W.X. analyzed the data; B.C., L.H. and Q.L. contributed reagents/materials/ analysis tools; D.Q. and B.C. wrote the paper. Authorship must be limited to those who have contributed substantially to the work reported.

Funding: This research received no external funding.

Acknowledgments: We thank the financial support of the National Natural Science Foundation of China (Nos. 21875075 and 61774077), Guangdong Province Natural Science Fund (Nos. 2018A0303130041), the Guangzhou Science and Technology Plan Project (201804010295) and National Undergraduate Innovative and Entrepreneurial Training Program (No. 201810561016, 201810561013).

Conflicts of Interest: The authors declare no conflict of interest.

References

1. Morales-Acevedo, A. Physical basis for the design of Cds/CdTe thin film solar cells. *Sol. Energy Mater. Sol. Cells* **2006**, *90*, 678–685. [CrossRef]
2. Woodhouse, M.; Goodrich, A.; Margolis, R.; James, T.; Dhere, R.; Gessert, T.; Barnes, T.; Eggert, R.; Albin, D. Perspectives on the pathways for cadmium telluride photovoltaic module manufacturers to address expected increases in the price for tellurium. *Sol. Mater. Sol. Cells* **2013**, *115*, 199–212. [CrossRef]
3. Green, M.A.; Emery, K.; Hishikawa, Y.; Warta, W.; Dunlop, E.D.; Levi, D.H.; Ho-Baillie, A.W.Y. Solar cell efficiency tables (version 49). *Prog. Photovolt. Res. Appl.* **2017**, *25*, 3–13. [CrossRef]
4. Fahrenbruch, A.L. Ohmic contacts and doping of CdTe. *Sol. Cells* **1987**, *21*, 399–412. [CrossRef]
5. Paudel, N.R.; Compaan, A.D.; Yan, Y. Sputtered CdS/CdTe solar cells with MoO_{3-X}/Au back contacts. *J. Electr. Mater.* **2013**, *113*, 26–30.
6. Zhang, M.; Qiu, L.; Li, W.; Zhang, J.; Wu, L.; Feng, L. Copper doping of MoO_x thin films for CdTe solar cells. *Mater. Sci. Semicond. Process.* **2018**, *86*, 49–57. [CrossRef]
7. Paudel, N.R.; Xiao, C.; Yan, Y. CdS/CdTe thin-film solar cells with Cu-free transition metal oxide/Au back contacts. *Prog. Photovolt. Res. Appl.* **2015**, *23*, 437–442. [CrossRef]
8. Paudel, N.R.; Yan, Y. Application of copper thiocyanate for high open-circuit voltages of CdTe solar cells. *Prog. Photovolt. Res. Appl.* **2016**, *24*, 94–101. [CrossRef]
9. Paudel, N.R.; Yan, Y. CdTe thin-film solar cells with cobalt-phthalocyanine back contacts. *Appl. Phys. Lett.* **2014**, *104*, 1–9. [CrossRef]
10. Fiederle, M.; Ebling, D.; Eiche, C.; Hug, P.; Joerger, W.; Laasch, M.; Schwarz, R.; Salk, M.; Benz, K. Studies of the compensation mechanism in CdTe grown from the vapour phase. *J. Cryst. Growth* **1995**, *146*, 142–147. [CrossRef]
11. Wang, T.; Du, S.; Li, W.; Liu, C.; Zhang, J.; Wu, L.; Li, B.; Zeng, G. Control of Cu doping and CdTe/Te interface modification for CdTe solar cells. *Mater. Sci. Semicond. Process.* **2017**, *72*, 46–51. [CrossRef]
12. Kuhn, T.A.; Ossau, W.; Waag, A.; Bicknell-Tassius, R.N.; Landwehr, G. Evidence of a deep donor in CdTe. *J. Cryst. Growth* **1992**, *117*, 660–665. [CrossRef]
13. Biglari, B.; Samimi, M.; Hageali, M.; Koebel, J.M.; Siffert, P. Effect of copper in high resistivity cadmium telluride. *J. Cryst. Growth* **1998**, *89*, 428–434. [CrossRef]
14. Desnica, U.V. Doping Limits in II–VI Compounds-Challenenges, Problems and Solutions. *Prog. Cryst. Growth Charact. Mater.* **1998**, *36*, 291–357. [CrossRef]
15. Jones, E.D.; Stewart, N.M.; Mullin, J.B. The diffusion of copper in cadmium telluride. *J. Cryst. Growth* **1992**, *117*, 244–248. [CrossRef]
16. Yang, B.; Ishikawa, Y.; Miki, T.; Doumae, Y.; Isshiki, M. Aging behavior of some residual impurities in CdTe single crystals. *J. Cryst. Growth* **1997**, *179*, 410–414. [CrossRef]
17. Ahn, B.T.; Yun, J.H.; Cha, E.S.; Park, K.C. Understanding the junction degradation mechanism in CdS/CdTe solar cells using a Cd-deficient CdTe layer. *Curr. Appl. Phys.* **2012**, *12*, 174–178. [CrossRef]
18. Han, J.; Fan, C.; Spanheimer, C.; Fu, G.; Zhao, K.; Klein, A.; Jaegermann, W. Electrical properties of the CdTe back contact: A new chemically etching process based on nitric acid/acetic acid mixtures. *Appl. Surf. Sci.* **2010**, *256*, 5803–5806. [CrossRef]
19. Durose, K.; Edwards, P.R.; Halliday, D.P. Materials aspects of CdTe/CdS solar cells. *J. Cryst. Growth* **1999**, *197*, 733–742. [CrossRef]

20. Uličná, S.; Isherwood, P.J.M.; Kaminski, P.M.; Walls, J.M.; Li, J.; Wolden, C.A. Development of ZnTe as a back contact material for thin film cadmium telluride solar cells. *Vacuum* **2017**, *139*, 159–163. [CrossRef]

21. Gul, Q.; Zakria, M.; Khan, T.M.; Mahmood, A.; Iqbal, A. Effects of Cu incorporation on physical properties of ZnTe thin films deposited by thermal evaporation. *Mater. Sci. Semicond. Process.* **2014**, *19*, 17–23. [CrossRef]

22. Park, K.C.; Cha, E.S.; Ahn, B.T. Sodium-doping of ZnTe film by close-spaced sublimation for back contact of CdTe solar cell. *Curr. Appl. Phys.* **2011**, *11*, S109–S112. [CrossRef]

23. Jin, G.; Wei, H.; Cheng, Z.; Sun, H.; Sun, H.; Yang, B. Aqueous-Processed Polymer/Nanocrystal Hybrid Solar Cells with Efficiency of 5.64%: The Impact of Device Structure, Polymer Content, and Film Thickness. *J. Phys. Chem.* **2017**, *121*, 2025–2034. [CrossRef]

24. Townsend, T.K.; Foos, E.E. Fully solution processed all inorganic nanocrystal solar cells. *Phys. Chem. Chem. Phys.* **2014**, *16*, 16458. [CrossRef] [PubMed]

25. Yoon, W.; Townsend, T.K.; Lumb, M.P.; Tischler, J.G.; Foos, E.E. Sintered CdTe Nanocrystal Thin Films: Determination of Optical Constants and Application in Novel Inverted Heterojunction Solar Cells. *IEEE Trans. Nanotechnol.* **2014**, *13*, 551–556. [CrossRef]

26. Panthani, M.G.; Kurley, J.M.; Crisp, R.W.; Dietz, T.C.; Ezzyat, T.; Luther, J.M.; Talapin, D.V. High Efficiency Solution Processed Sintered CdTe Nanocrystal Solar Cells: The Role of Interfaces. *Nano Lett.* **2014**, *14*, 670–675. [CrossRef]

27. Lin, H.; Xia, W.; Wu, H.N.; Tang, C.W. CdS/CdTe solar cells with MoOx as back contact buffers. *Appl. Phys. Lett.* **2010**, *97*, 69. [CrossRef]

28. Wen, S.; Li, M.; Yang, J.; Mei, X.; Wu, B.; Liu, X.; Heng, J.; Qin, D.; Hou, L.; Xu, W.; Wang, D. Rationally Controlled Synthesis of CdSe$_x$Te$_{1-x}$ Alloy Nanocrystals and Their Application in Efficient Graded Bandgap Solar Cells. *Nanomaterials* **2017**, *7*, 380. [CrossRef]

29. Du, X.; Chen, Z.; Liu, F.; Zeng, Q.; Jin, G.; Li, F.; Yao, D.; Yang, B. Improvement in Open-Circuit Voltage of Thin Film Solar Cells from Aqueous Nanocrystals by interface Engineering. *ACS Appl. Mater. Interfaces* **2016**, *8*, 900–907. [CrossRef] [PubMed]

30. Zeng, Q.; Hu, L.; Cui, J.; Feng, T.; Du, X.; Jin, G.; Liu, F.; Ji, T.; Li, F.; Zhang, H.; et al. High-Efficiency Aqueous-Processed Polymer/CdTe Nanocrystals Planar Heterojunction Solar Cells with Optimized Band Alignment and Reduced Interfacial Charge Recombination. ACS Appl. *Mater. Interfaces* **2017**, *9*, 31345–31351. [CrossRef] [PubMed]

31. Guo, X.; Tan, Q.; Liu, S.; Qin, D.; Mo, Y.; Hou, L.; Liu, A.; Wu, H.; Ma, Y. High-efficiency solution-processed CdTe nanocrystal solar cells incorporating a novel crosslinkable conjugated polymer as the hole transport layer. *Nano Energy* **2018**, *46*, 150–157. [CrossRef]

32. Crisp, R.W.; Panthani, M.G.; Rance, W.L.; Duenow, J.N.; Parilla, P.A.; Callahan, R.; Dabney, M.S.; Berry, J.J.; Talapin, D.V.; Luther, J.M. Nanocrystal Grain Growth and Device Architectures for High-Efficiency CdTe Ink-Based Photovoltaics. *ACS Nano* **2014**, *8*, 9063–9072. [CrossRef]

33. Liu, H.; Tian, Y.; Gao, K.; Lu, K.; Wu, R.; Qin, D.; Wu, H.; Peng, Z.; Hou, L.; Huang, W. Solution processed CdTe/CdSe nanocrystal solar cells with more than 5.5% efficiency by using an inverted device structure. *J. Mater. Chem.* **2015**, *3*, 4227–4234. [CrossRef]

34. Liu, S.; Liu, W.; Heng, J.; Zhou, W.; Chen, Y.; Wen, S.; Qin, D.; Hou, L.; Wang, D.; Xu, H. Solution-Processed Efficient Nanocrystal Solar Cells Based on CdTe and CdS Nanocrystals. *Coatings* **2018**, *8*, 26. [CrossRef]

35. Chen, Z.; Zhang, H.; Zeng, Q.; Wang, Y.; Xu, D.; Wang, L.; Wang, H.; Yang, B. In Situ Construction of Nanoscale CdTe-CdS Bulk Heterojunctions for Inorganic Nanocrystal Solar Cells. *Adv. Mater.* **2014**, *4*, 1400235. [CrossRef]

36. Qin, D.; Tan, Q.; Lu, K.; Li, M.; Hou, L.; Xie, Y.; Zhang, Z.; Xu, W.; Wu, H. Improving performance in CdTe/CdSe nanocrystals solar cells by using bulk nano-heterojunctions. *J. Mater. Chem. C* **2016**, *4*, 6483–6491.

37. Tian, Y.; Zhang, Y.; Lin, Y.; Gao, K.; Zhang, Y.; Liu, K.; Yang, Q.; Zhou, X.; Qin, D.; Wu, H.; et al. Solution-processed efficient CdTe nanocrystal/CBD-CdS hetero-junction solar cells with ZnO interlayer. *J. Nanopart.* **2013**, *15*, 1–9. [CrossRef]

38. Zeng, Q.; Chen, Z.; Zhao, Y.; Du, X.; Liu, F.; Jin, G.; Dong, F.; Zhang, H.; Yang, B. Aqueous-Processed Inorganic Thin-Film Solar Cells Based on CdSe$_x$Te$_{1-x}$ Nanocrystals: The Impact of Composition on Photovoltaic Performance. *ACS Appl. Mater. Interfaces* **2015**, *7*, 23223–23230. [CrossRef] [PubMed]

39. Chen, Z.; Zeng, Q.; Liu, F.; Jin, G.; Du, X.; Du, J.; Zhang, H.; Yang, B. Efficient inorganic solar cells from aqueous nanocrystals: The impact of composition on carrier dynamics. *RSC Adv.* **2015**, *5*, 74263. [CrossRef]

40. Rimmaudo, I.; Salavei, A.; Bing, L.X.; Mare, S.D.; Romeo, A. Superior stability of ultra thin CdTe solar cells with simple Cu/Au back contact. *Thin Solid Films* **2015**, *582*, 105–109. [CrossRef]
41. Demtsu, S.H.; Sites, J.R. Effect of back-contact barrier on thin-film CdTe solar cells. *Thin Solid Films* **2006**, *510*, 320–324. [CrossRef]
42. Yang, T.; Cai, W.; Qin, D.; Wang, E.; Lan, L.; Gong, X.; Peng, J.; Cao, Y. Solution-Processed Zinc Oxide Thin Film as a Buffer Layer for Polymer Solar Cells with an Inverted Device Structure. *J. Phys. Chem. C* **2010**, *114*, 6849–6853. [CrossRef]
43. Dobson, K.D.; Visoly-Fisher, I.; Hodes, G.; Cahen, D. Stability of CdTe/CdS thin-film solar cells. *Sol. Energy Mater. Sol. Cells* **2000**, *62*, 295–325. [CrossRef]

Article

Blue Electroluminescent Al$_2$O$_3$/Tm$_2$O$_3$ Nanolaminate Films Fabricated by Atomic Layer Deposition on Silicon

Yao Liu, Zhongtao Ouyang, Li Yang, Yang Yang * and Jiaming Sun *

School of Materials Science and Engineering, Tianjin Key Lab for Rare Earth Materials and Applications, Nankai University, Tianjin 300350, China; 15100299616@163.com (Y.L.); zhongtoe@163.com (Z.O.); materialyang@126.com (L.Y.)

Received: 18 February 2019; Accepted: 8 March 2019; Published: 11 March 2019

Abstract: Realization of a silicon-based light source is of significant importance for the future development of optoelectronics and telecommunications. Here, nanolaminate Al$_2$O$_3$/Tm$_2$O$_3$ films are fabricated on silicon utilizing atomic layer deposition, and intense blue electroluminescence (EL) from Tm^{3+} ions is achieved in the metal-oxide-semiconductor structured luminescent devices based on them. Precise control of the nanolaminates enables the study on the influence of the Tm dopant layers and the distance between every Tm$_2$O$_3$ layer on the EL performance. The 456 nm blue EL from Tm^{3+} ions shows a maximum power density of 0.15 mW/cm^2. The EL intensities and decay lifetime decrease with excessive Tm dopant cycles due to the reduction of optically active Tm^{3+} ions. Cross-relaxation among adjacent Tm$_2$O$_3$ dopant layers reduces the blue EL intensity and the decay lifetime, which strongly depends on the Al$_2$O$_3$ sublayer thickness, with a critical value of ~3 nm. The EL is attributed to the impact excitation of the Tm^{3+} ions by hot electrons in Al$_2$O$_3$ matrix via Poole–Frenkel mechanism.

Keywords: electroluminescence; nanolaminate; Al$_2$O$_3$; Tm$_2$O$_3$; atomic layer deposition

---

1. Introduction

Traditional electronic integrated circuits have been facing with a bottleneck in terms of power consumption, speed, and signal crosstalk as the communication frequency and bandwidth rise to a higher level. One possible solution is the optoelectronic integration which realizes photonic technologies on silicon chips [1–4]. However, applicable Si-based light sources have been unsolved for a long time. Rare earth (RE) ions are generally efficient luminescence centers in various matrixes. Nowadays diverse RE-doped insulating materials have been developed for the applications in solid state lasers and phosphors [5–9]. However, it has been widely known that the mismatch in the coordination structure and atomic size of silicon (tetrahedron) and RE ions (octahedron) limit the desired spectroscopic performance due to the clustering of RE ions in the Si host [10,11]. Aiming for the realization of compact Si-based optoelectronics, electroluminescence (EL) from RE^{3+} ions has been extensively reported in many compounds, such as SiN$_x$, TiO$_2$, and ZnO [12–15]. However, the efficiencies of the devices based on the aforementioned materials are far from practical utilization. One of the limitations is the large leakage current. RE-implanted SiO$_2$ MOS-structured light-emitting devices (MOSLEDs) have attracted much attention due to their notable EL efficiency and silicon compatibility [16,17]. In comparison, similar devices based on Al$_2$O$_3$ nanofilms present much lower working voltage, and comparable efficiency in our previous study, while their EL performance needs more exploration [18,19]. Blue emission, which has the highest photon energy (2.6–2.7 eV) of the three primary colors, is of great importance in display and lighting. Tm^{3+} ions have present efficient blue emissions in various matrixes including ZnS, ZnO, fluorophosphate, and many other oxides and fluorides [20–23]. The reported achievements are mostly focused on photoluminescence (PL), by virtue of upconversion to convert infrared photons to blue emission [20,24]. For practical application, electrically excited devices are urgently needed. Whether high-energy blue photons can be generated in this prototype device is still unknown. Using Tm-doped Al_2O_3 might exploit the merits of both oxides to realize efficient blue EL from Tm^{3+} ions.

Atomic layer deposition (ALD) is a monatomic vapor deposition technique achieved by alternating saturated gas–surface reactions, based on which the film can be deposited in a self-limited growth mode and exhibits superior homogeneity and excellent uniformity [25–29]. This technique supplies a convenient way to devise nanolaminates with optimal performance. In this work, we fabricate nanolaminate Al_2O_3/Tm_2O_3 films which function as blue EL layers in the Si-based MOSLEDs. The EL intensity and decay lifetime are compared by changing the Al_2O_3 or Tm_2O_3 sublayer cycles. The influence of the Tm clustering and interaction concerning the Al_2O_3 or Tm_2O_3 cycles are explored respectively. The 456 nm blue EL from Tm^{3+} ions shows a maximum power density of 0.15 mW/cm^2. The device characteristics are in good consistence with the previous reports on the excitation mechanism and the critical interlayer thickness for the cross-relaxation among adjacent dopant layers.

2. Materials and Methods

The nanolaminate Al_2O_3/Tm_2O_3 films were grown on <100>-oriented phosphorous–doped silicon (n-Si) substrates with the resistivity of 2–5 $\Omega \cdot$cm and a thickness of 500 μm (CETC-46 Ltd., Tianjin, China), which were cleaned through the standard RCA process before growth. The ALD equipment was a 4-inch chamber system (Nano Tech Savannah 100, Cambridge, MA, USA). Trimethylaluminum [TMA, $Al(CH_3)_3$, 99.999+%] and $Tm(THD)_3$ (THD = 2,2,6,6-teramethyl-3,5 heptanedionate, 99.9%, Strem Chemicals, Inc., Newburyport, MA, USA) were used as the metal precursors for Al_2O_3 and Tm_2O_3, while ozone was used as the oxidant. N_2 was used as the carrier and purge gas with a flow rate of 20 sccm. During the growth, the pulse time of TMA, $Tm(THD)_3$, and ozone was 0.015 s, 2 s, and 1.8 s, respectively. The TMA was maintained at room temperature while the Tm precursor was heated at 170 °C. The pipelines and the substrates were maintained at 190 °C and 325 °C. The growth rates for the Tm_2O_3 and Al_2O_3 films were 0.216 Å/cycle and 0.79 Å/cycle, respectively.

In order to investigate the luminescent characteristics of nanolaminate Al_2O_3/Tm_2O_3 films, a series of devices concerning the Tm_2O_3 dopant cycles and the Al_2O_3 interlayer cycles were fabricated as shown in Table 1. The total cycle numbers were adjusted correspondingly to obtain the luminescent films with a thickness of ~50 nm. The thickness of the film was measured by an homemade ellipsometer with a 632.8 nm He-Ne laser at an incident angle of 69.8°. As the thickness variation from the designed value for the nanolaminates are quite small (less than 3%), the nominal Tm concentrations are used to quantify the doping levels. All Al_2O_3/Tm_2O_3 films were subsequently annealed at 800 °C in N_2 atmosphere for 1 h to reduce defects and activate Tm^{3+} luminescence. Then, 120 nm TiO_2/Al_2O_3 nanolaminate films consisting of 2 nm Al_2O_3 and 8 nm TiO_2 sublayers were grown by ALD on Al_2O_3/Tm_2O_3 films as the protective layers. Afterwards, ~100 nm $ZnO:Al_2O_3$ films were grown by ALD as the transparent conductive electrodes, which were lithographically patterned into 0.5 mm circular dots. Finally, 100 nm Al electrodes were deposited on the back side of the Si substrates by thermal evaporation, and annealed afterwards in vacuum at 250 °C for 0.5 h to realize ohmic contact.

Table 1. The corresponding experimental parameters for clarity.

Sample Label	Tm^{3+} (at%)	Al_2O_3:Tm_2O_3 Cycle Number
AOT-1	0.69	13:1
AOT-2	1.37	13:2
AOT-4	2.64	13:4
AOT-6	3.83	13:6
AOT-8	4.95	13:8
AOT-d05	2.46	7:2
AOT-d1	1.37	13:2
AOT-d2	0.69	26:2
AOT-d3	0.45	40:2
AOT-d4	0.35	52:2
AOT-d6	0.23	78:2

The PL spectra from the luminescent nanolaminates were excited by a 355 nm laser. For EL and Current–Voltage (I–V) measurements, the devices were activated by means of a Keithley 2410 SourceMeter unit (Keithley Instruments Inc., Cleveland, OH, USA), with the negative voltage connecting to n-Si substrates. The PL and EL signals were detected by a monochromator (Zolix λ500, Zolix Instruments Co., Ltd, Beijing, China) and a Si photomultiplier connected to a Keithley 2010 multimeter (Keithley Instruments Inc., Cleveland, OH, USA). Photographic images were collected by a digital camera through a 20-fold objective microscope. Time-resolved photoluminescence (TRPL) was measured by a SR430 multi-channel scaler (Stanford Research Systems Inc., Sunnyvale, CA, USA) with a 355 nm laser working in the pulse mode. The decay lifetime of the EL emission was measured by the SR430 multichannel scaler, excited by a high-voltage amplifier equipped with a digital function signal generator (DG5072, RIGOL Technology Co., Ltd, Beijing, China). All the above measurements were performed at room temperature.

3. Results and Discussion

The Tm_2O_3 films deposited by ALD can be crystalized into Tm_2O_3 phase even without annealing, while the Al_2O_3 films are amorphous after annealing at 800 °C. However, the nanolaminate Al_2O_3/Tm_2O_3 film with the highest Tm content (AOT-8) is amorphous after annealing at 800 °C, therefore the nanolaminate structure restricts the grain growth of the dopant Tm_2O_3 layers. Figure 1a shows the PL spectra from the nanolaminate Al_2O_3/Tm_2O_3 films. The PL peaks at 456 nm are attributed to the transition of $^1D_2 \rightarrow {}^3F_4$ in Tm^{3+} ions [20–22]. The inset of Figure 1a presents the comparison of the PL intensities of all samples, which decrease with the Tm_2O_3 dopant layers. Due to the common cluttering characteristics of RE ions, with the increase of Tm content, the number of activated Tm^{3+} ions decreases and the cross relaxation between Tm^{3+} ions further reduce the radiative transitions [30,31]. For TRPL results shown in Figure 1b, the decay lifetime of these PL emissions from Tm^{3+} ions also decreases with the Tm content, which coincides with the PL intensities. The inset gives the fitting values of the PL decay lifetime, which are in the range of 0.13–1.25 μs. The PL decay lifetime decreases rapidly as the Tm dopant layers rise to 4. The cross relaxation and concentration quenching contribute to the nonradiative recombination and decrease the luminescence lifetime.

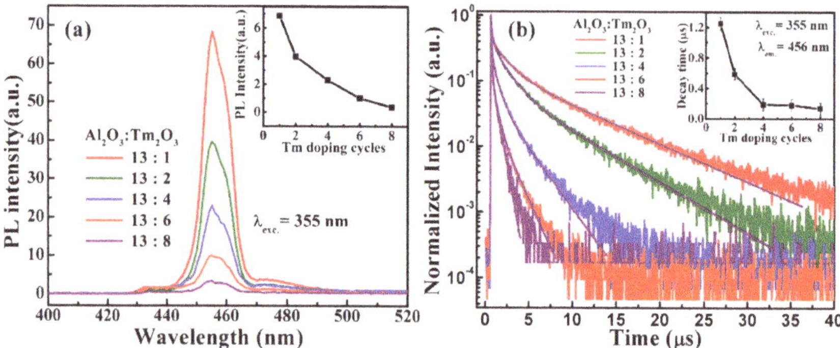

Figure 1. The (**a**) photoluminescence (PL) and (**b**) time-resolved photoluminescence (TRPL) spectra from the nanolaminate Al_2O_3/Tm_2O_3 films with different Tm dopant cycles excited by a 355 nm laser. The insets present the tendency of these PL intensities and PL decay lifetime with the Tm dopant cycles.

The schematic for the multilayered devices is shown in Figure 2a. The EL spectrum from the MOSLED based on the Al_2O_3/Tm_2O_3 nanolaminate with 2 cycles of Tm dopant (AOT-2) is presented in Figure 2b. The EL emissions mainly exhibit several peaks at the wavelengths of 368, 456, 474, and 802 nm, which originate from the radiative transitions from the $^1D_2$, $^3F_4$, $^1G_4$, and $^3H_4$ excited states to the $^3H_6$ ground state in Tm^{3+} ions, respectively, as sketched in the inset of Figure 2b [21–23]. It is noteworthy that the EL emissions at 456 nm and 474 nm are dominating and the blue light is easily seen by naked eyes, as shown in Figure 2c. These images were taken by a digital camera from this AOT-2 MOSLED at different injection currents. The blue EL emission gradually brightens with the increase of the injection current from 10 µA to 80 µA.

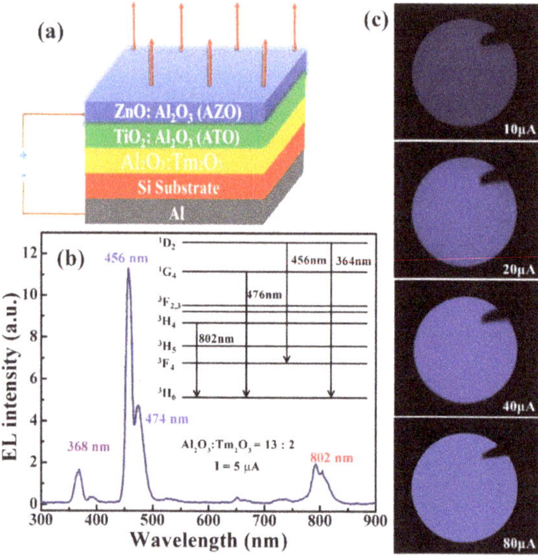

Figure 2. (**a**) The schematic for the luminescent devices based on the nanolaminate Al_2O_3/Tm_2O_3 films. (**b**) The EL spectrum from the device in which the Al_2O_3/Tm_2O_3 subcycle ratio is 13:2 (AOT-2), the inset shows the radiative transitions in the Tm^{3+} ions resulting in the EL emissions. (**c**) The images taken by a digital camera from this AOT-2 MOS-structured light-emitting device (MOSLED) at different injection currents.

Figure 3a shows EL spectra from the MOSLEDs based on the Al_2O_3/Tm_2O_3 films with different Tm dopant cycles at an injection current of 5 µA. The concentrations of Tm dopant are from 0.69% to 4.95%, respectively. The spectra exhibit four peaks at 368, 458, 474, and 802 nm as mentioned above. The inset shows that the 456 nm blue EL intensity increases with the Tm dopant cycles up to 2 and then decreases due to concentration quenching. The EL presents higher tolerance for Tm clustering than the PL performance. The dependence of the 456 nm EL power density on the injection current density are shown in Figure 3b. Generally, the EL intensity presents a linear relationship with the injection current density. A power density up to 0.15 mW/cm^2 was obtained from the optimal MOSLED at a current density of 2.87 A/cm^2. Initially, the EL output power density increases as the Tm dopant cycles increases to 2, due to the increase of the excitable Tm^{3+} ions. The further decline of the power density with the Tm dopant cycle is attributed to the clustering and cross relaxation which reduce the number of excited Tm^{3+} ions [30,31]. The efficiency and output power are lower than the previously reported devices based on the Tb and Yb doped Al_2O_3 nanolaminates [18,19]. As the energy of the blue photon is higher than that of the green EL from Tb^{3+} ions and the near-infrared one from Yb^{3+} ions, the excitation possibility of the radiative transitions within Tm^{3+} ions should be lower which leads to the limited efficiency and output power. In addition, the visible EL from the RE-doped SiO_2 is stronger than the devices in this work [32]. The higher working voltage needed for luminescence in SiO_2 evidences the necessity of high electrical field for excitation of the photon with higher energy, which is adverse to practical application. However, this EL output power density is superior to the EL devices based on the RE-doped ZnO as the leakage current is greatly restricted comparatively [13].

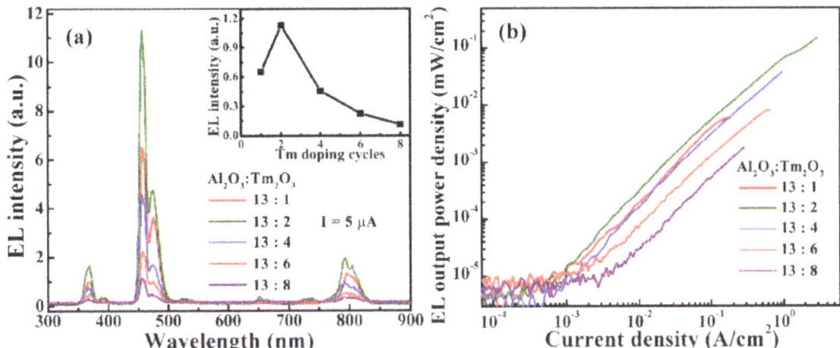

Figure 3. (a) EL spectra from the MOSLEDs based on the Al_2O_3/Tm_2O_3 films with different Tm dopant cycles at an injection current of 5 µA, the inset shows the tendency of this EL intensity with the Tm dopant layers. (b) The dependence of the 456 nm EL power density on the injection current density for the Al_2O_3/Tm_2O_3 MOSLEDs with different Tm dopant cycles.

Figure 4a,b shows the dependence of blue (456 nm) EL intensities, together with the injection current, on the applied voltages for the nanolaminate MOSLEDs based on different Al_2O_3/Tm_2O_3 films. All devices exhibit a typical I–V characteristic of the MOS structure, i.e., the current starts with a low background one under the low electric field, then exponentially increases with the voltage [16–19]. The difference on the leakage currents mainly depends on the process of device procedures, coming from the electrons hopping through the defects within the matrix. At this stage, no hot electrons are generated in the Al_2O_3/Tm_2O_3 conduction band with no EL emissions. Afterwards, the injection current increases exponentially with the applied voltage and the conduction mechanism is dominated by the Poole–Franked (P–F) mode until the device breakdown [18,19]. In the P–F conduction mode the plot of the $\ln(J/E)$ versus $E^{1/2}$ features a linear relationship (J is the current density and E is the electric field). As shown in Figure 4c, for all Al_2O_3/Tm_2O_3 MOSLEDs the P–F plots work in the EL-enabling voltages, with the threshold voltage of around 40 V (~3 MV/cm). The slopes of the linear plots of the P–F injections are similar while the little difference is caused by the slight variation of the injection current as mentioned above. Therefore, for the EL excitation, electrons are firstly injected into the conduction band of Al_2O_3 by trap-assisted tunneling and accelerated to gain energy under high electric field. These hot electrons excite the doped Tm^{3+} ions from the ground state to higher levels by inelastic collision. After the nonradiative relaxation, the radiative transitions in the Tm^{3+} ions from the excited state to ground state generate the characteristic EL emissions [20–22].

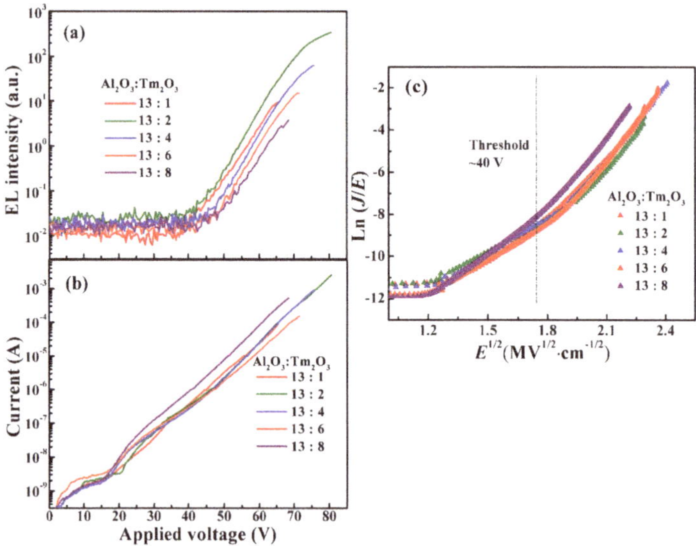

Figure 4. (**a**,**b**) The dependence of blue (456 nm) EL intensities, together with the injection current, on the applied voltages for the nanolaminate MOSLEDs based on different Al_2O_3/Tm_2O_3 films. (**c**) The plot of $\ln(J/E)$ versus $E^{1/2}$ (Poole–Frenkel conduction mode) for these MOSLEDs.

The EL decay lifetime of the 456 nm EL from different nanolaminate Al_2O_3/Tm_2O_3 MOSLEDs is measured under pulse excitation mode. The decay curves are shown in Figure 5a, which are close to the single exponential decay function. The decay lifetime decreases from 4.02 µs to 0.53 µs with the increase of Tm dopant cycles, as shown in Figure 5b. These values of EL decay lifetime are several times larger than that of PL decay lifetime shown in Figure 1b, and keep decreasing with the Tm doping concentration, which comes from the cross relaxation and concentration quenching caused by the excess Tm^{3+} ions. These phenomena again mean that the tolerance on the concentration quenching in EL performance is higher than that in PL.

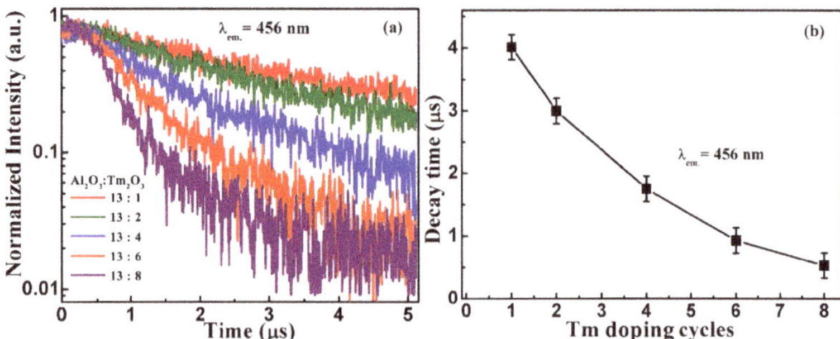

Figure 5. (**a**) The EL decay lifetime of the 456 nm EL from different nanolaminate Al_2O_3/Tm_2O_3 MOSLEDs and (**b**) the tendency of the EL decay lifetime with the Tm dopant cycles.

In the RE-doped Al_2O_3 MOSLEDs, the Al_2O_3 sublayer thickness affects the cross relaxation between excited RE ions, and the acceleration distance for injected electrons. In order to investigate the effect of the distance between Tm_2O_3 dopant layers, a series of MOSLEDs were fabricated in which the Al_2O_3 sublayer thickness varied from 0.5 nm to 6 nm while the Tm dopant cycles was fixed at 2. Figure 6a shows the dependence of the blue EL intensity on the injection current. Here, the EL intensities are divided by the cycle numbers to present the emissions from every Tm dopant cycle. With the increase of the thickness of Al_2O_3 sublayer, the contribution of a single Tm dopant cycle to the EL intensity firstly increases and then saturates as the Al_2O_3 interlayer thickness reaches 3 nm. Figure 6b presents the tendency. This phenomenon has been observed in our previous reports with a similar value, concerning the nonradiative interaction among excited RE^{3+} ions and the acceleration distance for the injected electrons [18,19]. Therefore, it is a common characteristic for the luminescent RE^{3+} ions in an Al_2O_3 matrix that the distance for the presence of nonradiative interaction and adequate electron acceleration is around 3 nm.

Furthermore, the decay lifetimes for these MOSLEDs are shown in Figure 6c, whose correlation with the Al_2O_3 interlayer thickness is summarized in Figure 6d. Similar to the EL intensity, the decay lifetime increases from 1.18 to 7.41 µs with the Al_2O_3 interlayer thickness increasing from 0.5 nm to 3 nm, and saturates at higher distances. The reduction of the decay lifetime at higher Tm doping concentrations is still ascribed to the increase of nonradiative cross relaxations between the two closely located Tm^{3+} dopant layers as mentioned above, with the similar critical Al_2O_3 interlayer thickness of 3 nm [19,33]. Considering the total EL intensities, the optimal Al_2O_3 interlayer thickness in these MOSLEDs is 2 nm. It should be noted that there is little difference between the total EL emission from nanolaminate Al_2O_3/Tm_2O_3 MOSLEDs with 1 nm and 2 nm Al_2O_3 interlayers. The effect of more dopant ions is offset by the relative lowered excitation efficiency. This optimal doping concentration is also consistent with previous reports (around 1 at%) on the RE doped luminescent materials [18,33].

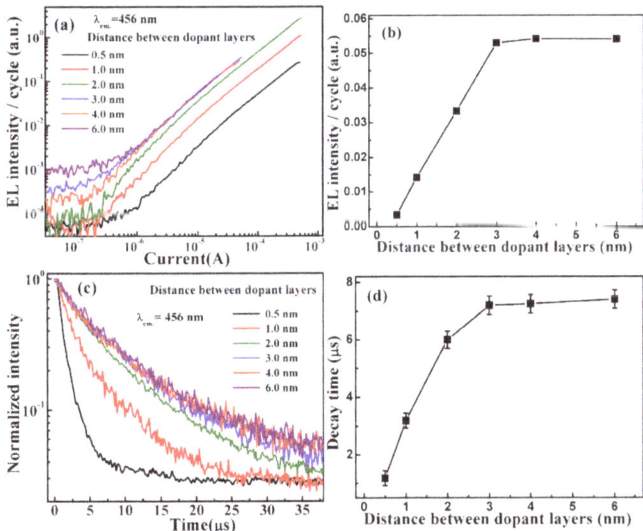

Figure 6. (a) The dependence of the blue EL intensity on the injection current from the Al_2O_3/Tm_2O_3 MOSLEDs with different Al_2O_3 sublayer thicknesses. Here, the EL intensities are divided by the cycle numbers to present the emissions from every Tm dopant cycle. (b) The relation of blue EL intensity with the Al_2O_3 sublayer thicknesses. (c) The EL decay lifetime for the Al_2O_3/Tm_2O_3 MOSLEDs and (d) the relation of lifetime to the Al_2O_3 sublayer thicknesses.

The blue EL intensities (output powers) from our prototype devices are quite low and incapable of practical application. This work confirms the potential to realize blue EL from Al_2O_3/Tm_2O_3 nanolaminates by ALD. Moreover, the devices are fabricated entirely by ALD, which is characterized by the precise control of the film deposition over large substrates, and the compatibility with Si-based CMOS technology. Therefore, MOSLEDs based on Al_2O_3/Tm_2O_3 nanolaminates can be easily expanded for mass-production. The challenging deficiencies are the low EL efficiency and output power, the high working voltage, and the limited injection current. Further optimization can be achieved by adopting a thicker Al_2O_3/Tm_2O_3 luminescent layer with more optimal dopant structure and a less resistant protective layer with higher dielectric constant, to obtain a higher emission intensity.

4. Conclusions

Blue EL is demonstrated from nanolaminate Al_2O_3/Tm_2O_3 MOSLEDs fabricated by ALD. The emission at 456 nm from Tm^{3+} ions exhibits a power density of 0.15 mW/cm^2. The decrease of the EL intensity and decay lifetime due to the clustering and cross-relaxation of the Tm^{3+} ions is observed by adjusting the Tm_2O_3 dopant cycles. The decay lifetime for the Tm^{3+} ions under optical excitation is in the range of 0.13–1.25 μs while under electrical excitation, the decay lifetime increases to 1.13–4.02 μs. The EL is attributed to the impact excitation of the Tm^{3+} ions by hot electrons in the Al_2O_3 matrix via the P–F mechanism. Consistent with the previous results, a critical Al_2O_3 interlayer thickness of ~3 nm for the nonradiative interaction among excited Tm^{3+} ions and the acceleration distance of the injected electrons works. This work could contribute to the development of Si-compatible RE-doped light sources by modifying the dopant structure in the nanolaminates to achieve efficient emissions.

Author Contributions: Conceptualization, J.S. and Y.Y.; methodology, Y.L. and Z.O.; validation, Y.L. and L.Y.; formal analysis, Y.L., Y.Y. and J.S.; writing—original draft preparation, Y.L.; writing—review and editing, Y.Y.; supervision, J.S.; funding acquisition, J.S. and Y.Y.

Funding: This research and APC were funded by National Natural Science Foundation of China, grant number 61674085 and 61705114, and China Postdoctoral Science Foundation, grant number 2017M611154.

Conflicts of Interest: The authors declare no conflict of interest.

References

1. Gan, X.; Shiue, R.J.; Gao, Y.; Meric, I.; Heinz, T.F.; Shepard, K.; Hone, J.; Assefa, S.; Englund, D. Chip-integrated ultrafast graphene photodetector with high responsivity. *Nat. Photonics* **2013**, *7*, 883–887. [CrossRef]
2. Marpaung, D.; Yao, J.; Capmany, J. Integrated microwave photonics. *Nat. Photonics* **2019**, *13*, 80–90. [CrossRef]
3. Atabaki, A.; Moazeni, S.; Pavanello, F.; Gevorgyan, H.; Notaros, J.; Alloatti, L.; Wade, M.T.; Sun, C.; Kruger, S.A.; Meng, H.; et al. Integrating photonics with silicon nanoelectronics for the next generation of systems on a chip. *Nature* **2018**, *556*, 349–354. [CrossRef] [PubMed]
4. Miritello, M.; Lo Savio, R.; Iacona, F.; Franzo, G.; Irrera, A.; Piro, A.M.; Bongiorno, C.; Priolo, F. Efficient luminescence and energy transfer in erbium silicate thin films. *Adv. Mater.* **2007**, *19*, 1582–1588. [CrossRef]
5. Muravyev, S.V.; Anashkina, E.A.; Andrianov, A.V.; Dorofeev, V.; Motorin, S.E.; Koptev, M.Y. Dual-band Tm^{3+}-doped tellurite fiber amplifier and laser at 1.9 μm and 2.3 μm. *Sci. Rep.* **2018**, *8*, 16164. [CrossRef] [PubMed]
6. Nilsson, J.; Payne, D.N. High-Power Fiber Lasers. *Science* **2011**, *332*, 921–922. [CrossRef] [PubMed]
7. Kim, J.H.; Holloway, P.H. Near-infrared-electroluminescent light-emitting planar optical sources based on gallium nitride doped with rare earths. *Adv. Mater.* **2005**, *17*, 91–96. [CrossRef]
8. Irrera, A.; Franzo, G.; Iacona, F.; Canino, A.; Di Stefano, G.; Sanfilippo, D.; Piana, A.; Fallica, P.G.; Priolo, F. Light emitting devices based on silicon nanostructures. *Physica E* **2007**, *38*, 181–187. [CrossRef]
9. Kenyon, A.J. Recent developments in rare-earth doped materials for optoelectronics. *Prog. Quantum Electron.* **2002**, *26*, 225–284. [CrossRef]
10. Gu, L.L.; Xiong, Z.H.; Chen, G.; Xiao, Z.S.; Gong, D.W.; Hou, X.Y.; Wang, X. Luminescent erbium-doped porous silicon bilayer structures. *Adv. Mater.* **2001**, *13*, 1402–1405. [CrossRef]

11. Eames, C.; Probert, M.I.J.; Tear, S.P. The structure and growth direction of rare earth silicide nanowires on Si(100). *Appl. Phys. Lett.* **2010**, *96*, 241903. [CrossRef]

12. Tatebayashi, J.; Yoshii, G.; Nakajima, T.; Kamei, H.; Takatsu, J.; Lebrun, D.M.; Fujiwaraa, Y. Control of the energy transfer between Tm^{3+} and Yb^{3+} ions in Tm,Yb-codoped ZnO grown by sputtering-assisted metalorganic chemical vapor deposition. *J. Appl. Phys.* **2018**, *123*, 161409. [CrossRef]

13. Yang, Y.; Li, Y.P.; Wang, C.X.; Zhu, C.; Lv, C.Y.; Ma, X.Y.; Yang, D.R. Rare earth doped ZnO films: A material platform to realize multicolor and near infrared electroluminescence. *Adv. Opt. Mater.* **2014**, *2*, 240–244. [CrossRef]

14. Zhu, C.; Lv, C.Y.; Jiang, M.M.; Zhou, J.W.; Li, D.S.; Ma, X.Y.; Yang, D.R. Green electroluminescence from Tb_4O_7 films on silicon: Impact excitation of Tb^{3+} ions by hot carriers. *Appl. Phys. Lett.* **2016**, *108*, 051113. [CrossRef]

15. Jambois, O.; Berencen, Y.; Hijazi, K.; Wojdak, M.; Kenyon, A.J.; Gourbilleau, F.; Rizk, R.; Garrido1, B. Current transport and electroluminescence mechanisms in thin SiO_2 films containing Si nanocluster-sensitized erbium ions. *J. Appl. Phys.* **2009**, *106*, 063526. [CrossRef]

16. Rebohle, L.; Braun, M.; Wutzler, R.; Liu, B.; Sun, J.M.; Helm, M.; Skorupa, W. Strong electroluminescence from SiO_2-Tb_2O_3-Al_2O_3 mixed layers fabricated by atomic layer deposition. *Appl. Phys. Lett.* **2014**, *104*, 251113. [CrossRef]

17. Sun, J.M.; Prucnal, S.; Skorupa, W.; Dekorsy, T.; Müchlich, A.; Helm, M.; Rebohle, L.; Gebel, T. Electroluminescence properties of the Gd^{3+} ultraviolet luminescent centers in SiO_2 gate oxide layers. *J. Appl. Phys.* **2006**, *99*, 103102. [CrossRef]

18. Ouyang, Z.; Yang, Y.; Sun, J. Near-infrared electroluminescence from atomic layer doped Al_2O_3:Yb nanolaminate films on silicon. *Scr. Mater.* **2018**, *151*, 1–5. [CrossRef]

19. Yang, Y.; Li, N.; Sun, J.M. Intense electroluminescence from Al_2O_3/Tb_2O_3 nanolaminate films fabricated by atomic layer deposition on silicon. *Opt. Express* **2018**, *26*, 9344–9352. [CrossRef] [PubMed]

20. Chen, Z.; Kang, S.; Zhang, H.; Wang, T.; Lv, S.; Chen, Q.; Dong, G.; Qiu, J. Controllable optical modulation of blue/green up-conversion fluorescence from Tm^{3+} (Er^{3+}) single-doped glass ceramics upon two-step excitation of two-wavelengths. *Sci. Rep.* **2017**, *7*, 45650. [CrossRef] [PubMed]

21. Lee, Y.W.; Chien, H.W.; Cho, C.H.; Chen, J.Z.; Chang, J.S.; Jiang, S. Heavily Tm^{3+}-Doped Silicate Fiber for High-Gain Fiber Amplifiers. *Fibers* **2013**, *1*, 82–92. [CrossRef]

22. Lee, D.S.; Steckl, A.J. Selective enhancement of blue electroluminescence from GaN:Tm. *Appl. Phys. Lett.* **2003**, *82*, 55–57. [CrossRef]

23. Guereñu, A.L.; Bastian, P.; Wessig, P.; John, L.; Kumke, M.U. Energy Transfer between Tm-doped upconverting nanoparticles and a small organic dye with large stokes shift. *Biosensors* **2019**, *9*, 9. [CrossRef] [PubMed]

24. Yamada, Y.; Kanemitsu, Y. Blue light emission from strongly photoexcited and electron-doped $SrTiO_3$. *J. Appl. Phys.* **2011**, *109*, 102410. [CrossRef]

25. Chawla, V.; Ruoho, M.; Weber, M.; Chaaya, A.A.; Taylor, A.A.; Charmette, C.; Miele, P.; Bechelany, M.; Michler, J.; Utke, I. Fracture mechanics and oxygen gas barrier properties of Al_2O_3/ZnO nanolaminates on PET deposited by atomic layer deposition. *Nanomaterials* **2019**, *9*, 88. [CrossRef] [PubMed]

26. Bouriaux, L.F.; Rosamond, M.C.; Williams, D.A.; Davies, A.G.; Wälti, C. Field-enhanced direct tunneling in ultrathin atomic-layer-deposition-grown Au-Al_2O_3-Cr metal-insulator-metal structures. *Phys. Rev. B* **2017**, *96*, 115435. [CrossRef]

27. Zhang, F.; Sun, G.; Zheng, L.; Liu, S.; Liu, B.; Dong, L.; Wang, L.; Zhao, W.; Liu, X.; Yan, G.; et al. Interfacial study and energy-band alignment of annealed Al_2O_3 films prepared by atomic layer deposition on 4H-SiC. *J. Appl. Phys.* **2013**, *113*, 044112. [CrossRef]

28. Elam, J.W.; Routkevitch, D.; Mardilovich, P.P.; George, S.M. Conformal coating on ultrahigh-aspect-ratio nanopores of anodic alumina by atomic layer deposition. *Chem. Mater.* **2003**, *15*, 3507–3517. [CrossRef]

29. Bosund, M.; Mizohata, K.; Hakkarainen, T.; Putkonen, M.; Söderlund, M.; Honkanen, S.; Lipsanen, H. Atomic layer deposition of ytterbium oxide using β-diketonate and ozone precursors. *Appl. Surf. Sci.* **2009**, *256*, 847–851. [CrossRef]

30. Desirena, H.; De la Rosa, E.; Diaz-Torres, L.A.; Kumar, G.A. Concentration effect of Er^{3+} ion on the spectroscopic properties of Er^{3+} and Yb^{3+}/Er^{3+} co-doped phosphate glasses. *Opt. Mater.* **2006**, *28*, 560–568. [CrossRef]

31. Goldner, P.; Pellé, F. Size dependence of the luminescence spectra and dynamics of Eu^{3+}: Y_2O_3 nanocrystals. *J. Lumin.* **1999**, *83*, 297–300.
32. Sun, J.M.; Prucnal, S.; Skorupa, W.; Helm, M.; Rebohle, L.; Gebel, T. Increase of blue electroluminescence from Ce-doped SiO_2 layers through sensitization by Gd^{3+} ions. *Appl. Phys. Lett.* **2006**, *89*, 091908. [CrossRef]
33. Ouyang, Z.; Yang, Y.; Sun, J.M. Electroluminescent Yb_2O_3:Er and $Yb_2Si_2O_7$:Er nanolaminate films fabricated by atomic layer deposition on silicon. *Opt. Mater.* **2018**, *80*, 209–215. [CrossRef]

 *nanomaterials*

Article

Preliminary Study of Ge-DLC Nanocomposite Biomaterials Prepared by Laser Codeposition

Miroslav Jelinek [1,2,*], Tomáš Kocourek [1,2], Karel Jurek [1], Michal Jelinek [3], Barbora Smolková [1], Mariia Uzhytchak [1] and Oleg Lunov [1]

[1] Institute of Physics of the Czech Academy of Sciences, Na Slovance 2, 182 21 Prague 8, Czech Republic; kocourek@fzu.cz (T.K.); jurek@fzu.cz (K.J.); smolkova@fzu.cz (B.S.); uzhytchak@fzu.cz (M.U.); lunov@fzu.cz (O.L.)

[2] Faculty of Biomedical Engineering, Czech Technical University in Prague, nam. Sitna 3105, 27 201 Kladno, Czech Republic

[3] Faculty of Nuclear Science and Physical Engineering, Czech Technical University in Prague, Brehova 7, 115 19 Prague 1, Czech Republic; michal.jelinek@fjfi.cvut.cz

* Correspondence: jelinek@fzu.cz; Tel.: +420-266-052-733

Received: 4 February 2019; Accepted: 13 March 2019; Published: 18 March 2019

Abstract: This paper deals with the synthesis and study of the properties of germanium-doped diamond-like carbon (DLC) films. For deposition of doped DLC films, hybrid laser technology was used. Using two deposition lasers, it was possible to arrange the dopant concentrations by varying the laser repetition rate. Doped films of Ge concentrations from 0 at.% to 12 at.% were prepared on Si (100) and fused silica (FS) substrates at room temperature. Film properties, such as growth rate, roughness, scanning electron microscopy (SEM) morphology, wavelength dependent X-ray spectroscopy (WDS) composition, VIS-near infrared (IR) transmittance, and biological properties (cytotoxicity, effects on cellular morphology, and ability to produce reactive oxygen species (ROS)) were studied in relation to codeposition conditions and dopant concentrations. The analysis showed that Ge-DLC films exhibit cytotoxicity for higher Ge doping.

Keywords: germanium; DLC; doped biomaterials; pulsed laser deposition; reactive oxygen species; apoptosis; cytotoxicity

---

1. Introduction

Recent advances in the application of body implantable medical devices have created a demand for studies developing new robust biocompatible materials with improved physicochemical properties [1]. Materials that are biocompatible and/or have biocompatible end degradation products are critically important for such devices. Diamond-like carbon (DLC) films represent a promising material for the modification of medical implants, providing high mechanical and chemical stability and a high degree of biocompatibility [2–4].

DLC chemical modifications with silver, germanium, and copper have been utilized to further improve the surface properties of DLC coatings in terms of biocompatibility and optical, electronic, and mechanical properties [4–8]. Among the modifications, germanium (Ge)-based DLCs and alloys have received growing interest in studies relating to biomedical application due to their improved physicochemical properties [9–13]. For instance, a recent study showed that Ge-DLC films had better mechanical properties (high hardness, high durability, low stress, low absorption, broad band infrared (IR) transparency, variable refraction index from 1.7 to 4) and good adhesion to many IR substrates compared with pure DLC films [14]. Furthermore, germanium carbide (GeC) thin films were shown to be easily fabricated by pulsed laser deposition (PLD) at room temperature [15]. Another study revealed that, for concentrations below about 10 at.%, carbon atoms were incorporated in the germanium lattice and created stabile GeC alloy [16]. Moreover, changes to band gap were observed in region from 1.6 to 2.8 eV for GeC films [16]. Radio frequency plasma-enhanced chemical vapor deposition (PECVD) was shown to be effective in creating protective antireflective DLC coating on germanium [17]. Indeed, enhancement in maximum transmission above 90% in the 3–6 µm wavelength range was found [17]. Additionally, one can generate multilayer DLC and germanium-doped DLC films [18]. Such films possess hardness above 48.1 GPa, which is almost the same as that of pure DLC film [18]. Compared with the pure DLC film, the critical load of Ge-DLC film on the germanium substrate increases from 71.6 mN to 143.8 mN [18]. Moreover, Ge-DLC film showed no change after fastness tests [18]. Importantly, germanium-doped DLC films with different germanium concentrations showed similar properties [19]. Furthermore, germanium-carbon (GexC1-x) alloy coatings have been shown to have low infrared optical absorption [20]. The Ge-DLC films prepared by pulsed vacuum arc deposition and electron beam evaporation showed that the Ge-doping significantly improved the optical and mechanical properties of the films [21]. Hollow-cathode plasma-enhanced chemical vapor deposition was shown to be effective in generation of DLC and Ge-DLC coatings [22]. Such coatings showed a significant anti-biofouling effect on *Pseudomonas aeruginosa* [22]. In contrast, neither modified DLC nor Ge-DLC showed any significant inhibitory effect on *Staphylococcus aureus* [22]. The overall surface energy of the Ge-DLC coating was approximately 40% larger than that of undoped modified DLC [22]. The resultant wettability was also higher, and the polar component of surface energy for Ge-DLC was significantly larger [22].

From the above literature review, we can conclude that Ge-DLC films represent a promising material for a wide range of applications, including in optics and biomedicine. However, from a material point of view, the results and data presented to date have sometimes been contradictory and neither fully nor systematically related to deposition parameters and Ge concentrations. No study of the properties on a larger and finer scale of Ge dopant concentrations has been performed. Indeed, there is a potential applicability of Ge alloys as biocompatible thin films due to their improved physicochemical properties [10–12]. Specifically, studies of biocompatibility testing of Ge alloys have shown that such alloys have presumably low cytotoxicity profile and are well tolerated by the organism [10–12]. Nonetheless, from a biological point of view, the number of such studies is limited. Indeed, only a few works have addressed the long-term cytotoxicity of germanium-based alloys so far [10–12]. However, it is difficult to make any reasonable conclusion out of these papers, because they lack crucial data on positive and negative controls during toxicity assessment [10–12]. It is worth noting here that long-term administration of high-dose germanium products presents a potential human health hazard [23]. Thus, it is of great importance to study the cytotoxic effects of Ge alloys on sensitive biological systems.

This study aims to present and discuss preliminary results obtained on Ge-doped DLC layers. We conducted a physical and biocompatibility study of Ge-doped DLC layers directly connected with dopant concentrations. Furthermore, we investigated the in vitro adhesion, proliferation, and toxicity of one of the very sensitive cell lines (hepatic, Huh7) upon culturing on Ge-doped DLC layers using various bioassays. We combined physicochemical analysis (the growth rate, roughness, morphology, composition, transmission) of Ge-doped DLC layers with their biological properties (cytotoxicity, effects on cellular morphology, and ability to produce reactive oxygen species (ROS)) for Ge concentrations ranging from zero up to 12 at.%. Here, we show a preliminary study of the cytotoxicity of Ge-based substrates with a wide range of germanium concentrations. However, long-term acute effects of such substrates on physiological functions of the cells have not been thoroughly investigated. Additionally, with our study, we aim to show that Ge-based films might have significant adverse effects and might not be as safe as they are considered.

2. Experimental

Deposition: Ge-DLC films were deposited by dual PLD using two krypton fluoride (KrF) excimer lasers (λ = 248 nm, τ = 20 ns), as shown in Figure 1. The flows of two target material were directed to rotating substrate. Our arrangement allowed, in a simple way, to prepare doped layers with a large scale of dopant concentrations by changing the laser fluence on targets and lasers' repetition rates. The first laser Compex Pro (Coherent Inc., Santa Clara, CA, USA) beam was focused onto a high-purity graphite target with the energy density of 8 J cm^{-2}. The repetition rate was set in the range from 18 Hz to 26 Hz. The second laser Lumonics PM-800 (Coherent Inc., Santa Clara, CA, USA) was focused onto a Ge target with the energy density of 1.5 J cm^{-2} and the repetition rate of 1 Hz to 23 Hz. The number of pulses was adjusted to reach approximately the same layer thickness. The substrate was in a distance of 40 mm away from the targets. The targets were rotated (0.5 Hz). To increase the films' adhesion, the substrates were cleaned by radiofrequency (RF) discharge before deposition for 2 min at 100 W power. The base vacuum of the coating system was 5×10^{-4} Pa. The films were codeposited in argon ambient (0.25 Pa) at room substrate temperature. Substrates of Si (100) were used for study of morphology and composition, and FS substrates were used for biotesting. Deposition conditions are summarized in Table 1.

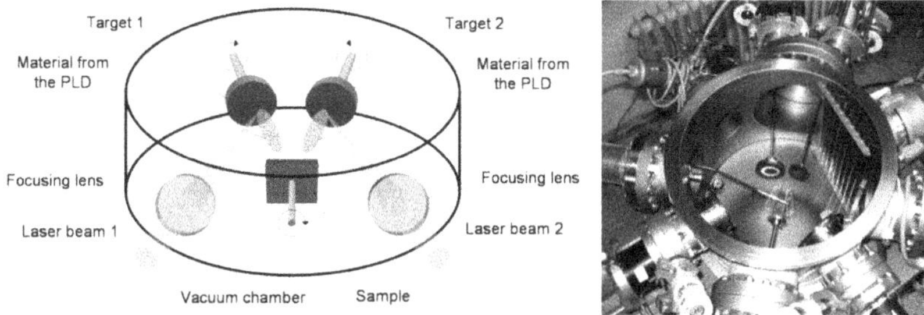

Figure 1. Dual beam laser deposition (schema—left, photo of chamber—right).

Table 1. Deposition conditions, Ge concentration, and film roughness (Ra) for germanium-doped diamond-like carbon (Ge-DLC) films prepared by double pulsed laser deposition (PLD) arrangement. Film thickness—cca 160 nm. Laser 1—energy density 8 J cm^{-2}, spot size 2 × 1 mm^{2}. Laser 2—energy density 1.5 J cm^{-2}, spot size 4 × 1.5 mm^{2}. WDS—wavelength dependent X-ray spectroscopy.

Sample No.	Laser 1—Compex (Graphite Target)		Laser 2—Lumonics (Ge Target)		Ge in DLC (WDS) [at.%]	Roughness Ra [nm]
	No. of Pulses	Rep. Rate [Hz]	No. of Pulses	Rep. Rate [Hz]		
5	21,685	14	29,429	19	12	120.1
6	29,083	30	17,450	18	9	83.6
7	31,891	27	12,992	11	5	63.6
8	36,216	30	6036	5	2.5	61.7
9	38,680	37	2091	2	1	14.4
10	40,000	30	-	-	0	2.8

Surface, thickness, roughness: The surface of the samples was analyzed by mechanical profilometer Alphastep IQ—KLA Tencor (KLA Corporation, Milpitas, CA, USA) and by SEM. The profilometer scan was 2 mm long and 50 μm s^{-1} fast; sampling frequency was 50 Hz, sensor range 13 μm, and stylus force was 4.5 mg. A cut-off value of 250 μm was used for calculation of the roughness parameters. A stylus with a 5 μm tip radius and 60° angle was used. Two scans were performed in two perpendicular directions. The roughness average (Ra for line) was adapted from ISO 4287/1 and calculated by either Alphastep or Gwyddion software.

Morphology, composition (SEM, WDS): The SEM morphology and composition of Ge-DLC thin layers was determined using an electron microprobe JEOL JXA 733 (JEOL Inc., Peabody, MA, USA) employing a wavelength dispersive X-ray spectrometer (WDS). The energy of primary electrons was kept at 3 keV to minimize their penetration depth and the absorption of emitted X-rays. For this energy, an electron spot diameter was estimated to be in the range 1–2 μm with the information depth of about 0.2 μm so that the contribution of the substrate could be neglected. In this case, the Ge Lα line was used for analysis. The accuracy of the measurement of Ge and C was better than 5%.

Transmittance of the Ge-DLC: films was measured using a spectrophotometer Shimadzu UV-3600 (Shimadzu, Kyoto, Japan) in a wavelength range from 200 to 3200 nm. A fused silica plate was used as a reference sample, and therefore the transmittance characteristics of the deposited films were measured.

Cell culture: Human hepatocellular carcinoma cell line (Huh7) obtained from the Japanese Collection of Research Bioresources (JCRB) was cultured in Eagle's minimum essential medium (EMEM; ATCC) supplemented with 10% fetal bovine serum (FBS, Thermo Fisher Scientific) and 0.1% (v/v) penicillin/streptomycin (Sigma, St. Louis, MO) as recommended by the supplier. Cultures were kept in a humidified 5% CO_2 atmosphere at 37 °C, and the medium was changed once a week [24].

Cell viability assay: Cytotoxic activity of the compounds was determined quantitatively by a fluorimetric assay utilizing propidium iodide (PI), as described earlier [25].

Prior to cell seeding, all experimental Ge-based substrates were sterilized by treatment with 70% ethanol for 20 min, followed by ultraviolet (UV) exposure for one hour. For all cell experiments in this study, the cells were seeded on the sterilized substrates at an initial density of 25,000 cells/cm^2 and were maintained under standard cell culture conditions (37 °C, 5% CO_2). Cells were allowed to grow on Ge-based substrates for 5 days under standard cell culture conditions [24]. When the cells were in the exponential growth phase, the monolayer was washed once in phosphate-buffered saline (PBS), then stained by the addition of 1 mL PI (50 μg/mL) to each dish for 5 min in the dark at room temperature. The nuclei were counterstained with Hoechst 33,342 (Thermo Fisher Scientific, Waltham, MA, USA). Fluorescence images were recorded with epifluorescent microscope IM-2FL (Optika Microscopes, Ponteranica, Italy). ImageJ (NIH) software was used for image processing and fluorescent micrograph quantification. Cell counting was carried out for five fields of view per sample. Three independent samples per each surface were assessed for cell proliferation measurements, and the reported values are the mean ± SEM (standard error of the mean). PI positive cells were considered to be dead cells. The survival rate was subsequently calculated using the following equation:

$$\text{Survival rate (\%)} = (\text{Hoechst positive cells} - \text{PI positive cells}) \times 100/\text{Hoechst positive cells}. \quad (1)$$

As a positive control, the cells were treated with 20% ethanol for 60 min.

Additionally, cell viability was analyzed by alamarBlue assay (Thermo Fisher Scientific, Waltham, MA, USA). AlamarBlue is a resazurin-based solution that functions as a cell health indicator using the reducing power of living cells to quantitatively measure viability. Resazurin, the active ingredient of alamarBlue reagent, is a nontoxic, cell-permeable compound that is blue in color and virtually nonfluorescent. Upon entering living cells, resazurin is reduced to resorufin, a compound that is red in color and highly fluorescent [26]. Cell growth and seeding were the same as for PI measurements. Cells were allowed to grow on Ge-based substrates for 5 days. Afterward, alamarBlue reagent was added to each dish and incubated for 0.5 h at 37 °C to form resorufin. The red fluorescence was measured using a TECAN microplate reader SpectraFluor Plus (TECAN, Mannedorf, Switzerland) at 590 nm. Three independent experiments were performed for each measurement. Cell viability was normalized to control values (no Ge exposure) and expressed as mean ± SEM.

Furthermore, we assessed viability using calcein AM method (Thermo Fisher Scientific, Waltham, MA, USA). Cells were grown for 5 days on germanium substrates as for PI or alamarBlue assays. Calcein AM provides a simple, rapid, and accurate method to measure cell viability and/or cytotoxicity. Calcein AM is a nonfluorescent, hydrophobic compound that easily permeates intact, live cells. The hydrolysis of calcein AM by intracellular esterases produces calcein, a hydrophilic, strongly fluorescent compound that is well retained in the cell cytoplasm [27]. After growing on substrates, cells were stained with calcein-AM (1 μM) for 30 min. The green fluorescence was measured using a TECAN microplate reader SpectraFluor Plus (TECAN, Mannedorf, Switzerland) at 488 nm. Three independent experiments were performed for each measurement. Cell viability was normalized to control values (no Ge exposure) and expressed as mean $\pm$ SEM. In order to confirm the validity of the live/dead staining in all cases, cells were also treated with 20% ethanol for 60 min and subsequently analyzed.

Detection of intracellular reactive oxygen species: ROS levels were measured using the Cellular ROS/Superoxide Detection Assay Kit (Abcam, Cambridge, UK). ROS levels were assessed as described previously [28]. After growing on Ge substrates, the cells were labeled with fluorescent reporter dyes, which are oxidized by ROS with high specificity, according to the manufacturer's instruction (Abcam). For total ROS detection, we used the cell permeant reagent $2',7'$-dichlorofluorescein diacetate (DCFDA), a fluorogenic dye that measures hydroxyl, peroxyl, and other ROS activity within the cell. Dihydroethidium (hydroethidine or DHE) was used for superoxide detection. After diffusion in to the cell, DCFDA is deacetylated by cellular esterases to a nonfluorescent compound, which is later oxidized by ROS into $2',7'$-dichlorofluorescin (DCF). DCF is a highly fluorescent compound, which can be detected by fluorescence spectroscopy and/or microscopy with maximum excitation and emission spectra of 495 nm and 529 nm, respectively [29]. DHE has been shown to be oxidized by superoxide to form 2-hydroxyethidium (2-OH-E$^+$) (ex 500–530 nm/em 590–620 nm) or by nonspecific oxidation by other sources of ROS to form ethidium (E$^+$) (ex 480 nm/em 576 nm) [29]. Thus, one can discriminate specifically the total ROS and superoxide. The nuclei were counterstained with Hoechst 33,342 (Thermo Fisher Scientific, Waltham, MA, USA). Fluorescence images were recorded with epifluorescent microscope IM-2FL (Optika Microscopes, Ponteranica, Italy). ImageJ (NIH) software was used for image processing and fluorescent micrograph quantification. Cell counting was carried out for five fields of view per sample. Three independent samples per each surface were assessed for cell proliferation measurements, and the reported values are the mean $\pm$ SEM. Cells treated with H_2O_2 (1 mM for 30 min) were used as a ROS positive control.

Detection of apoptosis: The Dead Cell Apoptosis Kit (Thermo Fisher Scientific, Waltham, MA, USA) was used to measure early apoptosis by detecting phosphatidylserine expression and membrane permeability. Huh7 were grown on different Ge substrates for 5 days. Afterward, cells were stained with Dead Cell Apoptosis Kit according to the manufacturer's instructions. Phosphatidylserine expression as an early sign of apoptosis was determined by the binding of Alexa Fluor 488 annexin V. Propidium iodide was used to differentiate necrotic cells. Hoechst 33,342 was used as nucleus staining. After staining, labeled cells were then imaged using spinning disk confocal microscopy IXplore SpinSR10 (Olympus, Tokyo, Japan). ImageJ software (NIH) was used for image processing and quantification. As positive control, 20 µM staurosporine for 2 h was used. ImageJ software was used for image processing and fluorescent micrograph quantification.

Spinning disk confocal microscopy: In order to visualize in great detail the morphological changes of Huh7 upon their growth on Ge substrates, we utilized brand new high-resolution spinning disk confocal microscopy (Spin SR, Olympus). Huh7 were grown on different Ge substrates for 5 days and labeled with propidium iodide—red dye, Hoechst 33,342 nuclear stain—blue. The merging of blue and red gives magenta color. Cell membranes were labeled with CellMask™ Green (green). Labeled cells were then imaged using high-resolution spinning disk confocal microscopy (SpinSR, Olympus). Fluorescence images were taken with the acquisition software cellSens (Olympus). ImageJ software (NIH) was used for image processing.

Statistical analysis: Data obtained from independent experiments are presented as the mean $\pm$ SEM. Statistical analysis was performed using one-way analysis of variance and the Newman–Keuls test. Differences were considered statistically significant at $p < 0.05$.

3. Results and Discussion

3.1. Growth Rate, Roughness, Morphology

From the profilometer measurements, we calculated the growth rate. With Ge doping changing from 12 at.% to zero at.%, the growth rate moved from 7.4×10^{-3} nm/pulse to 4.0×10^{-3} nm/pulse.

Roughness measurements showed that Ra changed from about 3 nm to 120 nm. From the SEM photos, we could see that the layer consisted of small "grain" droplets, and the surface was covered with grains of sizes up to several micrometers. For lower dopant concentrations, the surface was smoother (see Figure 2).

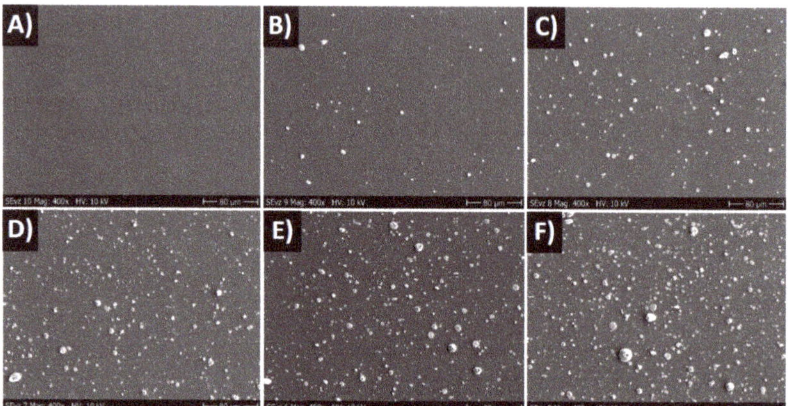

Figure 2. SEM images of Ge-doped DLC layers, 400× magnification. (**A**) 0 at.%, (**B**) 1 at.%, (**C**) 2.5 at.%, (**D**) 5 at.%, (**E**) 9 at.%, and (**F**) 12 at.% of Ge in DLC.

3.2. Composition (WDS)

The measured Ge dopant concentrations moved with deposition conditions from 0.0 at.% to 12 at.% (see Table 1). The WDS analyses showed that the Ge content in the grains was about 10 times higher compared with the smooth parts of the film.

3.3. Transmittance

The film transmittance in the UV-VIS and near-infrared regions generally decreased with increasing germanium content (Figure 3). For example, at 1700 nm, the transmittance decreased from 59% for pure DLC (0 at.% of Ge) to 34% for 12 at.% of Ge. The absorption peak around 2700 nm might have been caused by residual OH absorption of fused silica substrates. Our Ge-doped DLC films changed only the value of transparency in UV-VIS but were still transparent compared with pure germanium wafers, which exhibited transparency after ~1.7 micrometers.

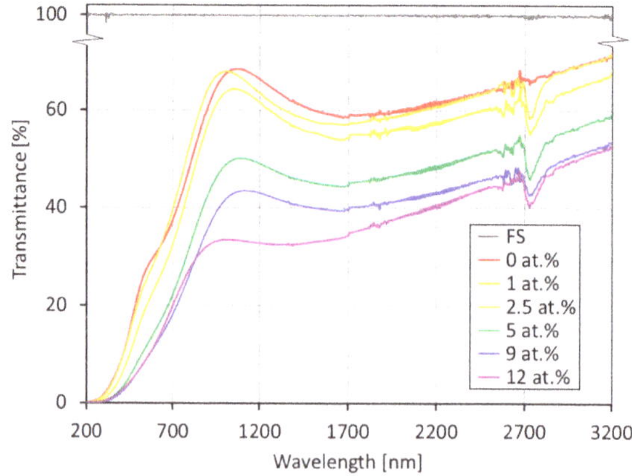

Figure 3. Transmission curves of DLC and Ge-doped DLC films in region up to 3.2 micrometers.

3.4. In Vitro Biocompatibility Evaluations

Biocompatibility of the materials themselves is a crucial initial step in the research devoted to biomedical applications of any tested material. We first studied how the substrates with different amounts of doped Ge influence the overall growth rate of the cells. Our research was focused on a specific cell type—hepatocytes. These cells are polarized, specialized, and species-specific, making them uniquely susceptible to infections [30]. Therefore, we utilized Huh7 cells for the studies. Huh7 is one of the most widely used in vitro model systems for the study of human hepatocytes [31].

It is worth noting here that for quite some time, germanium and germanium compounds were considered to be safe and used in dietary supplements [32]. Indeed, numerous reports have shown that germanium supplements present a potential hazard to humans at high doses [23]. However, till now, Ge-containing healthcare products are still available [10]. This fact has created a perception that Ge-based implants would be of high biocompatibility and well tolerated by the human body [10–12]. Thus, no systemic toxicity of Ge implants in humans has been reported. There is a lack of rigorous analysis of Ge-induced cytotoxicity on relevant cellular models. In fact, detailed studies in the past two decades have shown that redox active metals and metalloids (like Ge) undergo redox cycling reactions and possess the ability to produce reactive oxygen species [23,33,34]. Increased formation of ROS overwhelms cellular antioxidant protection and subsequently induces DNA damage, lipid peroxidation, and cell death [33]. Recently, we showed that among widely used hepatic cell lines, Huh7 are the most susceptible to redox imbalance and oxidative damage and proposed Huh7 a as fragile hepatic cell line [35–37]. Thus, to assess acute toxicity of Ge-based substrates, we used Huh7 cell line in our experiments.

Huh7 cells were cultured on substrates with different percentages of doped Ge for 5 days. After 5 days of culturing, the cell viability was assessed utilizing propidium iodide labeling. Propidium iodide (PI) is a well-known membrane impermeant dye that is generally excluded from viable cells. It binds to double-stranded DNA by intercalating between base pairs. Thus, an increase in PI-labeled cells reflects the cytotoxic response. Figure 4A shows the cytotoxic behavior of the cells grown on different Ge substrates as observed by phase-contrast microscopy (PhC) and dead assay by fluorescence imaging. For all samples, most Huh7 cells attached tightly to the surface of the materials and spread effectively during the culture period. Figure 4B summarizes the viability determined by a live/dead cell count from five different areas on each sample. Overall, we were able to subdivide substrates in three categories in accordance with their cytotoxic effects (Figure 4). FS, Ge O%, and Ge 1% substrates had low (nearly no) cytotoxicity; Ge 2.5% and Ge 5% had moderate toxicity; and Ge 9% and Ge 12% had highly toxic (Figure 4). The results gave a substantial indication of the biocompatibility of the investigated substrates. Indeed, propidium iodide staining shows dead cells with already permeabilized membrane. Thus, to assess in more detail the prerequisites of Ge-induced cytotoxicity, we utilized two metabolic activity assays (Figure 4C,D). Consistent with the PI viability results, the viability of Huh7 was concentration-dependently decreased after 5 days of exposure to substrates with different Ge concentrations. Both calcein AM (Figure 4D) and alamarBlue (Figure 4C) assays showed similar results for Ge-induced cytotoxicity. Moreover, viability results from these metabolic activity assays were in line with PI viability results (Figure 4).

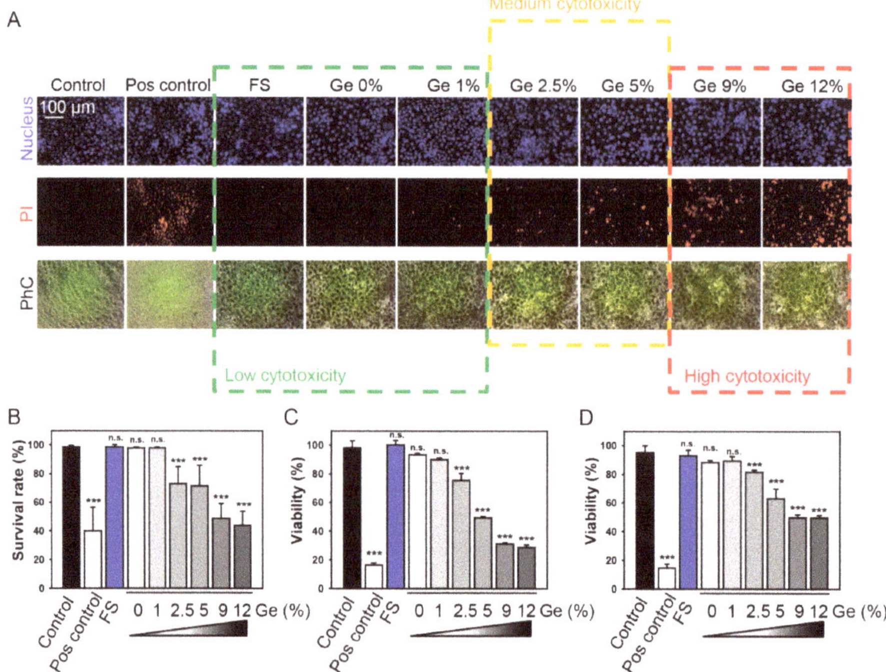

Figure 4. Cell viability of Huh7 cells on different Ge substrates after 5 days of cultivation. (**A**) Cells were stained with propidium iodide (PI, red) to assess cell viability. The nuclei were counterstained with Hoechst blue. Cells were imaged using epifluorescent microscope IM-2FL (OPTIKA Italy). Representative images are shown. As a positive control, cells were treated with 20% ethanol for 60 min. (**B**) Quantitative analysis of survival rate of cells grown on different Ge substrates. ImageJ software (NIH) was used for image processing and quantification. One-way ANOVA with Newman–Keuls multiple comparison test was used. Data are expressed as means $\pm$ SEM ($n = 3$), *** $p < 0.001$. (**C**) Cell viability as detected by the alamarBlue assay of Huh7 grown on different Ge substrates for 5 days, $n = 3$ each. The data were normalized to control values (no Ge exposure), which were set as 100% cell viability. As a positive control, cells were treated with 20% ethanol for 60 min. One-way ANOVA with Newman–Keuls multiple comparison test was used. Data are expressed as means $\pm$ SEM ($n = 3$), *** $p < 0.001$. (**D**) Cell viability as detected by the calcein AM assay of Huh7 grown on different Ge substrates for 5 days, $n = 3$ each. The data were normalized to control values (no Ge exposure), which were set as 100% cell viability. As a positive control, cells were treated with 20% ethanol for 60 min. One-way ANOVA with Newman–Keuls multiple comparison test was used. Data are expressed as means $\pm$ SEM ($n = 3$), *** $p < 0.001$.

The growth of Huh7 on Ge substrates for 5 days did not induce early signs of apoptosis [38,39], namely, translocation of phosphatidylserine to the outer cell membrane leaflet, as measured by the binding of Alexa488-labeled annexin V with concomitant increase in membrane permeability, as shown by propidium iodide staining (Figure 5A). Indeed, there was minor number of cells positive on annexin V in Huh7 grown on Ge substrates (Figure 5A). In contrast, there was a massive Ge concentration-dependent increase in membrane permeability, as shown by propidium iodide incorporation (Figure 5A). In fact, Huh7 treatment with the well-known compound staurosporine resulted in the formation of distinguished apoptotic hallmarks (Figure 5A). This constellation suggested that Ge substrates resulted in either late stage of apoptotic cell death or some necrotic events.

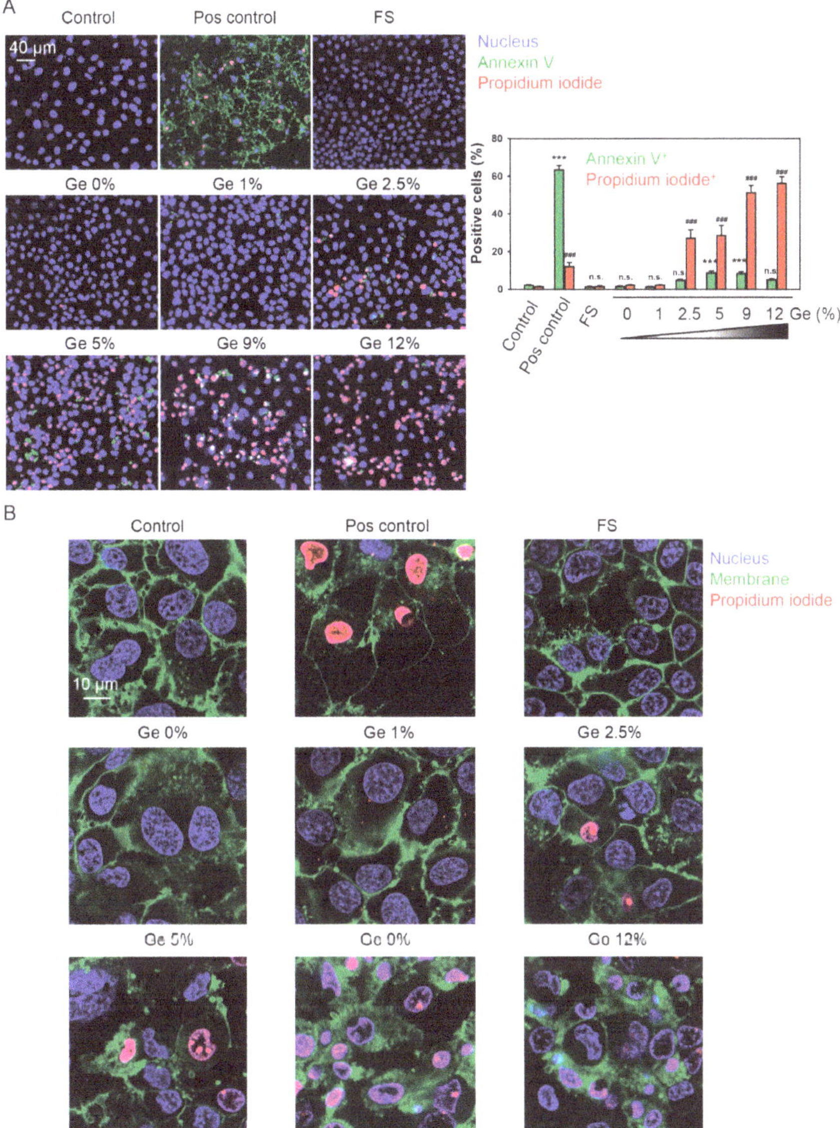

Figure 5. (**A**) Analysis of apoptotic cell death. Huh7 were grown on different Ge substrates for 5 days and labeled with annexin V—green dye, propidium iodide—red dye, Hoechst 33,342 nuclear stain—blue. The merging of blue and red gives magenta color. Labeled cells were then imaged using spinning disk confocal microscopy. ImageJ software (NIH) was used for image processing and quantification. One-way ANOVA with Newman–Keuls multiple comparison test was used. Data are expressed as means ± SEM (*n* = 3), *** *p* < 0.001. As positive control, 20 µM staurosporine for 2 h was used. (**B**) Huh7 were grown on different Ge substrates for 5 days and labeled with propidium iodide—red dye, Hoechst 33,342 nuclear stain—blue. The merging of blue and red gives magenta color. Cell membranes were labeled with CellMask™ Green (green). Labeled cells were then imaged using high-resolution spinning disk confocal microscopy. ImageJ software (NIH) was used for image processing. As a positive control, cells were treated with 20% ethanol for 60 min.

We further assessed how the morphology of cells changed upon culturing on Ge substrates. We stained cellular membrane with CellMask Green plasma membrane stain. As can be clearly see from Figure 5B and consistent with cell viability and annexin V staining, the cells grown on Ge substrates with 0%–1% germanium showed no significant changes in size, shape, and membrane morphology. However, starting from 2.5% germanium, cells exhibited significant morphological changes (Figure 5B). Huh7 exposure to substrates with high Ge concentrations ($\geq$2.5%) concentration-dependently resulted in massive CellMask intracellular incorporation, indicating membrane permeabilization (Figure 5B). Additionally, culturing on Ge substrates with high Ge concentrations ($\geq$2.5%) resulted in vesicular shedding (Figure 5B). Indeed, multivesicular bodies and microvesicles are shed from the plasma membrane during the cell death process [38]. Such vesicles can be exosomes, apoptotic, or necrotic bodies. To clearly assess the difference, one needs to do a more in-depth study on this matter. In general, vesicular shedding has been implicated in an increasing number of physiological and pathological contexts as mediators of local and systemic intercellular communication [40–42]. Such vesicles might augment immune and inflammatory responses [40–42]. Importantly, positive control (20% ethanol) treatment showed distinct morphologic changes in comparison with Ge substrates (Figure 5B). No vesicle shedding was observed, and cells showed ballooning morphology, indicating accidental necrotic cell death induced by ethanol (Figure 5B).

The next logical step was to check the ROS accumulation. Excess ROS results in oxidative stress and subsequent cell death. It is well known that ROS are emerging as key effectors in signal transduction [43]. Accumulating evidence also suggests that ROS play a major role in the mediation of metal-induced cellular responses [34]. It is worth noting that the ability of Ge-based materials to produce ROS in cells has not been tested previously [10–12].

We found that Ge substrates possess a dose-dependent ROS production in Huh7 cells (Figure 6A–C). As can be clearly seen from Figure 6B,C, Ge substrates induced dose-dependent ROS accumulation in cells with the highest amount of ROS, produced after cell culturing on substrates with higher (12%) Ge amount (Figure 6B,C). We used two distinct fluorescent probes. One probe was indicative of cellular production of different ROS types, while the other was superoxide ($O_2{}^-$)-specific. This allowed us to monitor changes in the total ROS level as well as specifically verify the level of superoxide. Cell culturing on Ge substrates triggered a dose-dependent accumulation of superoxide as well (Figure 6A–C). Indeed, substrates showed dose-dependent ROS and superoxide accumulation, which correlates with cytotoxicity.

To confirm the role for ROS in the induction of Ge cytotoxicity, we used the ROS scavenger *N*-acetyl-L-cysteine (NAC, a potent ROS scavenger [44–46]). As expected, the ROS scavenger was able to antagonize the cytotoxic effects elicited by Ge substrates on Huh7 cells (Figure 7). Importantly, cell death induced by 20% ethanol (used as a positive control) was not antagonized by NAC (Figure 7). High concentrations of ethanol induce accidental necrosis by the direct rupturing of cell membrane without concomitant accumulation of ROS [47]. These data confirmed a pivotal role of ROS accumulation in Ge-induced cytotoxicity.

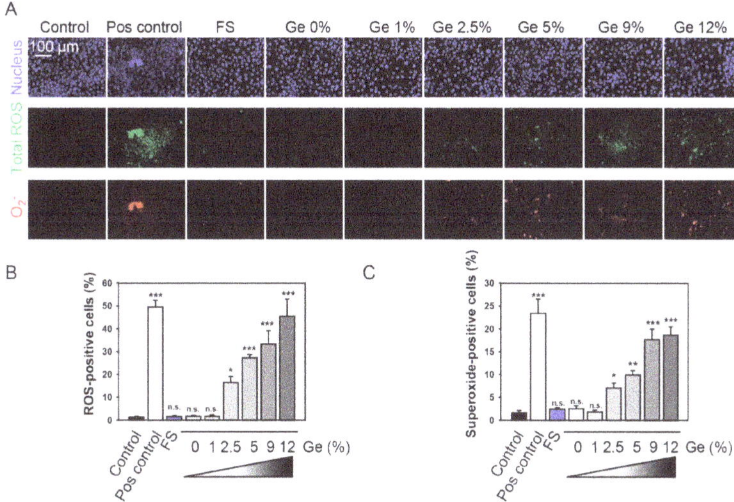

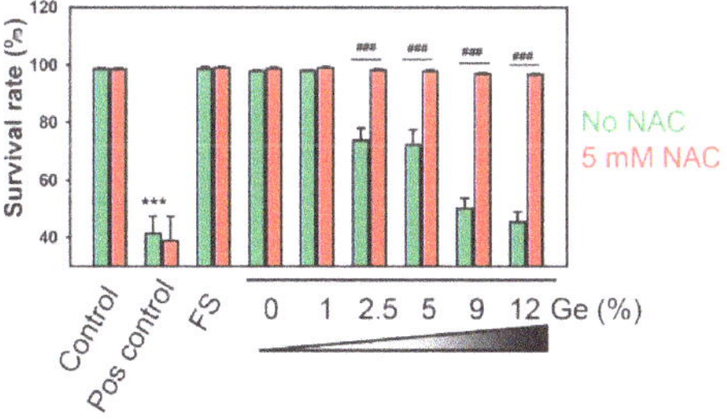

Figure 6. Cell growth on different Ge substrates results in reactive oxygen species (ROS) accumulation. (**A**) Cells were cultured on different Ge substrates for 5 days. Cells were labeled with the ROS-sensitive fluorescent dyes using the cellular ROS/superoxide detection kit (Abcam, Cambridge, United Kingdom). Total ROS were labeled with green dye and superoxide anion with red dye. The nuclei were counterstained with Hoechst blue. Cells were imaged using epifluorescent microscope IM-2FL (OPTIKA Italy). Cells treated with H_2O_2 (1 mM for 30 min) were used as a ROS positive control. Representative images out of three independent experiments are shown. (**B**) Quantitative analysis of total ROS-positive cells cultured on different Ge substrates for 5 days. ImageJ software (NIH) was used for image processing and quantification. One-way ANOVA with Newman–Keuls multiple comparison test was used. Data are expressed as means ± SEM ($n = 3$), * $p < 0.05$ ** $p < 0.01$ *** $p < 0.001$. (**C**) Quantitative analysis of total superoxide-positive cells cultured on different Ge substrates for 5 days. ImageJ software (NIH) was used for image processing and quantification. One-way ANOVA with Newman–Keuls multiple comparison test was used. Data are expressed as means ± SEM ($n = 3$), * $p < 0.05$ ** $p < 0.01$ *** $p < 0.001$.

Figure 7. Treatment with ROS scavenging agent *N*-acetyl-L-cysteine (NAC) completely abolished the cytotoxicity of Ge substrates. Huh7 cells were grown on different Ge substrates for 5 days. To inhibit ROS accumulation, Huh7 were supplemented with 5 mM NAC. Cells were stained with propidium iodide to assess cell viability. The nuclei were counterstained with Hoechst blue. Afterward, cells were imaged using epifluorescent microscope IM-2FL (OPTIKA Italy), and quantitative analysis of survival rate of cells was then done using ImageJ software (NIH). One-way ANOVA with Newman–Keuls multiple comparison test was used. Data are expressed as means ± SEM ($n = 3$), *** $p < 0.001$. As a positive control, cells were treated with 20% ethanol for 60 min.

To summarize this part, we used a hepatic cell line to assess cytotoxicity mediated by Ge substrates and found that concentration of Ge up to 1% in the substrates is not toxic for cell culture. In contrast, concentrations of Ge higher than 5% showed substantial degree of cytotoxicity. Furthermore, we identified the source of Ge-mediated toxicity. Indeed, high Ge concentrations (>2.5%) in the substrates resulted in intracellular ROS production, the accumulation of which leads to cell death.

4. Conclusions

Ge-doped DLC layers with Ge doping up to 12 at.% were prepared using two-laser codeposition from Ge and graphite targets. Film properties, such as growth rate, roughness, SEM morphology, WDS composition, VIS-near IR transmittance, and biological properties (cytotoxicity, effects on cellular morphology, and ability to produce ROS) were studied in relation to dopant concentrations. The growth rate of the films was low (from 4.0×10^{-3} nm/pulse to 7.4×10^{-3} nm/pulse). Roughness (Ra) increased with germanium doping. Transparency decreased with Ge doping, but the shape of the curves was similar to that for pure (undoped) DLC films. Analysis showed that Ge-DLC films exhibit cytotoxicity for higher Ge doping.

Generally, systemic studies on toxicity of Ge are still rather limited [23]. A number of studies have shown that Ge supplements present a potential hazard to humans at high doses [23]. Ge compounds are relatively less toxic compared with other metalloids and metals. However, relatively high doses of germanium dioxide and other inorganic Ge compounds can cause severe poisoning, including death [23]. In the present study, we identified a threshold for Ge concentration in cell culture substrate to avoid severe toxic reaction. We found that Ge concentrations higher than 2.5% induced signs of cell death in hepatic cells. Interestingly, Ge-based materials with Ge concentration as low as 1.5% have already been shown to exhibit cytotoxic effects [10–12]. In this regard, our Ge 1% material showed less toxicity in comparison to 1.5 Ge described in [10]. Although ROS produced during physiological processes are rapidly inactivated by antioxidant enzymes, the excess of ROS can induce apoptotic cell death. ROS have been identified as a major reason for metalloids- and metals-induced cytotoxicity [34]. We revealed that Ge concentrations higher than 2.5% resulted in ROS production and cellular accumulation. Excess of ROS production mediated by Ge substrates results in cell death. Therefore, the cytotoxicity of Ge-based materials should be carefully considered when transitioning to clinical practice. Our results imply that the cytotoxic effects of Ge substrates require more intensive study and that they should be considered in biomedical applications.

Author Contributions: O.L., and M.J. designed experiments and interpreted data. M.J., T.K., K.J. and M.J. carried out synthesis and physicochemical characterization of germanium-doped diamond-like carbon. M.J. provided critical input to the overall research direction. B.S., M.U. and O.L. performed epifluorescent and confocal imaging analysis. B.S. and M.U. cultured cells, performed cytotoxicity analysis. All authors discussed and analyzed the data. O.L. and M.J. wrote the manuscript with input from all co-authors.

Funding: We would like to thank CTU Grant Agency for their support by grant SGS18/157/OHK4/2T/17. The work was also supported by Operational Program Research, Development, and Education financed by European Structural and Investment Funds and the Czech Ministry of Education, Youth, and Sports (Project No. SOLID21 - CZ.02.1.01/0.0/0.0/16_019/0000760).

Conflicts of Interest: The authors declare no conflict of interest.

References

1. Joung, Y.H. Development of implantable medical devices: From an engineering perspective. *Int. Neurourol. J.* **2013**, *17*, 98–106. [CrossRef] [PubMed]
2. Thorwarth, G.; Falub, C.V.; Muller, U.; Weisse, B.; Voisard, C.; Tobler, M.; Hauert, R. Tribological behavior of DLC-coated articulating joint implants. *Acta Biomater.* **2010**, *6*, 2335–2341. [CrossRef] [PubMed]
3. Thomson, L.A.; Law, F.C.; Rushton, N.; Franks, J. Biocompatibility of diamond-like carbon coating. *Biomaterials* **1991**, *12*, 37–40. [CrossRef]

4. Gorzelanny, C.; Kmeth, R.; Obermeier, A.; Bauer, A.T.; Halter, N.; Kumpel, K.; Schneider, M.F.; Wixforth, A.; Gollwitzer, H.; Burgkart, R.; et al. Silver nanoparticle-enriched diamond-like carbon implant modification as a mammalian cell compatible surface with antimicrobial properties. *Sci. Rep.* **2016**, *6*, 22849. [CrossRef] [PubMed]

5. Harrasser, N.; Jussen, S.; Obermeir, A.; Kmeth, R.; Stritzker, B.; Gollwitzer, H.; Burgkart, R. Antibacterial potency of different deposition methods of silver and copper containing diamond-like carbon coated polyethylene. *Biomater. Res.* **2016**, *20*, 17. [CrossRef] [PubMed]

6. Dwivedi, N.; Kumar, S.; Carey, J.D.; Tripathi, R.K.; Malik, H.K.; Dalai, M.K. Influence of silver incorporation on the structural and electrical properties of diamond-like carbon thin films. *ACS Appl. Mater. Interfaces* **2013**, *5*, 2725–2732. [CrossRef] [PubMed]

7. Palyanov, Y.N.; Kupriyanov, I.N.; Borzdov, Y.M.; Surovtsev, N.V. Germanium: A new catalyst for diamond synthesis and a new optically active impurity in diamond. *Sci. Rep.* **2015**, *5*, 14789. [CrossRef] [PubMed]

8. Wei, Q.; Sharma, A.K.; Sankar, J.; Narayan, J. Mechanical properties of diamond-like carbon composite thin films prepared by pulsed laser deposition. *Compos. Part B-Eng.* **1999**, *30*, 675–684. [CrossRef]

9. Evans, A.C.; Franks, J.; Revell, P.J. Diamond-like carbon applied to bioengineering materials. *Surf. Coat. Tech.* **1991**, *47*, 662–667. [CrossRef]

10. Bian, D.; Zhou, W.; Deng, J.; Liu, Y.; Li, W.; Chu, X.; Xiu, P.; Cai, H.; Kou, Y.; Jiang, B.; et al. Development of magnesium-based biodegradable metals with dietary trace element germanium as orthopaedic implant applications. *Acta Biomater.* **2017**, *64*, 421–436. [CrossRef]

11. Hwang, S.W.; Park, G.; Edwards, C.; Corbin, E.A.; Kang, S.K.; Cheng, H.; Song, J.K.; Kim, J.H.; Yu, S.; Ng, J.; et al. Dissolution chemistry and biocompatibility of single-crystalline silicon nanomembranes and associated materials for transient electronics. *ACS Nano* **2014**, *8*, 5843–5851. [CrossRef]

12. Kang, S.K.; Park, G.; Kim, K.; Hwang, S.W.; Cheng, H.; Shin, J.; Chung, S.; Kim, M.; Yin, L.; Lee, J.C.; et al. Dissolution chemistry and biocompatibility of silicon- and germanium-based semiconductors for transient electronics. *ACS Appl. Mater. Interfaces* **2015**, *7*, 9297–9305. [CrossRef]

13. Adiga, S.P.; Jin, C.; Curtiss, L.A.; Monteiro-Riviere, N.A.; Narayan, R.J. Nanoporous membranes for medical and biological applications. *Int. Rev. Nanomed. Nanobiotechnol.* **2009**, *1*, 568–581. [CrossRef]

14. Sousani, F.; Jamali, H.; Mozafarinia, R.; Eshaghi, A. Thermal stability of germanium-carbon coatings prepared by a RF plasma enhanced chemical vapor deposition method. *Infrared Phys. Techn.* **2018**, *93*, 255–259. [CrossRef]

15. Mahmood, A.; Iqbal, M.; Ali, Z.; Shafi, H.Z.; Shah, A.; Batani, D. Optical analysis of germanium carbide thin films deposited by reactive pulsed laser ablation. *J. Laser Micro Nanoen.* **2010**, *5*, 204–208. [CrossRef]

16. Mahmood, A.; Shah, A.; Castillon, F.F.; Araiza, L.C.; Heiras, J.; Raja, M.Y.A.; Khizar, M. Surface analysis of GeC prepared by reactive pulsed laser deposition technique. *Curr. Appl. Phys.* **2011**, *11*, 547–550. [CrossRef]

17. Varade, A.; Krishna, A.; Reddy, K.N.; Chellamalai, M.; Shashikumar, P.V. Diamond-like carbon coating made by RF plasma enhanced chemical vapour deposition for protective antireflective coatings on germanium. *Proc. Mat. Sci.* **2014**, *5*, 1015–1019. [CrossRef]

18. Cheng, Y.; Lu, Y.M.; Guo, Y.L.; Huang, G.J.; Wang, S.Y.; Tian, F.T. Multilayers diamond-like carbon film with germanium buffer layers by pulsed laser deposition. *Surf. Rev. Lett.* **2017**, *24*, 02. [CrossRef]

19. Ankit, K.; Varade, A.; Reddy, K.N.; Dhan, S.; Chellamalai, M.; Balashanmugam, N.; Krishna, P. Synthesis of high hardness IR optical coating using diamond-like carbon by PECVD at room temperature. *Diam. Relat. Mater.* **2017**, *78*, 39–43.

20. Martin, P.M.; Johnston, J.W.; Bennett, W.D. Properties of reactively-deposited Sic and Gec alloys. *P. Soc. Photo-Opt. Ins.* **1990**, *1323*, 291–298.

21. Lu, Y.M.; Huang, G.J.; Guo, Y.L.; Wang, S.Y. Diamond-like carbon film with gradient germanium-doped buffer layer by pulsed laser deposition. *Surf. Coat. Tech.* **2018**, *337*, 290–295. [CrossRef]

22. Robertson, S.N.; Gibson, D.; MacKay, W.G.; Reid, S.; Williams, C.; Birney, R. Investigation of the antimicrobial properties of modified multilayer diamond-like carbon coatings on 316 stainless steel. *Surf. Coat. Tech.* **2017**, *314*, 72–78. [CrossRef]

23. Tao, S.H.; Bolger, P.M. Hazard assessment of germanium supplements. *Regul. Toxicol. Pharmacol.* **1997**, *25*, 211–219. [CrossRef]

24. Lunova, M.; Zablotskii, V.; Dempsey, N.M.; Devillers, T.; Jirsa, M.; Sykova, E.; Kubinova, S.; Lunov, O.; Dejneka, A. Modulation of collective cell behaviour by geometrical constraints. *Integr. Biol.* **2016**, *8*, 1099–1110. [CrossRef] [PubMed]

25. Wrobel, K.; Claudio, E.; Segade, F.; Ramos, S.; Lazo, P.S. Measurement of cytotoxicity by propidium iodide staining of target cell DNA. Application to the quantification of murine TNF-alpha. *J. Immunol. Methods* **1996**, *189*, 243–249. [CrossRef]

26. Back, S.A.; Khan, R.; Gan, X.; Rosenberg, P.A.; Volpe, J.J. A new Alamar Blue viability assay to rapidly quantify oligodendrocyte death. *J. Neurosci. Methods* **1999**, *91*, 47–54. [CrossRef]

27. Braut-Boucher, F.; Pichon, J.; Rat, P.; Adolphe, M.; Aubery, M.; Font, J. A non-isotopic, highly sensitive, fluorimetric, cell-cell adhesion microplate assay using calcein AM-labeled lymphocytes. *J. Immunol. Methods* **1995**, *178*, 41–51. [CrossRef]

28. Lynnyk, A.; Lunova, M.; Jirsa, M.; Egorova, D.; Kulikov, A.; Kubinova, S.; Lunov, O.; Dejneka, A. Manipulating the mitochondria activity in human hepatic cell line Huh7 by low-power laser irradiation. *Biomed. Opt. Express* **2018**, *9*, 1283–1300. [CrossRef]

29. Buettner, G.R. Moving free radical and redox biology ahead in the next decade(s). *Free Radic. Biol. Med.* **2015**, *78*, 236–238. [CrossRef]

30. March, S.; Ramanan, V.; Trehan, K.; Ng, S.; Galstian, A.; Gural, N.; Scull, M.A.; Shlomai, A.; Mota, M.M.; Fleming, H.E.; et al. Micropatterned coculture of primary human hepatocytes and supportive cells for the study of hepatotropic pathogens. *Nat. Protoc.* **2015**, *10*, 2027–2053. [CrossRef]

31. Treyer, A.; Musch, A. Hepatocyte polarity. *Compr. Physiol.* **2013**, *3*, 243–287.

32. Schauss, A.G. Nephrotoxicity and neurotoxicity in humans from organogermanium compounds and germanium dioxide. *Biol. Trace Elem. Res.* **1991**, *29*, 267–280. [CrossRef]

33. Jomova, K.; Valko, M. Advances in metal-induced oxidative stress and human disease. *Toxicology* **2011**, *283*, 65–87. [CrossRef]

34. Tchounwou, P.B.; Yedjou, C.G.; Patlolla, A.K.; Sutton, D.J. Heavy metals toxicity and the environment. *Mol. Clin. Environ. Toxicol.* **2012**, *101*, 133–164.

35. Lunova, M.; Smolkova, B.; Lynnyk, A.; Uzhytchak, M.; Jirsa, M.; Kubinova, S.; Dejneka, A.; Lunov, O. Targeting the mTOR signaling pathway utilizing nanoparticles: A critical overview. *Cancers* **2019**, *11*, 82. [CrossRef]

36. Smolkova, B.; Lunova, M.; Lynnyk, A.; Uzhytchak, M.; Churpita, O.; Jirsa, M.; Kubinova, S.; Lunov, O.; Dejneka, A. Non-thermal plasma, as a new physicochemical source, to induce redox imbalance and subsequent cell death in liver cancer cell lines. *Cell. Physiol. Biochem.* **2019**, *52*, 119–140.

37. Lunova, M.; Prokhorov, A.; Jirsa, M.; Hof, M.; Olzynska, A.; Jurkiewicz, P.; Kubinova, S.; Lunov, O.; Dejneka, A. Nanoparticle core stability and surface functionalization drive the mTOR signaling pathway in hepatocellular cell lines. *Sci. Rep.* **2017**, *7*, 16049. [CrossRef]

38. Fink, S.L.; Cookson, B.T. Apoptosis, pyroptosis, and necrosis: Mechanistic description of dead and dying eukaryotic cells. *Infect. Immun.* **2005**, *73*, 1907–1916. [CrossRef]

39. Zhang, Y.; Chen, X.; Gueydan, C.; Han, J. Plasma membrane changes during programmed cell deaths. *Cell Res.* **2018**, *28*, 9–21. [CrossRef]

40. Lynch, C.; Panagopoulou, M.; Gregory, C.D. Extracellular vesicles arising from apoptotic cells in tumors: Roles in cancer pathogenesis and potential clinical applications. *Front. Immunol.* **2017**, *8*, 1174. [CrossRef]

41. Caruso, S.; Poon, I.K.H. Apoptotic cell-derived extracellular vesicles: More than just debris. *Front. Immunol.* **2018**, *9*, 1486. [CrossRef]

42. Gyorgy, B.; Szabo, T.G.; Pasztoi, M.; Pal, Z.; Misjak, P.; Aradi, B.; Laszlo, V.; Pallinger, E.; Pap, E.; Kittel, A.; et al. Membrane vesicles, current state-of-the-art: Emerging role of extracellular vesicles. *Cell. Mol. Life Sci.* **2011**, *68*, 2667–2688. [CrossRef]

43. Winterbourn, C.C. Reconciling the chemistry and biology of reactive oxygen species. *Nat. Chem. Biol.* **2008**, *4*, 278–286. [CrossRef]

44. Sun, S.Y. N-acetylcysteine, reactive oxygen species and beyond. *Cancer Biol. Ther.* **2010**, *9*, 109–110. [CrossRef]

45. Halasi, M.; Wang, M.; Chavan, T.S.; Gaponenko, V.; Hay, N.; Gartel, A.L. ROS inhibitor N-acetyl-L-cysteine antagonizes the activity of proteasome inhibitors. *Biochem. J.* **2013**, *454*, 201–208. [CrossRef]

46. Smolkova, B.; Uzhytchak, M.; Lynnyk, A.; Kubinova, S.; Dejneka, A.; Lunov, O. A critical review on selected external physical cues and modulation of cell behavior: Magnetic nanoparticles, non-thermal plasma and lasers. *J. Funct. Biomater.* **2019**, *10*, 2. [CrossRef]

47. Castilla, R.; Gonzalez, R.; Fouad, D.; Fraga, E.; Muntane, J. Dual effect of ethanol on cell death in primary culture of human and rat hepatocytes. *Alcohol Alcohol.* **2004**, *39*, 290–296. [CrossRef]

 *nanomaterials* 

Article

Surface Characteristics and Biological Evaluation of Si-DLC Coatings Fabricated Using Magnetron Sputtering Method on Ti6Al7Nb Substrate

Dorota Bociaga [1,*], Anna Sobczyk-Guzenda [1], Piotr Komorowski [1,2], Jacek Balcerzak [3], Krzysztof Jastrzebski [1], Karolina Przybyszewska [1] and Anna Kaczmarek [1]

[1] Faculty of Mechanical Engineering, Institute of Materials Science and Engineering, Lodz University of Technology, 1/15 Stefanowskiego St., 90-924 Lodz, Poland; anna.sobczyk-guzenda@p.lodz.pl (A.S.-G.); piotr.komorowski@p.lodz.pl (P.K.); kj.drakkainen@gmail.com (K.J.); przybyszewska.karolina94@gmail.com (K.P.); anna.m.kaczmarek90@gmail.com (A.K.)

[2] Bionanopark Ltd., Molecular and Nanostructural Biophysics Laboratory, 114/116 Dubois St., 93-465 Lodz, Poland

[3] Faculty of Process and Environmental Engineering, Department of Molecular Engineering, Lodz University of Technology, 213 Wolczanska St., 90-924 Lodz, Poland; jacek.balcerzak@p.lodz.pl

*   Correspondence: dorota.bociaga@p.lodz.pl; Tel.: +48-513-155-177

Received: 4 May 2019; Accepted: 27 May 2019; Published: 29 May 2019

Abstract: Diamond-like carbon (DLC) coatings are well known as protective coatings for biomedical applications. Furthermore, the incorporation of different elements, such as silicon (Si), in the carbon matrix changes the bio-functionality of the DLC coatings. This has also been proven by the results obtained in this work. The Si-DLC coatings were deposited on the Ti6Al7Nb alloy, which is commonly used in clinical practice, using the magnetron sputtering method. According to the X-ray photoelectron spectroscopy (XPS) analysis, the content of silicon in the examined coatings varied from ~2 at.% up to ~22 at.%. Since the surface characteristics are key factors influencing the cell response, the results of the cells' proliferation and viability assays (live/dead and XTT (colorimetric assays using tetrazolium salt)) were correlated with the surface properties. The surface free energy (SFE) measurements, Fourier transform infrared spectroscopy (FTIR), and X-ray photoelectron spectroscopy (XPS) analysis demonstrated that the polarity and wettability of the surfaces examined increase with increasing Si concentration, and therefore the adhesion and proliferation of cells was enhanced. The results obtained revealed that the biocompatibility of Si-doped DLC coatings, regardless of the Si content, remains at a very high level (the observed viability of endothelial cells is above 70%).

Keywords: DLC bio-functionality; silicon doping; diffusion barrier; biocompatibility; proliferation improvement; endothelial cells

---

1. Introduction

Implants are introduced into the body in order to recreate different mechanical and biological functions, and therefore improve the quality of human life or even save it. Since implants remain in constant contact with the living tissues and body fluids, they must fulfill strict requirements concerning their mechanical and biological properties. Currently, the most frequently implanted biomaterials are metals, which may be further classified into three main groups, i.e., stainless steels, titanium-based alloys, and cobalt-based alloys [1–3]. Metallic biomaterials possess high strength, resistance to fracture, as well as good corrosion resistance. However, biomaterials, in order to be safely inserted into the living organism, should also exhibit high biocompatibility, which means they must not cause any adverse tissue reactions such as acute and chronic inflammation or irritation of the surrounding tissues. Despite good mechanical properties, even metals with good corrosion resistance do not show full chemical stability in the highly aggressive environment of the human body. Therefore, metallic biomaterials may release harmful degradation products, which includes metal ions that can cause a negative biological response. The released degradation products not only accumulate in the surrounding tissues and organs, but also induce inflammation [4] and allergic reactions [5]. As a consequence, post-implantation infection may occur and lead to the revision or removal of the implant [6–9].

Hence, today, growing attention has been given to the modification of the surface of metallic biomaterials in order to enhance their properties, also in the context of a biological response [10]. One of the most commonly investigated solutions is the use of diamond-like carbon (DLC) coatings [11,12]. According to the numerous studies carried out by different scientific groups, DLC films exhibit high wear resistance [13], very good physiochemical properties [14], superior corrosion resistance [15], as well as excellent bio- and hemocompatibility [16–21]. Therefore, diamond-like carbon coatings can act as protecting coatings for implants and reduce the risk of adverse tissue reactions due to the implantation [22].

Moreover, the incorporation of different elements into the carbon matrix may result in even further improvement of certain properties of DLC films. Depending on the dopant used, different features may be achieved, including varying bio-functionality which determines the potential biomedical applications of the doped diamond-like carbon films. However, according to the literature, the biological response towards the doped DLC coatings depends not only on the element used, but also on the physiochemical properties of the surface. Those are in turn controlled by the synthesis method, different parameters of the deposition process, the form and amount of the incorporated dopant, as well as the substrate itself.

One of the most vastly investigated elements applied as a dopant for DLC coatings is silicon (Si), which according to the literature improves hemocompatibility, cell proliferation, and antibacterial resistance [23–27]. De Scheerder et al. also proved that the inflammatory reactions may be inhibited by the incorporation of SiO_x in the diamond-like carbon films [28]. Furthermore, the addition of Si into the DLC matrix enhances its chemical, mechanical, as well as tribological characteristics, causing the increase of hardness along with the reduction of the friction coefficient and residual stress [29–32]. As a result, the Si-incorporated carbon films may serve as a barrier which effectively prevents the diffusion of metal ions from the substrate, and therefore improves the biocompatibility of metallic biomaterials.

Despite numerous studies concerning the Si-DLC coatings, there are only a few reports concerning the evaluation of Si-DLC coatings deposited on titanium-based alloys using the magnetron sputtering technique. Therefore, the aim of this work was to perform a complex study concerning the physiochemical and biological characteristics of Si-doped DLC coatings deposited on Ti6Al7Nb substrates via co-sputtering of silicon and graphite targets.

2. Materials and Methods

2.1. Deposition of the Diamond-Like Carbon (DLC) DLC and Si-DLC Coatings

The DLC and Si-DLC coatings were both fabricated on Ti6Al7Nb substrates (Bibus Metals Sp. Z o.o. [Ltd.], Dabrowa, Poland). The applied discoidal samples had a diameter of 16 mm and a thickness of 6 mm. Before the deposition process, all substrates were ground and polished using the OP-S silica suspension (StruersApS). This was followed by ultrasonic cleaning in acetone for 10 min.

For the deposition of the examined DLC and Si-DLC films the multi-target magnetron sputtering system (PREVAC Sp. z o.o. [Ltd.], Rogow, Poland) was used. The apparatus consisted of a chamber with peripheral accessories, a control cabinet, and a computer enabling remote control of individual components. The chamber was equipped with three magnetron guns ION'X®(Thin Film Consulting, Grafenberg, Germany) with individual cooling systems. Two of them worked in a constant current mode, Pinnacle series power supplies (Advanced Energy, Fort Collins, CO, USA), while the third one was operated using a high frequency RF generator, Cesar RF Power Generator (Advanced Energy, Fort

Collins, CO, USA). The working pressure was ensured by a pump system consisting of a SCROLLVAC CS 30D primary pump (Leybold GmbH, Köln, Germany) and Turbovac SL700 turbo pump (Leybold GmbH, Köln, Germany). The rotary pump was used both to obtain the initial vacuum in the working chamber and to preserve the low operating pressure of the turbo pump. The system was equipped with two graphite (DC—direct current) and one silicon (RF—radio frequency) sputter targets (Kurt J. Lesker Company, Jefferson Hills, PA, USA) with a purity of 99.999%. In order to assure the uniform thickness of the coatings the samples were placed on a rotary table moving at a constant velocity of 0.33 rpm. Prior to the deposition, the plasma etching in argon atmosphere was carried out for 30 minutes at the bias of −700 V and pressure equal to 1 Pa. The input power of both graphite targets during the synthesis process of the DLC and Si-DLC coatings was 200 W each, while the pressure value was kept constant at 0.6 Pa with the Ar flow of 10 sccm. The silicon was introduced by co-sputtering of the silicon target with the input power varying from 0 to 80 W in order to achieve different Si concentrations. The deposition time was equal to 60 min.

2.2. Surface Morphology by Scanning Electron Microscopy (SEM)

The surface morphology of the deposited DLC and Si-DLC coatings was analyzed using the scanning electron microscope JSM-6610LV (JEOL). The secondary (SE) electron imaging mode was applied in order to obtain topographical contrast, while the backscattered (BSE) electrons imaging mode was used for compositional contrast. The applied accelerating voltage was equal to 20 kV. All of the performed SEM observations were carried out under high vacuum.

2.3. Chemical Composition and Bonding by X-ray Photoelectron Spectroscopy (XPS)

The surface chemical composition of the examined DLC and Si-DLC films was investigated by X-ray photoelectron spectroscopy. For this purpose, the Kratos AXIS Ultra spectrometer was used. The system was equipped with a monochromatic Al Kα X-ray source with an excitation energy of 1486.6 eV. High resolution measurements were performed with the power of the anode equal to 150 W and the pass energy of the hemispherical electron energy analyzer set to 20 eV. The XPS spectra were collected from the analysis areas of 300 μm × 700 μm. The obtained data were used to establish a quantitative elemental composition of the coatings. Moreover, the atomic bonds between the elements were studied in detail.

2.4. Chemical Structure by Fourier Transform Infrared Spectroscopy (FTIR)

The analysis of the chemical structure of the deposited DLC and Si-DLC coatings was carried out using Fourier transform infrared spectroscopy (FTIR). Investigations were performed using a Nicolet iS50 spectrophotometer (Thermo Scientific) operated in the absorbance mode in the range of 1700–500 cm^{-1}. The resolution of spectral measurements was equal to 1 cm^{-1} and a single measurement cycle consisted of 120 scans.

2.5. Surface Free Energy (SFE) and Wettability

The wettability of the DLC and Si-DLC coatings was measured using the sessile drop technique for two liquids of different polarity and known surface tension, i.e., distilled water and diiodomethan (Sigma Aldrich). The FM40 Easy Drop system with Drop Shape Analysis software (Krüss GmbH, Hamburg, Germany) was applied for that purpose. Prior to the measurements, steam sterilization of all of the examined samples was conducted in order to reflect the state of the surface in contact with the biological material.

The wettability measurements were followed by the assessment of the surface free energy. This was based on the Owens-Wendt theoretical model, and accordingly, the SFE of a solid consists of two components, dispersive and polar [33,34]. The following equations were used for the determination of the SFE values:

$$\gamma L \times (1 + \cos\theta)/2 = (\gamma Sd \times \gamma Ld)^{0.5} + (\gamma Sp \times \gamma Lp)^{0.5} \tag{1}$$

$$\gamma S = \gamma Sd + \gamma Sp \tag{2}$$

where, γL—liquid's surface free energy; γS—solid's free energy; γSd, γSp—dispersive (d) and polar (p) component of the surface energy γS; γLd, γLp—dispersive (d) and polar (p) component of the surface energy γL; θ—contact angle.

2.6. Endothelial Cells' Viability and Proliferation

The biocompatibility of the DLC and Si-DLC films was evaluated using human endothelial cell line, EA.hy926, (ATCC - American Type Culture Collection). All of the samples were ultrasonically cleaned in ethanol and then in ultrapure water (0.055 µS/cm) for 15 min. Next, steam sterilization (121 °C, 15 min) was performed using the autoclave (J.P. Selecta Autoclave 401731).

Afterwards, the endothelial cells were seeded onto the surface of the examined samples at a density of 6×10^4 cells per well. The cultures were carried out in DMEM medium (Biowest, Nuaillé, France) for 48 h in standard conditions (i.e. 37 °C) and humidified atmosphere of 5% CO_2 in air. The cells with no contact with any biomaterial were used as a control, as well as the Ti6Al7Nb substrates.

2.6.1. Live/Dead Assay

The proliferation and viability of the endothelial cells on the surface of the examined samples were evaluated by means of live/dead assay. The IN Cell Analyzer 2000 (GE Healthcare, Chicago, IL, USA) automated microscope was used in order to visualize live and dead cells. Prior to the microscopic observations, the samples were incubated in Hank's Balanced Salt Solution containing fluorescent dyes at room temperature for 15 min. The applied mixture of fluorescent dyes included: 5 µg/mL of Hoechst 33342 (Molecular Probes, Eugene, OR, USA), 1–5 µM calcein-AM (Santa Cruz Biotechnology, Dallas, TX, USA), and 1 µg/mL of propidium iodide (Molecular Probes). The analysis of the obtained series of microscopic images was performed using the IN Cell Analyzer software (GE Healthcare, Chicago, IL, USA). The classification of EA.hy926 cells into two subpopulations, (i.e. live and dead) was based on the observed fluorescent signal. The cells stained with the green-fluorescent calcein-AM (retained in the cytoplasm) were labelled as live, while cells stained with the red-fluorescent propidium iodide were counted as dead. The percentage of live and dead cells on the surface allowed assessment of its cytotoxicity. The proliferation of the cells, understood as the total cells' count, was evaluated on the basis of the blue-fluorescent signal coming from cells' nuclei stained with Hoechst 33342 dye which intercalates to the DNA.

2.6.2. XTT Assay

The XTT assay, based on the mitochondrial activity of the cells, was used in order to assess the viability of cells which were in, both, direct and indirect contact with the surface of the deposited DLC and Si-DLC coatings. For that purpose, the medium was removed after 48 h of culture and the XTT solution (XTT Cell Viability Assay Kit, Biotium Inc., Fremont, CA, USA) was added in accordance with the procedure described by the manufacturer. The samples were then incubated in the XTT solution for 4 h in standard conditions (37 °C, humidified atmosphere of 5% CO_2 in air). After that, the absorbance was measured using a Multiskan GO microplate spectrophotometer (Thermo Fisher Scientific, Waltham, MA, USA) at two different wavelengths, i.e., 450 nm and 620 nm (reference). The following formula was applied to calculate the viability of the endothelial cells:

$$\text{Viability [\%]} = (A/Ac) \times 100\% \tag{3}$$

where, A—absorbance measured for the investigated sample and Ac—absorbance measured for the control (cells with no contact with any biomaterial at all).

3. Results and Discussion

3.1. Surface Morphology by Scanning Electron Microscopy (SEM)

The SEM images of the surface morphology of the DLC and Si-DLC films are presented in Figure 1. No significant changes in the morphology of the examined coatings due to the addition of silicon were observed in both the topographic (Figure 1A) as well as the compositional (Figure 1B) images. The SEM examination revealed that the surface of the fabricated coatings was smooth and uniform without any defects or delamination, which proves the high quality of the obtained magnetron sputtered films. Furthermore, no silicon conglomerates were observed. However, due to the very low thickness of the deposited coatings (~250 nm), the contrast arising from the two-phase Ti6Al7Nb alloy may be observed.

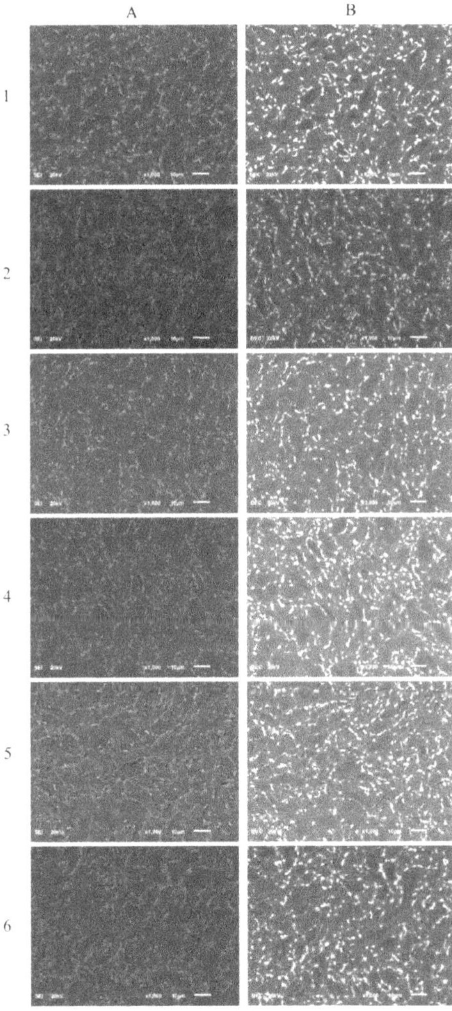

Figure 1. The scanning electron microscopy (SEM) images of the deposited coatings: (**A**) secondary electron imaging; (**B**) backscattered electron imaging, (**1**) diamond-like carbon (DLC), (**2**) Si-DLC 0, (**3**) Si-DLC 1, (**4**) Si-DLC 2, (**5**) Si-DLC 3, (**6**) Si-DLC 4.

3.2. Chemical Composition and Bonding by X-ray Photoelectron Spectroscopy (XPS)

The comparative XPS wide scans of exemplary surfaces of samples, Si-DLC 0 and Si-DLC 4, are depicted in the Figure 2. The spectra revealed that the examined Si-DLC coatings are only composed of oxygen (O 1s and O 2s band), carbon (C 1s band), and silicon (Si 2s and Si 2p band), without any impurities.

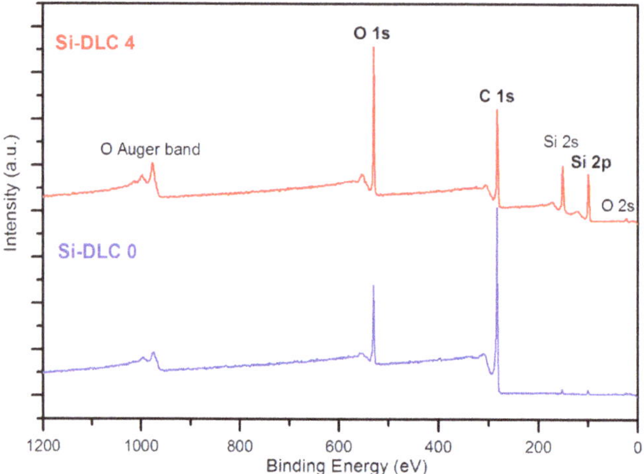

Figure 2. The X-ray photoelectron spectroscopy (XPS) wide scans of the surface of the Si-DLC 0 and Si-DLC 4 samples.

The results of the quantitative XPS analyses of all of the samples' surfaces are presented in the Table 1. Data shown confirms that with increasing magnetron sputtering power, the relative atomic content of silicon also increases.

Table 1. Chemical composition of the deposited DLC and Si-DLC coatings according to the XPS analysis.

Sample	Si Target Power [W]	Si [at.%]	O [at.%]	C [at.%]	N [at.%]
DLC	0	-	9.84 ± 0.11	89.48 ± 0.24	0.68 ± 0.21
Si-DLC 0	10	1.83 ± 0.01	13.72 ± 0.16	84.45 ± 0.41	-
Si-DLC 1	20	3.79 ± 0.04	21.93 ± 0.34	74.28 ± 0.30	-
Si-DLC 2	40	5.58 ± 0.03	18.80 ± 0.07	75.62 ± 0.10	-
Si-DLC 3	60	14.34 ± 0.10	28.90 ± 0.30	56.76 ± 0.19	-
Si-DLC 4	80	22.15 ± 0.37	24.85 ± 0.18	53.00 ± 0.54	-

In terms of the chemical structure of the deposited films, high resolution XPS analysis of oxygen O 1s, carbon C 1s, and silicon Si 2p bands was conducted. Figure 3 presents a C 1s band for each of the Si-DLC coatings divided into components representing specific atomic bonds.

The spectra of the samples from the Si-DLC 3 to the Si-DLC 0 (Figure 3B–E) were deconvoluted, according to the literature [35,36], into six components assigned to: (SiO_xC_y) silicon oxycarbides (at 283.60 eV), sp^2 hybridized carbon (at 284.50 eV), sp^3 hybridized carbon (at 285.30–285.50 eV), C-O bonds (at 286.50 eV), C = O bonds (at 287.60–287.80 eV), and ester bonds COO-R (at 288.70–288.90 eV). Additionally, for the Si- DLC 4 film, also the Si-C bond was distinguished in the position of 282.80 eV (Figure 3A), which was reported in the literature [36]. The relative content of the abovementioned components, identified in carbon C 1s band, is presented in the Table 2 for every investigated sample.

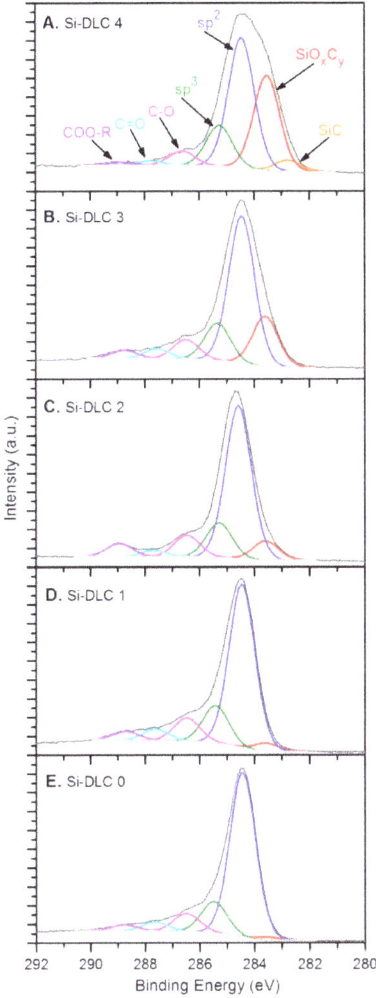

Figure 3. The deconvolution of the C 1s band for the deposited Si-DLC coatings: (**A**) Si-DLC 4, (**B**) Si-DLC 3, (**C**) Si-DLC 2, (**D**) Si-DLC 1, (**E**) Si-DLC 0.

Table 2. Decomposition of the C 1s peak for the examined DLC and Si-DLC coatings.

Sample	Si-C/SiO$_x$C$_y$	C sp^2	C sp^3	C-O	C=O	COO-R	sp^2/sp^3
DLC	-	59.28 ± 0.69	17.98 ± 0.65	11.86 ± 0.07	5.95 ± 0.04	4.93 ± 0.01	3.30 ± 0.16
Si-DLC 0	1.50 ± 0.00	71.70 ± 1.00	13.50 ± 0.10	7.80 ± 0.20	3.65 ± 0.25	2.10 ± 0.50	5.30 ± 0.11
Si-DLC 1	3.30 ± 0.10	63.95 ± 0.75	14.85 ± 0.05	9.50 ± 0.20	4.70 ± 0.20	3.70 ± 0.20	4.31 ± 0.07
Si-DLC 2	9.30 ± 1.60	57.15 ± 4.65	16.95 ± 2.95	8.95 ± 0.15	1.70 ± 0.20	5.50 ± 0.30	3.53 ± 0.89
Si-DLC 3	18.10 ± 0.40	53.55 ± 0.65	14.30 ± 0.80	7.45 ± 0.25	3.65 ± 0.05	2.95 ± 0.05	3.76 ± 0.26
Si-DLC 4	36.00 ± 0.50	42.85 ± 0.75	13.65 ± 0.05	4.70 ± 0.20	1.35 ± 0.05	0.95 ± 0.05	3.14 ± 0.07

The analysis of the data presented in Table 2 and the comparison of the C 1s spectra depicted in Figure 3 led to the conclusion that the increase of the magnetron sputtering power results in the significant evolution of silicon oxycarbide and eventually the formation of silicon carbide bonds. Additionally, a slight decrease of the sp^2/sp^3 hybridization ratio may be also observed as a result of the increasing sputtering power and higher Si content (Figure 4).

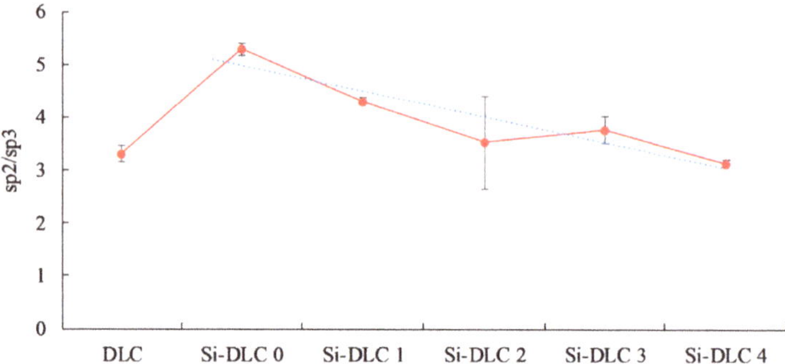

Figure 4. The sp^2/sp^3 bonding ratio of the deposited DLC and Si-DLC coatings.

Decomposition of the O 1s band, presented in Table 3, confirmed the oxycarbide bonds relative content depended on the magnetron sputtering power and the Si content in the structure of the Si-DLC films.

Table 3. Decomposition of the O 1s peak for the obtained Si-DLC coatings.

Sample	C=O	Si-O-C/SiO$_x$C$_y$	SiO$_x$
Si-DLC 0	11.85 ± 0.95	67.45 ± 1.35	20.70 ± 2.30
Si-DLC 1	7.20 ± 1.10	69.25 ± 1.05	23.55 ± 2.15
Si-DLC 2	7.15 ± 2.25	70.15 ± 0.15	22.70 ± 2.10
Si-DLC 3	1.25 ± 0.25	83.15 ± 0.25	15.60 ± 0.20
Si-DLC 4	-	84.85 ± 0.15	15.15 ± 0.15

3.3. Chemical Structure by Fourier Transform Infrared Spectroscopy (FTIR)

Figure 5 presents the FTIR spectra of the deposited Si-DLC coatings in the range of 1700–500 cm^{-1}. In the analyzed region, three wide bands can be distinguished at 1650–1230 cm^{-1}, 1160–1000 cm^{-1}, and 1000–540 cm^{-1}. In the range of 1650–1230 cm^{-1} the maxima corresponding to the C = O bonds (1580 cm^{-1}), the – CH$_3$ and/or –OH groups (1412–1370 cm^{-1}), the SiOCOCH$_3$ groups (1350 cm^{-1}), as well as the C = O (1305 cm^{-1}) can be observed [37–39]. The presence of the – CH$_3$ groups in the deposited coatings may be due to the contamination of the graphite targets used as a source of carbon during the deposition process. The C = O bonds are most probably present at the surface of the deposited coatings due to the oxidation of the surface after the deposition process. According to Ong et al. [40], the absorbance of the oxygen atoms from the atmosphere may occur as a result of the unfused bonds present on the surface of the deposited coatings.

In the range of 1160–1000 cm^{-1}, vibrations of bonds belonging to the Si-O groups with a high-intensity peak at 1080 cm^{-1} are noticed. This maximum may be assigned to the stretching modes of the Si-O-C bonds [39,41,42]. Thus, in order to evaluate the difference in the content of the Si-O bonds in the deposited films, the peak at 1080 cm^{-1} was used. The lowest intensity of this peak was revealed for the Si-DLC 0 coatings with the lowest content of Si. With the increasing concentration of Si in the examined coatings, the intensity of the analyzed peak also increased. The Si-O bond may be responsible for the change in the surface wettability (Figure 6). With the increasing content of the Si-O bonds, the surface polarity increases and, as a consequence, also the surface wettability is raised, which may influence the biological response.

In the range of 1000–500 cm^{-1}, the peaks corresponding to the stretching modes of the Si-CH$_3$ bonds at (905 cm^{-1}) and the Si-O-C groups at 830 cm^{-1}, as well as in the non-hydrogenated Si-C groups (800 cm^{-1}) are observed. Moreover, the peaks originating from bending vibrations of the C = O bonds are noticed at 670 and 568 cm^{-1}, respectively [39,41,42].

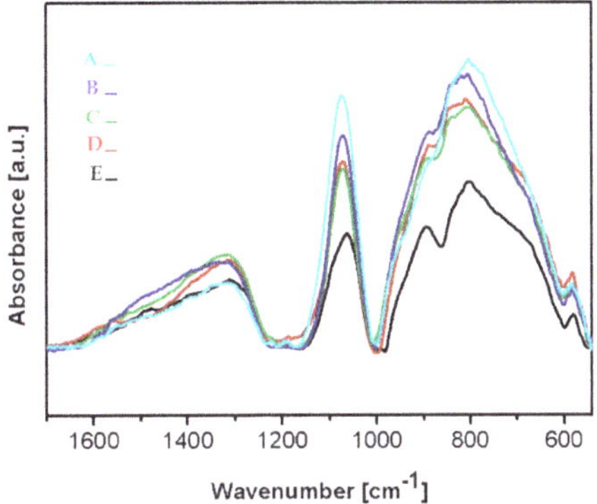

Figure 5. The Fourier transform infrared spectroscopy (FTIR) spectra of the deposited coatings Si-DLC coatings: (**A**) Si-DLC 4, (**B**) Si-DLC 3, (**C**) Si-DLC 2, (**D**) Si-DLC 1, (**E**) Si-DLC 0.

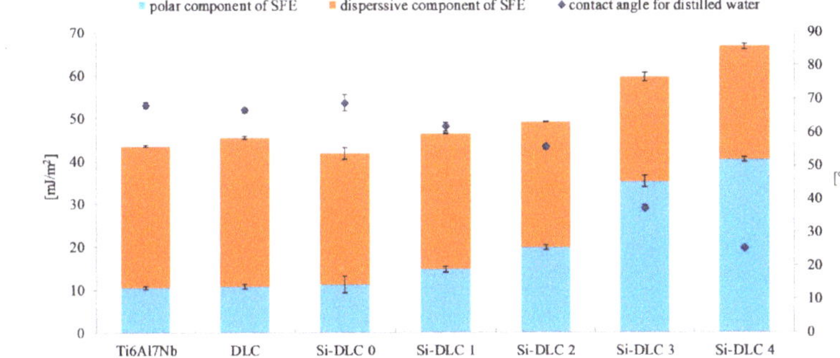

Figure 6. The surface free energy and wettability of the deposited DLC and Si-DLC coatings after sterilization.

It is observed that as the Si concentration in the investigated coatings increases, the amount of the Si-CH$_3$ bonds also increases, except for the Si-DLC 4 sample. In this coating, the content of the Si-CH$_3$ bonds is reduced, but the amount of the Si C bonds increases. The peak corresponding to the vibrations of the Si-C bonds in the spectrum of the coating with the highest content of Si (Si-DLC 4) is also shifted towards the lower wave numbers, which may be connected to the fact that silicon in the structure of this coating is mostly bonded to the non-hydrogenated carbon. The increase in the nonpolar Si-C bonds does not result in the increase in the dispersive component of the SFE. On the contrary, the opposite effect is observed (Figure 6) and the increasing content of Si resulted in the higher surface polarity and wettability. This may be due to the fact that the Si-O and the C=O bonds, which are polar in nature, are localized mainly at the surface of the fabricated films which are directly responsible for the increasing polar component of the SFE with the increasing content of Si.

3.4. Surface Free Energy (SFE) and Wettability

Figure 6 presents the results of the surface free energy and wettability assessment of the deposited coatings. It is observed that the wettability of the Ti6Al7Nb substrate, DLC, and Si-DLC 0 coatings is similar. However, for the higher concentrations of Si, the wettability of the surface significantly increases. At the same time, it must be noted that all the examined surfaces are hydrophilic. The higher the content of Si, the higher the wettability of the surface. The results obtained confirmed the findings presented by Ong et al. [40] and Okpalugo et al. [43] who found that the polar component of the surface free energy is higher as Si content increases, especially when it exceeds 16 at.%. Similar results were also reported in our previous work [27].

The changes in the SFE and wettability are explained by the differences in the chemical structure of the deposited coatings. The O 1s region of the XPS spectra of the investigated films is deconvoluted into the following components: $C = O$, Si-O-C, and SiO_x bonds. For higher magnetron sputtering powers, the content of the SiO_x bonds decreases despite the increasing concentration of Si. Moreover, the contribution of the $C = O$ bonds is reduced from 10.9%, for the Si-DLC 0 coatings with 1.83 at.%. of Si, to zero, for the Si-DLC 4 coating containing 22.15 at.%. of Si. At the same time, the content of the Si-O-C groups increases which is especially significant for the Si-DLC 4 coating with the highest amount of Si. The most notable growth of the surface polarity is observed for the Si-DLC 3 and the Si-DLC 4 coatings, as seen in the Figure 6. This may be explained by the fact that in the Si-O-C groups two different bonds are present, i.e., the covalent C-O bond and the ionic Si-O bond. Due to the fact that the ionic bond is stronger than the covalent bond, the breakdown of the C-O bonds is much more possible. The newly created unfused bonds may be saturated with water vapour. As a result, the C-OH and Si-OH groups are formed, which causes the increase of hydrophilicity of the surface. In addition, in the case of the Si-DLC 4 sample, the CH_3 groups (responsible for the dispersion component of the SFE) disappear, which also leads to the increase of the surface wettability.

3.5. Endothelial Cells' Viability and Proliferation

In general, cells show good spreading, proliferation, and differentiation on hydrophilic surfaces. Nevertheless, the major factor determining the nature of the cells' interaction with biomaterials is the composition and conformation of the proteins adsorbed on the surface. Due to the fact that the adhesion of the cells to the surface of the material requires a series of cytoplasmic, transmembrane, and extracellular proteins which assemble into the stable contact sites [44], the adsorption of serum and extracellular matrix proteins is likely affecting the adhesion and behavior of cells [45]. Therefore, the observed difference in the proliferation and adhesion of endothelial cells may be caused by the difference in the absorption of proteins responsible for the cell colonization process. This is especially important in the case of proteins involved in the formation of the extracellular matrix (ECM), such as proteoglycans and glycoproteins, i.e., fibronectin, laminin, and collagen. In addition to the amount of proteins adsorbed on the surface, their biological activity, which is connected, for example, with their conformation, may be changed [46,47]. Moreover, the size of the biological molecules and the time of contact with the biomaterial's surface can determine the tissue–biomaterial interaction [48].

The adsorption of proteins responsible for the cell colonization and their activity may be affected by the physiochemical properties of the surface, i.e., the surface free energy and the associated surface wettability, as well as the charge and chemical composition of the biomaterial's surface. According to the literature, the higher surface wettability results in better adhesion and proliferation of the eukaryotic cells [46,49–51] This was confirmed by the results obtained. The life/dead assay showed that with the increasing content of Si, the cells' viability on the surface of the deposited coatings is enhanced as a result of the increased surface hydrophilicity. The biocompatibility evaluation of the deposited Si-DLC coatings demonstrated slightly enhanced proliferation of the endothelial (EA.hy926) cells, and adhesion on the surface of the Si-free DLC coating as compared with the Ti6Al7Nb substrate (Figure 7). Moreover, the incorporation of Si further improves the viability of endothelial cells on the surface of the Si-DLC coatings, especially for the Si-DLC 3 sample with 14.34 at.%. of Si. The higher the sputtering power and the resulting Si content, the more hydrophilic surface is obtained. This is due to the higher content of the Si-O bonds, which increase the polar component of the SFE and change the surface wettability. The higher the concentration of Si, the higher the wettability of the surface observed. In the case of the Si-DLC 4 coating where the amount of Si was above 16 at.%., the increase in the polar component of the SFE and wettability of the surface was still observed [40,43], but their effect on the behavior of the cells was suppressed and the reduction of proliferation of the cells occurred.

Nevertheless, the biocompatibility of Si-doped DLC coatings, regardless of the Si content, remains at a very high level.

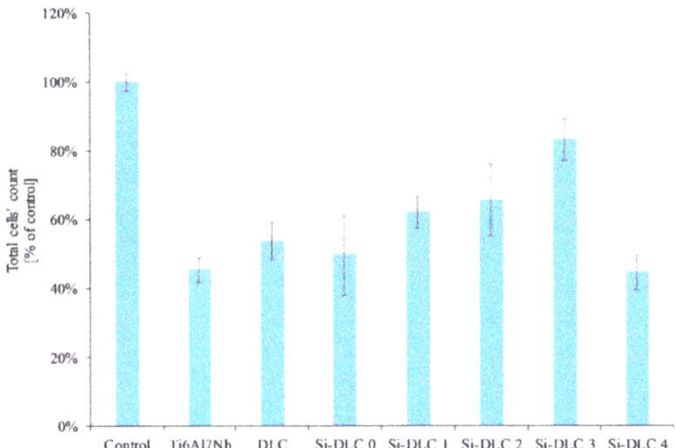

Figure 7. Endothelial cells' proliferation on the surface of the deposited Si-DLC coatings (live/dead assay).

Another important factor influencing the cell colonization of the surface is the structure of the DLC coatings, i.e., the sp^2/sp^3 ratio [52]. According to the literature, the Si element increases the sp^3 content. As the silicon content increases, the sp^2/sp^3 ratio decreases and as a result the proliferation of the cells is enhanced. According to T.T Liao el at. [50], the sp^2 bonding, characteristic of the graphite phase, results in lower adsorption of proteins and cells on the surface, while the sp^3 bonding, typical for the diamond phase, improves cell colonization. This is explained by the presence of free electrons on the surface of the diamond-like carbon coatings (the sp^2 phase) which suppress the adhesion of proteins and cells. A similar tendency was revealed for the examined coatings. With an increasing amount of Si, the content of the sp^3 also increased (Table 2), which, in turn, resulted in a higher proliferation of the endothelial cells. The only exception was the Si-DLC 4 film with the highest concentration of Si (22.15 at.%.). Despite the highest surface wettability and the lowest content of the sp^2, the proliferation of cells on the surface of the Si-DLC 4 coatings was lowered as compared with other Si-DLC coatings (similar to the Ti6Al7Nb sample without the DLC coating). This may be due to the fact that the content of Si in the Si-DLC 4 coating is too high and it affects the process of cells' division. The critical concentration of Si may be different for different cell lines and it may be associated with cellular adaptation [48,53]. Moreover, according to some authors, the high hydrophilicity of surface may inhibit the adhesion of cells due to the preferential adsorption of water molecules [54]. In this case, it can also be connected to the influence of surface charge on the behavior of cells [55]. According to Thevenot et al., the presence of negative charges may facilitate the adsorption of proteins promoting adhesion of the cells [44]. Moreover, Keselowsky et al. reported that surfaces with differently charged functional groups ($–CH_3$, $–OH$, $–COOH$, and $–NH_2$ groups) positively influenced the adsorption of fibrynogen as well as direct integrin binding and specificity [55]. In addition, Schmidt et al. pointed out that the functional groups such as $–CH_3$, $–OH$, $–COOH$ affect the proteins responsible for cells' behavior [56]. In the case of the sample Si-DLC 4, the content of the $Si-CH_3$ bonds is reduced, while the amount of the SiO_xC_y increases significantly. According to the Lagonegro et al. [57] the silicon oxycarbide reduces the cells' proliferation ability. Therefore, for the amount of Si above 20 at.%., the chemical and structural changes in the Si-DLC coatings caused the deterioration of biocompatibility.

Furthermore, the live/dead assay (Figure 8) demonstrated that no significant increase in the number of dead cells for any of the examined samples was observed. This proves that none of the deposited coatings exhibit cytotoxicity towards the endothelial cells.

As far as the XTT assay was concerned (Figure 9), it revealed that the Si incorporation into the DLC matrix had no significant influence on viability of EA.hy926 cells in indirect contact with the investigated surfaces. The overall change in the mitochondrial activity of endothelial cells in contact with the deposited Si-DLC coatings is negligible. Only a slight raise in the mean value of cells' viability may be noticed with the increasing Si content up to 14.34 at.%., followed by a small decrease for the coating with 22.15 at.%. of Si. This may be explained by the changes in the cells' proliferation and viability on the surface of the fabricated coatings as indicated by the results of the live/dead assay.

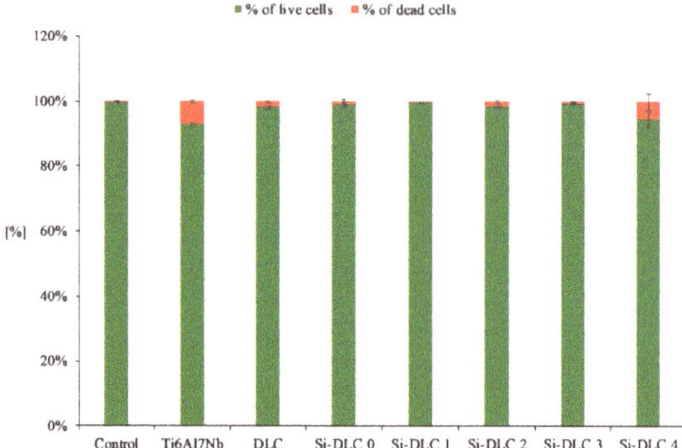

Figure 8. Cytotoxicity of the deposited Si-DLC coatings towards the endothelial cells (live/dead assay).

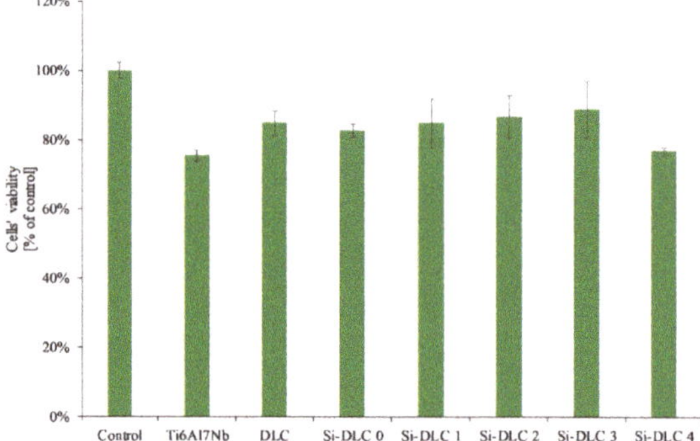

Figure 9. Endothelial cells' viability in direct and indirect contact with the deposited Si-DLC coatings (XTT assay).

To sum up, according to the ISO 10993-5, a biomaterial may be considered biocompatible if the cells' viability is higher than 70% [58] and, according to the results presented in this work, this criterion was fulfilled by all of the analyzed Si-DLC coatings, as the observed cells' viability was above 70%.

4. Conclusions

The results presented in this work showed that the magnetron sputtering is an effective method to produce homogenous and biocompatible Si-doped DLC coatings. The use of varying sputtering powers allows a different content of silicon in the coatings to be obtained, which affects both the chemical composition and structure, and therefore the surface properties and biological response. The addition of silicon to the DLC coating deposited on the Ti6Al7Nb alloy has a very positive effect on the proliferation and viability of endothelial cells.

The higher the concentration of Si, the higher the wettability of the surface observed. The increase of the polar component of the SFE and wettability of the coatings with the highest amount of Si is connected with the high content of the Si-O bonds. Additionally, for the Si-DLC 4 sample the lower content of the $-CH_3$ groups (responsible for the dispersive component of the SFE) is observed and the C-OH and Si-OH groups are formed which causes the increase of the polar component. According to the literature, the increase in the surface wettability promotes the cells' proliferation. Our research has shown that this relationship exists, but only to a certain level. A change in the biological response observed for the Si-DLC 4 coating may indicate that Si is tolerated by the endothelial cells up to a limit, which lies between 14 and 22 at.%. Above that limit the proliferation of cells decreases despite the increase in the surface wettability. This indicates that the biological response is not determined by the surface wettability alone. A balance must be maintained between the polar and dispersive components of the SFE, so that the extracellular matrix proteins are firstly attached to the surface. Otherwise, the preferential attachment of water to the surface may occur and limit the adhesion of cells.

Further research in order to precisely determine the optimal content of Si for cells is required and planned by the authors. An in-depth study of the interactions of cells and biomaterials' surface will allow a better understanding of the mechanisms underlying the cells' adhesion and proliferation.

Author Contributions: Conceptualization, D.B.; data curation, D.B. and K.J.; formal analysis, A.K.; investigation, A.S.-G., K.J., P.K., J.B., K.P., and A.K.; methodology, D.B., A.S.-G., and P.K.; project administration, D.B.; supervision, D.B.; visualization, K.P. and A.K.; writing—original draft, A.S.-G., J.B., and A.K.; writing—review and editing, D.B.

Funding: This research was funded by the National Centre for Research and Development, grant number LIDER/040/707/L-4/12/NCBR/2013.

Conflicts of Interest: The authors declare no conflict of interest.

References

1. Chen, W.; Oh, S.; Ong, A.P.; Oh, N.; Liu, Y.; Courtney, H.S.; Appleford, M.; Ong, J.L. Antibacterial and osteogenic properties of silver-containing hydroxyapatite coatings produced using a sol gel process. *J. Biomed. Mater. Res. Part A* **2007**, *82A*, 899–906. [CrossRef]

2. Metzler, P.; von Wilmowsky, C.; Stadlinger, B.; Zemann, W.; Schlegel, K.A.; Rosiwal, S.; Rupprecht, S. Nano-crystalline diamond-coated titanium dental implants–A histomorphometric study in adult domestic pigs. *J. Cranio-Maxillofacial Surg.* **2013**, *41*, 532–538. [CrossRef] [PubMed]

3. Jin, W.; Chu, P.K. Orthopedic Implants. In *Encyclopedia of Biomedical Engineering*; Elsevier: Amsterdam, Netherlands, 2019; pp. 425–439. ISBN 9780128051443.

4. Morais, S.; Pereira, M.C. Application of stripping voltammetry and microelectrodes in vitro biocompatibility and in vivo toxicity tests of AISI 316L corrosion products. *J. Trace Elem. Med. Biol.* **2000**, *14*, 48–54. [CrossRef]

5. Kręcisz, B.; Pałczyński, C. Uczulenie na metale a implanty medyczne. *Alergia* **2012**, *4*, 17–18.

6. Gruendemann, B.J.; Mangum, S.S. *Infection Prevention in Surgical Settings*; W.B. Saunders: Philadelphia, PA, USA, 2001; ISBN 9780721690353.

7. Wilson, J. (Nurse); *Infection Control in Clinical Practice*; Baillière Tindall/Elsevier: Amsterdam, Netherlands, 2006; ISBN 0702027618.

8. Darouiche, R.O. Treatment of Infections Associated with Surgical Implants. *N. Engl. J. Med.* **2004**, *350*, 1422–1429. [CrossRef] [PubMed]

9. Antoci, V.; Chen, A.F.; Parvizi, J. Orthopedic Implant Use and Infection. In *Comprehensive Biomaterials II*; Elsevier: Amsterdam, Netherlands, 2017; pp. 133–151. ISBN 9780081006924.

10. Li, Z.; Aik Khor, K. Preparation and Properties of Coatings and Thin Films on Metal Implants. In *Encyclopedia of Biomedical Engineering*; Elsevier: Amsterdam, Netherlands, 2019; pp. 203–212.

11. Love, C.A.; Cook, R.B.; Harvey, T.J.; Dearnley, P.A.; Wood, R.J.K. Diamond like carbon coatings for potential application in biological implants—A review. *Tribol. Int.* **2013**, *63*, 141–150. [CrossRef]

12. Okpalugo, T.I.T.; Ogwu, A.A. DLC thin films for implantable medical devices. *Thin Film Coatings Biomater. Biomed. Appl.* **2016**, 261–287.

13. Grill, A. Tribology of diamondlike carbon and related materials: an updated review. *Surf. Coatings Technol.* **1997**, *94–95*, 507–513. [CrossRef]

14. Dorner, A.; Schürer, C.; Reisel, G.; Irmer, G.; Seidel, O.; Müller, E. Diamond-like carbon-coated Ti6Al4V: influence of the coating thickness on the structure and the abrasive wear resistance. *Wear* **2001**, *249*, 489–497. [CrossRef]

15. Hang, R.; Ma, S.; Chu, P.K. Corrosion behavior of DLC-coated NiTi alloy in the presence of serum proteins. *Diam. Relat. Mater.* **2010**, *19*, 1230–1234. [CrossRef]

16. Cui, F.; Li, D. A review of investigations on biocompatibility of diamond-like carbon and carbon nitride films. *Surf. Coatings Technol.* **2000**, *131*, 481–487. [CrossRef]

17. Hauert, R. An overview on the tribological behavior of diamond-like carbon in technical and medical applications. *Tribol. Int.* **2004**, *37*, 991–1003. [CrossRef]

18. Dearnaley, G.; Arps, J.H. Biomedical applications of diamond-like carbon (DLC) coatings: A review. *Surf. Coatings Technol.* **2005**, *200*, 2518–2524. [CrossRef]

19. Tang, X.S.; Wang, H.J.; Feng, L.; Shao, L.X.; Zou, C.W. Mo doped DLC nanocomposite coatings with improved mechanical and blood compatibility properties. *Appl. Surf. Sci.* **2014**, *311*, 758–762. [CrossRef]

20. Joska, L.; Fojt, J. The effect of porosity on barrier properties of DLC layers for dental implants. *Appl. Surf. Sci.* **2012**, *262*, 234–239. [CrossRef]

21. Regan, E.M.; Uney, J.B.; Dick, A.D.; Zhang, Y.; Nunez-Yanez, J.; McGeehan, J.P.; Claeyssens, F.; Kelly, S. Differential patterning of neuronal, glial and neural progenitor cells on phosphorus-doped and UV irradiated diamond-like carbon. *Biomaterials* **2010**, *31*, 207–215. [CrossRef]

22. Grill, A. Diamond-like carbon coatings as biocompatible materials—an overview. *Diam. Relat. Mater.* **2003**, *12*, 166–170. [CrossRef]

23. Hauert, R. A review of modified DLC coatings for biological applications. *Diam. Relat. Mater.* **2003**, *12*, 583–589. [CrossRef]

24. Bendavid, A.; Martin, P.J.; Comte, C.; Preston, E.W.; Haq, A.J.; Magdon Ismail, F.S.; Singh, R.K. The mechanical and biocompatibility properties of DLC-Si films prepared by pulsed DC plasma activated chemical vapor deposition. *Diam. Relat. Mater.* **2007**, *16*, 1616–1622. [CrossRef]

25. Zhao, Q.; Liu, Y.; Wang, C.; Wang, S. Evaluation of bacterial adhesion on Si-doped diamond-like carbon films. *Appl. Surf. Sci.* **2007**, *253*, 7254–7259. [CrossRef]

26. Ren, D.W.; Zhao, Q.; Bendavid, A. Anti-bacterial property of Si and F doped diamond-like carbon coatings. *Surf. Coatings Technol.* **2013**, *226*, 1–6. [CrossRef]

27. Bociaga, D.; Kaminska, M.; Sobczyk-Guzenda, A.; Jastrzebski, K.; Swiatek, L.; Olejnik, A. Surface properties and biological behaviour of Si-DLC coatings fabricated by a multi-target DC–RF magnetron sputtering method for medical applications. *Diam. Relat. Mater.* **2016**, *67*, 41–50. [CrossRef]

28. De Scheerder, I.; Szilard, M.; Yanming, H.; Ping, X.B.; Verbeken, E.; Neerinck, D.; Demeyere, E.; Coppens, W.; Van de Werf, F. Evaluation of the biocompatibility of two new diamond-like stent coatings (Dylyn) in a porcine coronary stent model. *J. Invasive Cardiol.* **2000**, *12*, 389–394.

29. Kim, M.-G.; Lee, K.-R.; Eun, K.Y. Tribological behavior of silicon-incorporated diamond-like carbon films. *Surf. Coatings Technol.* **1999**, *112*, 204–209. [CrossRef]

30. Lee, K.-R.; Kim, M.-G.; Cho, S.-J.; Yong Eun, K.; Seong, T.-Y. Structural dependence of mechanical properties of Si incorporated diamond-like carbon films deposited by RF plasma-assisted chemical vapour deposition. *Thin Solid Films* **1997**, *308–309*, 263–267. [CrossRef]

31. Gangopadhyay, A.K.; Willermet, P.A.; Tamor, M.A.; Vassell, W.C. Amorphous hydrogenated carbon films for tribological applications I. Development of moisture insensitive films having reduced compressive stress. *Tribol. Int.* **1997**, *30*, 9–18. [CrossRef]

32. Soum-Glaude, A.; Rambaud, G.; Grillo, S.E.; Thomas, L. Investigation of the tribological behavior and its relationship to the microstructure and mechanical properties of a-SiC:H films elaborated by low frequency plasma assisted chemical vapor deposition. *Thin Solid Films* **2010**, *519*, 1266–1271. [CrossRef]

33. Mittal, K.L. *Contact Angle, Wettability and Adhesion. Volume 5*; VSP, Koninklijke Brill NV: Leiden, Netherlands, 2008; ISBN 9004158642.

34. Owens, D.K.; Wendt, R.C. Estimation of the surface free energy of polymers. *J. Appl. Polym. Sci.* **1969**, *13*, 1741–1747. [CrossRef]

35. Önneby, C.; Pantano, C.G. Silicon oxycarbide formation on SiC surfaces and at the SiC/SiO$_2$ interface. *J. Vac. Sci. Technol. A Vac. Surfaces Film.* **1997**, *15*, 1597–1602. [CrossRef]

36. Socha, R.P.; Laajalehto, K.; Nowak, P. Oxidation of the silicon carbide surface in Watts' plating bath. *Surf. Interface Anal.* **2002**, *34*, 413–417. [CrossRef]

37. Silverstein, R.M.; Bassler, G.C. Spectrometric identification of organic compounds. *J. Chem. Educ.* **1962**, *39*, 546. [CrossRef]

38. Lambert, J.B.; Claridge, T.D.W. *Introduction to Organic Spectroscopy*; Oxford Science Publication: Oxford, UK, 1987; ISBN 0023673001.

39. Arkles, B.; Larson, G.L. Infrared Analysis of Organosilicon Compounds. In *Silicon Compd. Silanes Silicones*; Gelest Inc.: Morrisville, PA, USA, 2013.

40. Ong, S.-E.; Zhang, S.; Du, H.; Too, H.-C.; Aung, K.-N. Influence of silicon concentration on the haemocompatibility of amorphous carbon. *Biomaterials* **2007**, *28*, 4033–4038. [CrossRef] [PubMed]

41. Ucovsky, G.I.; Nemanich, R.J.; Knights, J.C. Structural interpretation of the vibrational spectra of a-Si:H alloys. *Phys. Rev. B* **1979**, *19*, 2064. [CrossRef]

42. Smith, A.L.; Albert, L. *Analysis of Silicones*; Wiley: Hoboken, NJ, USA, 1974; ISBN 0471800104.

43. Okpalugo, T.I.T.; Ogwu, A.A.; Maguire, P.D.; McLaughlin, J.A.D. Platelet adhesion on silicon modified hydrogenated amorphous carbon films. *Biomaterials* **2004**, *25*, 239–245. [CrossRef]

44. Thevenot, P.; Hu, W.; Tang, L. Surface chemistry influences implant biocompatibility. *Curr. Top. Med. Chem.* **2008**, *8*, 270–280. [PubMed]

45. Keselowsky, B.G.; Collard, D.M.; García, A.J. Surface chemistry modulates fibronectin conformation and directs integrin binding and specificity to control cell adhesion. *J. Biomed. Mater. Res. Part A* **2003**, *66A*, 247–259. [CrossRef] [PubMed]

46. Chai, F.; Mathis, N.; Blanchemain, N.; Meunier, C.; Hildebrand, H.F. Osteoblast interaction with DLC-coated Si substrates. *Acta Biomater.* **2008**, *4*, 1369–1381. [CrossRef]

47. Ahmed, M.H.; Byrne, J.A.; McLaughlin, J.; Ahmed, W.; Ahmed, M.H.; Byrne, J.A.; McLaughlin, J.; Ahmed, W. Study of Human Serum Albumin Adsorption and Conformational Change on DLC and Silicon Doped DLC Using XPS and FTIR Spectroscopy. *J. Biomater. Nanobiotechnol.* **2013**, *04*, 194–203. [CrossRef]

48. Walkowiak-Przybyło, M.; Komorowski, P.; Walkowiak, B. Differences in the expression of cell cycle genes in osteoblasts and endothelial cells cultured on the surfaces of Ti6Al4V and Ti6Al7Nb alloys. *J. Biomed. Mater. Res. Part A* **2017**, *105*, 1607–1617. [CrossRef]

49. Filova, E.; Vandrovcova, M.; Jelinek, M.; Zemek, J.; Houdkova, J.; Jan Remsa; Kocourek, T.; Stankova, L.; Bacakova, L. Adhesion and differentiation of Saos-2 osteoblast-like cells on chromium-doped diamond-like carbon coatings. *J. Mater. Sci. Mater. Med.* **2017**, *28*, 17. [CrossRef] [PubMed]

50. Goddard, J.M.; Hotchkiss, J.H. Polymer surface modification for the attachment of bioactive compounds. *Prog. Polym. Sci.* **2007**, *32*, 698–725. [CrossRef]

51. Xu, L.-C.; Siedlecki, C.A. Effects of surface wettability and contact time on protein adhesion to biomaterial surfaces. *Biomaterials* **2007**, *28*, 3273–3283. [CrossRef] [PubMed]

52. Liao, T.T.; Zhang, T.F.; Li, S.S.; Deng, Q.Y.; Wu, B.J.; Zhang, Y.Z.; Zhou, Y.J.; Guo, Y.B.; Leng, Y.X.; Huang, N. Biological responses of diamond-like carbon (DLC) films with different structures in biomedical application. *Mater. Sci. Eng. C* **2016**, *69*, 751–759. [CrossRef] [PubMed]

53. Komorowski, P.; Walkowiak-Przybyło, M.; Walkowiak, B. Early cell response to contact with biomaterial's surface. *J. Biomed. Mater. Res. Part B Appl. Biomater.* **2016**, *104*, 880–893. [CrossRef]

54. Ogwu, A.A.; Okpalugo, T.I.T.; Ali, N.; Maguire, P.D.; McLaughlin, J.A.D. Endothelial cell growth on silicon modified hydrogenated amorphous carbon thin films. *J. Biomed. Mater. Res. Part B Appl. Biomater.* **2008**, *85B*, 105–113. [CrossRef]

55. Ishikawa, J.; Tsuji, H.; Sato, H.; Gotoh, Y. Ion implantation of negative ions for cell growth manipulation and nervous system repair. *Surf. Coatings Technol.* **2007**, *201*, 8083–8090. [CrossRef]

56. Schmidt, D.R.; Waldeck, H.; Kao, W.J. Protein Adsorption to Biomaterials. In *Biological Interactions on Materials Surfaces*; Springer US: New York, NY, USA, 2009; pp. 1–18.

57. Lagonegro, P.; Rossi, F.; Galli, C.; Smerieri, A.; Alinovi, R.; Pinelli, S.; Rimoldi, T.; Attolini, G.; Macaluso, G.; Macaluso, C.; et al. A cytotoxicity study of silicon oxycarbide nanowires as cell scaffold for biomedical applications. *Mater. Sci. Eng. C* **2017**, *73*, 465–471. [CrossRef]

58. ISO 10993-5:2009-Biological evaluation of medical devices–Part 5: Tests for in vitro cytotoxicity. Available online: https://www.iso.org/standard/36406.html (accessed on 3 May 2019).

 *nanomaterials*

Article

Biomimetic Collagen/Zn^{2+}-Substituted Calcium Phosphate Composite Coatings on Titanium Substrates as Prospective Bioactive Layer for Implants: A Comparative Study Spin Coating vs. MAPLE

Ionela Andreea Neacsu [1], Laura Vasilica Arsenie [1], Roxana Trusca [1], Ioana Lavinia Ardelean [1], Natalia Mihailescu [2], Ion Nicolae Mihailescu [2], Carmen Ristoscu [2], Coralia Bleotu [3], Anton Ficai [1,4] and Ecaterina Andronescu [1,4,*]

[1] Faculty of Applied Chemistry and Materials Science, University Politehnica of Bucharest, Polizu Street No.1, 011061 Bucharest, Romania; neacsu.a.ionela@gmail.com (I.A.N.); arsenielaura@yahoo.com (L.V.A.); roxanatrusca@yahoo.com (R.T.); dy4_ioana@yahoo.com (I.L.A.); anton_ficai81@yahoo.com (A.F.)

[2] National Institute for Laser, Plasma and Radiation Physics, Atomiştilor Street No. 409, 077125 Măgurele, Romania; natalia.serban@gmail.com (N.M.); ion.mihailescu@inflpr.ro (I.N.M.); carmen.ristoscu@inflpr.ro (C.R.)

[3] Stefan S. Nicolau' Institute of Virology, Romanian Academy, 011061 Bucharest, Romania; cbleotu@yahoo.com

[4] Academy of Romanian Scientists, Splaiul Independentei Street No. 54, 011061 Bucharest, Romania

[*] Correspondence: ecaterina.andronescu@upb.ro

Received: 8 April 2019; Accepted: 26 April 2019; Published: 3 May 2019

Abstract: Synthesis of biomimetic materials for implants and prostheses is a hot topic in nanobiotechnology strategies. Today the major approach of orthopaedic implants in hard tissue engineering is represented by titanium implants. A comparative study of hybrid thin coatings deposition was performed by spin coating and matrix-assisted pulsed laser evaporation (MAPLE) onto titanium substrates. The Collagen-calcium phosphate (Coll-CaPs) combination was selected as the best option to mimic natural bone tissue. To accelerate the mineralization process, Zn^{2+} ions were inserted by substitution in CaPs. A superior thin film homogeneity was assessed by MAPLE, as shown by scanning electron microscopy (SEM) and Fourier transform infrared (FTIR) microscopy. A decrease of P-O and amide absorbance bands was observed as a consequence of different Zn^{2+} amounts. A variety of structural modifications of the apatite layer are then generated, which influenced the confinement process towards the collagen template. The in-vitro Simulated Body Fluid (SBF) assay demonstrated the ability of Coll/Zn^{2+}-CaPs coatings to stimulate the mineralization process as a result of synergic effects in the collagen-Zn^{2+} substituted apatite. For both deposition methods, the formation of droplets associated to the growth of CaPs particulates inside the collagen matrix was visualized. This supports the prospective behavior of MAPLE biomimetic coatings to induce mineralization, as an essential step of fast implant integration with vivid tissues.

Keywords: Zn^{2+} substituted Coll-CaPs biomimetic layers; MAPLE; spin coating

---

1. Introduction

The prevention of post-surgical inflammation or rejection after the implantation of metallic devices is of growing interest in tissue engineering. In particular, Ti prostheses exhibit a pronounced tendency of corrosion, releasing abrasive metallic particles (as a result of redox processes) into the physiological medium [1,2]. Then, the corrosion leads to undesirable consequences, such as inflammation and infections that inhibit the artificial bone reconstruction. Another inconvenience in using common Ti orthopaedic implants resides in the low capacity to promote the interaction between the natural tissue and device [3]. The most important objective to overcome these disadvantages is therefore oriented towards the fabrication of smart biomaterials with enhanced biocompatibility, osseointegration, high surface properties and superior chemical stability.

An intelligent solution to address these limitations resorts to the deposition of thin coatings on Ti substrates. The aim of our work reflects a comparative study of two deposition methods—spin coating and matrix-assisted pulsed laser evaporation (MAPLE)—involved in the production of biocompatible collagen-Zn^{2+} substituted calcium phosphate thin coatings.

The materials developed are able to display a biomimetic behavior similar to the bone tissue. Materials selection was inspired by the composition of mammalian bone, which consists of a mixture of proteins and minerals [4]. The main inorganic component is hydroxyapatite (HAp) and is in a proportion of ~50 wt% and ~70 vol%, depending on sex and age. The rest is water and matrix and is pre-existing serving as a scaffold for mineralization. Collagen, the most abundant protein in the body, reaches about 90 wt% in bone matrix. Composite materials based on collagen and calcium phosphates (Coll-CaPs) are usually considered a good choice for hard tissue engineering because of the structural and compositional similarity with the natural bone. The most promising systems are therefore bone-like apatite layers designed by calcium phosphates (CaP) particles (including hydroxyapatite-HAp and tricalcium phosphate-TCP as major components) embedded in a collagen template [5–7]. Loading the biomimetic organic-inorganic matrix with platelet-rich plasma (PRP), bone morphogenetic proteins (BMPs) or bisphosphonates facilitates the osseous regeneration [8–10]. Moreover, complementary functional motifs dependent on supramolecular interactions (e.g., hydrogen bonds, electrostatic interactions, etc.) define the advantageous selection of these two components [7,11]. An advanced biocompatibility could be envisaged by using HAp, which in association with TCP improves the properties of the final implant in respect with the mechanical characteristics and bioactivity [12,13]. The apatite network enables the cationic exchanges via substitution processes, without altering the final features of the implant. Some studies highlighted the substantial benefits of using Ag^+ and Zn^{2+} cations as substitution agents in preventing the post-infection due to the enhanced antimicrobial activity [14–17]. Ag^+ is susceptible to be more toxic than Zn^{2+} even at ppm level concentrations, the ionic charge imbalance between Ca^{2+} and Ag^+ cations leading to structural modifications of the calcium-phosphate network [18]. On the other side, Zn^{2+} ions exhibit not only advanced antimicrobial behavior, but also promote the osteointegration processes [19].

Besides the importance of organic and inorganic synthons involved in the formation of the coating, the deposition method plays a major role in obtaining the desired chemical and biological properties. Requirements such as biocompatibility, osteoregeneration, or resorbability are generally related to an improved interaction of the implant with the natural tissue and physiological medium.

Generally, the coatings obtained via spin coating are not uniformly distributed on the metallic substrate, the random distribution of the composite generating differences in the properties of the implant, minimizing a complete regeneration (e.g., an imbalanced mineralization process in some areas that are characterized by an excess of composite material) and also areas strongly exposed to corrosion and inflammatory pitting points [20]. Nevertheless, the spin coating method could be properly balanced by easily monitoring some parameters (e.g., time and rotation speed, acceleration and deceleration), combined with chemical modifications of the deposited organic material (e.g., cross-linking with imines or aldehydes to preserve the collagen structure) or inorganic precursors. Also, the modifications of the metallic substrate are developed prior to the deposition method (e.g., chemical etching to offer an advanced adherence of the biocomposite to the disk). In contrast to spin coating, the matrix-assisted pulsed laser evaporation (MAPLE) technique promises a significant advantage by preserving the collagen/CaP ratio, due to the prevention of collagen denaturation during deposition process [21–23]. One therefore expects significantly improvement of both collagen-apatite interaction, as well as adherence to the Ti substrate. In MAPLE, the uniformity of the coating is easily achieved due to the additive character of the process [24]. More exactly, one can stop the deposition as soon as the desired thickness and uniformity are reached.

One should stress that, in contrast to other well-recognized deposition methods, like plasma (magnetron sputtering, electron/ion bombardment, pulsed laser deposition, laser induced forward transfer) or other techniques (Molecular Beam Epitaxy, Atomic Layer Deposition), both spin coating and MAPLE are conforming to requirements of safe transfer of organic materials. Indeed, they either do not use high temperature processing or protect via ice or other shields the delicate compounds like collagen against irreversible damage by temperature and/or laser radiation. We consider therefore that a comparison between the relative performances of the two techniques is relevant for potential bioactive layer deposition on implants.

Spin Coating and MAPLE were comparatively applied in the present study for thin films deposition on chemically etched Ti substrates. The Coll/CaPs biocomposite coatings with different Zn^{2+} contents (within 0–2 wt%) were investigated concerning morphological and structural characteristics by infrared mapping (IR), scanning electron microscopy (SEM), Fourier transform infrared spectroscopy (FTIR), energy dispersive spectrometry (EDAX), but also in correlation with their in-vitro functional properties in terms of an artificial mineralization process in simulated body fluid (SBF).

2. Materials and Methods

2.1. Materials

Calcium nitrate tetrahydrate ($Ca(NO_3)_2 \cdot 4H_2O$), zinc nitrate hexahydrate($Zn(NO_3)_2 \cdot 6H_2O$), ammonium hydroxide (NH_3 aq.) (30 wt% solution), hydrofluoric acid (HF), hydrogen peroxide (H_2O_2) (30 wt% solution), glutaraldehyde ($C_5H_8O_2$) (25 wt% solution), sodium chloride (NaCl), potassium chloride (KCl), potassium phosphate dibasic trihydrate ($K_2HPO_4 \cdot 3H_2O$), magnesium chloride hexahydrate ($MgCl_2 \cdot 6H_2O$), calcium chloride ($CaCl_2$), sodium sulphate (Na_2SO_4) and triethanolamine ($C_6H_{15}NO_3$) were purchased from Sigma-Aldrich, Darmstadt, Germany. Sodium phosphate dibasic dihydrate ($Na_2HPO_4 \cdot 2H_2O$) and nitric acid (HNO_3) were supplied by Fluka, Riedel-de-Haen, Germany. Sodium bicarbonate ($NaHCO_3$) was obtained from Riedel-de-Haen, Seelze, Germany. The collagen hydrogel (1.46 wt% anhydrous substance, pH = 2.5) was supplied by Sanimed International Impex SRL, Bucharest, Romania. HeLa cell culture was provided by American Tissue Culture Collection, Manassas, VA, USA, while Dulbecco's Modified Eagle's Medium and bovine serum were obtained from Sigma-Aldrich, St. Louis, MO, USA. Ti substrates (disk-shaped; 12 mm diameter, 0.5 mm thick) were provided by NextMaterials SRL, Milano, Italy [25].

2.2. Coll/Zn^{2+}-CaPs Hydrogels: Preparation

Coll/Zn^{2+}-CaPs hydrogels were synthesized by the in-situ generation of Zn^{2+}-substituted calcium phosphates in the collagen gel, the ratio between the collagen and CaPs was set at 1:1 (wt:wt). A Ca^{2+}-Zn^{2+} suspension was added dropwise in a flask containing 10 g of collagen hydrogel, kept on an ice bath. The mixture was mechanically stirred for 1 h at 500 rpm. Next, a sodium phosphate dibasic solution was added, the final mixture being stirred for 3 h at 500 rpm until complete homogenization. The co-existence of HAp and TCP as major components in the prepared coatings was further abbreviated as CaPs.

2.3. Deposition Methods

2.3.1. Spin Coating

The hydrogels prepared were deposited by spin coating on Ti substrates, with a Laurell WS-650 apparatus (3000 rpm, deposition time = 2 s/per layer, North Wales, PA, USA). Prior to deposition, Ti substrates were mechanically polished and modified by chemical etching in order to assure a suitable roughness benefic for an advanced adherence of the composite. The etching was performed by consecutively immersing the Ti substrates in HF-HNO_3 solution (ratio = 1:3), and in an

H_2O_2 15 wt% solution, respectively. After deposition, the Coll/Zn^{2+}-CaPs biocomposite coatings were cross-linked with glutaraldehyde 2.5 wt% solution and maintained at 4 °C for 24 h.

2.3.2. Matrix-Assisted Pulsed Laser Evaporation (MAPLE)

10 mL of Coll/CaPs hydrogel with different Zn^{2+} contents (0.5, 1 and 2 wt%) of Zn^{2+} have been used for the preparation of one MAPLE target. Before deposition, the target was frozen in a special copper holder at 77 K in liquid nitrogen and maintained at this temperature during deposition, using a cryogenic rotating setting (Figure 1) [26].

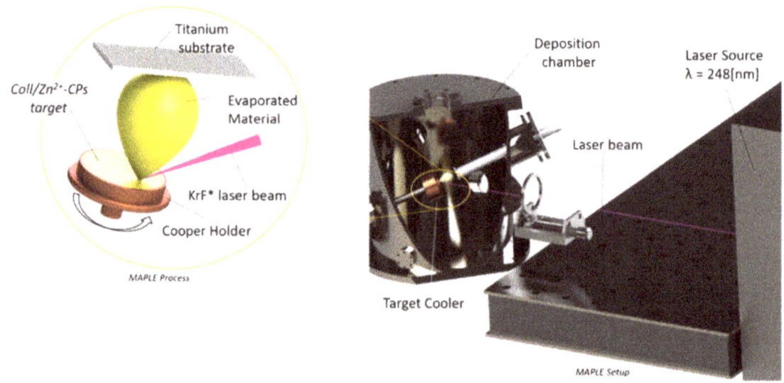

Figure 1. Schematic of matrix-assisted pulsed laser evaporation (MAPLE) set-up.

The chemically etched Ti substrates were successively cleaned in an ultrasonic bath for 15 min in acetone, alcohol and deionized H_2O and blown dry with high purity nitrogen before use as substrates for deposition. The experiments were conducted using a COMPEX Pro 205 KrF * (λ = 248 nm, $\tau_{FWHM} \leq$ 25 ns) excimer laser source (Coherent) which was operated at a fluence of 0.5 J/cm^{-2} at a repetition rate of 10 Hz. The laser beam was focused with MgF_2 cylindrical lens at an angle of 45° on the surface of the cryogenic target. Ti substrate was placed inside the reaction chamber parallel to the Coll/Zn^{2+}-CaPs target at a separation distance of 5 cm. For the growth of one film, 50,000 subsequent laser pulses have been applied. In order to obtain a uniform layer and to avoid drilling, the target and substrate were continuously rotated at 50 rpm, while the background pressure inside the deposition chamber was of 2×10^{-2} mbar.

2.4. Structural and Morphological Analyses

IR mapping analysis was applied to study the homogeneity of Coll/Zn^{2+}-CaPs films in correlation with different changes after Zn^{2+} substitutions in the hybrid coating. The analysis was done with an IR Thermo Scientific Nicolet iN10MX microscope (Waltham, MA USA) in reflection mode, at a spectral resolution of 4 cm^{-1} and a spatial resolution of 100 μm × 100 μm. Each spectrum cumulates the co-adding of 16 scans. For higher resolution a cooled MCT detector was used. FTIR spectra were acquired with a spectrophotometer Nicolet iS50R equipped with 3 beam splitters in the (12,500–50) cm^{-1} range. Spectra between 3800–500 cm^{-1} were recorded by co-adding 64 scans at a spectral resolution of 4 cm^{-1}. The morphology of the biocomposite coatings was studied via scanning electron microscopy (SEM), with a Quanta Inspect F50 microscope coupled with an energy dispersive spectrometer (EDAX) (Oregon, OR, USA).

2.5. In-Vitro Testing in Simulated Body Fluids (SBF)

The in-vitro mineralization process was assessed by the immersion of Coll/Zn^{2+}-CaPs composite coatings in SBF (10 mL of SBF/per disk) at 37 °C in a thermostatic bath. After 14 days, the coated Ti disks were characterized by FTIR and SEM analysis. 1 L of SBF solution was prepared according to

Kokubo methodology [27]. The concentrations of the ionic species involved in SBF preparation are listed in Table 1. The pH of the final buffer solution is 7.25.

Table 1. Concentrations of the ionic species used in simulated body fluids (SBF) preparation.

Ionic Species	Na^+	K^+	Mg^{2+}	Ca^{2+}	Cl^-	HCO_3^-	HPO_4^{2-}	SO_4^{2-}
Concentration of the ionic species (mmol/L)	142.0	5.0	1.5	2.5	148.8	4.2	1.0	0.5

2.6. Cell Viability

The cell viability assay was effectuated on simple and SBF immersed Coll/Zn^{2+}-CaPs composited deposited via MAPLE technology. The eukaryotic HeLa cell culture was maintained in Dulbecco's Modified Eagle's Medium supplemented with 10 wt% heat-inactivated fetal bovine serum at 37 °C in a 5 wt% CO_2 humid atmosphere. Coll/Zn^{2+}-CaPs were plated in 24-well plates and 1×10^5 cells were added onto each MAPLE sample. After 24 h, the Coll/Zn^{2+}-CaPs materials were moved to other wells, fixed in 70 wt% ethanol, stained with 50 µg/mL propidium iodide (PI) and washed with phosphate buffer solution (PBS). The Observer D1 Zeiss microscope (Jena, Germany) was used for the fluorescence assay and samples imaging. Remaining cells were collected, fixed in 70 wt% ethanol overnight at −20 °C and washed twice with PBS. Then, the cells were incubated with 100 µg/mL PI for 30 min and shielded against light at room temperature before flow cytometric analysis. DNA content and cell cycle distribution were monitored using a XML Beckman Coulter cytometer and measured using FlowJo software (Indianapolis, IN, USA).

3. Results and Discussion

3.1. Coll/Zn^{2+}-CaPs Biocomposite Coatings: Structural and Morphological Characterization

3.1.1. FTIR

Structural modifications in the collagen-calcium phosphate association were monitored by FTIR analysis (Figure 2). The use of one or another of the methods for thin films deposition did not significantly influenced the functional changes in the hybrid assembly (at 1032 cm^{-1} specific to phosphate units and 1242 cm^{-1} specific to amide collagen bonds).

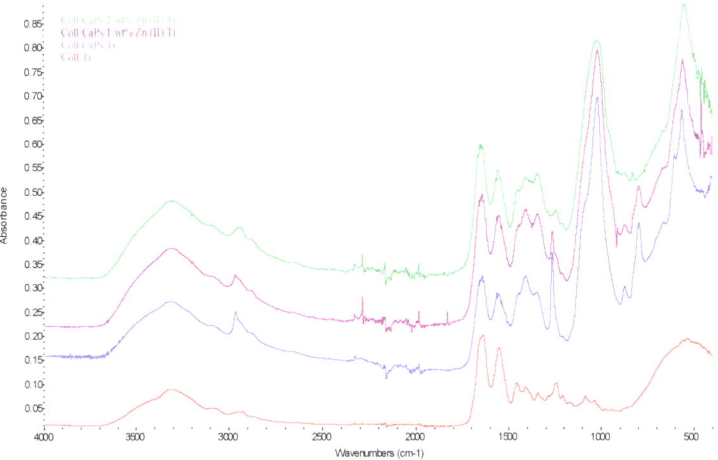

Figure 2. FTIR spectra of the Coll/Zn(II)-CPs biocomposites deposited on Ti substrates.

The Zn^{2+} content stands for a key factor in determining the molecular synergism between the collagen and inorganic network. An increase of the Zn^{2+} amount induces a gradual decrease of the absorbance in phosphate and amide regions. A large substitution of Ca^{2+} by Zn^{2+} generates structural modifications of apatite pattern, in particular for phosphate groups. These permutations of the P-O and/or P–OH functions stay at the origin of weak interactions and complementarity functional effects between the organic biopolymer (e.g., –OH, –COOH, –NH_2 groups of collagen) and phosphate moieties in Zn^{2+}-substituted CaPs.

3.1.2. IR Microscopy

The high homogeneity of thin films deposited on biocompatible Ti substrates is a mandatory condition in the development of an implant prototype with enhanced biological behavior [28]. IR mapping analysis of the biocomposite films deposited by spin coating is illustrated in Figure 3.

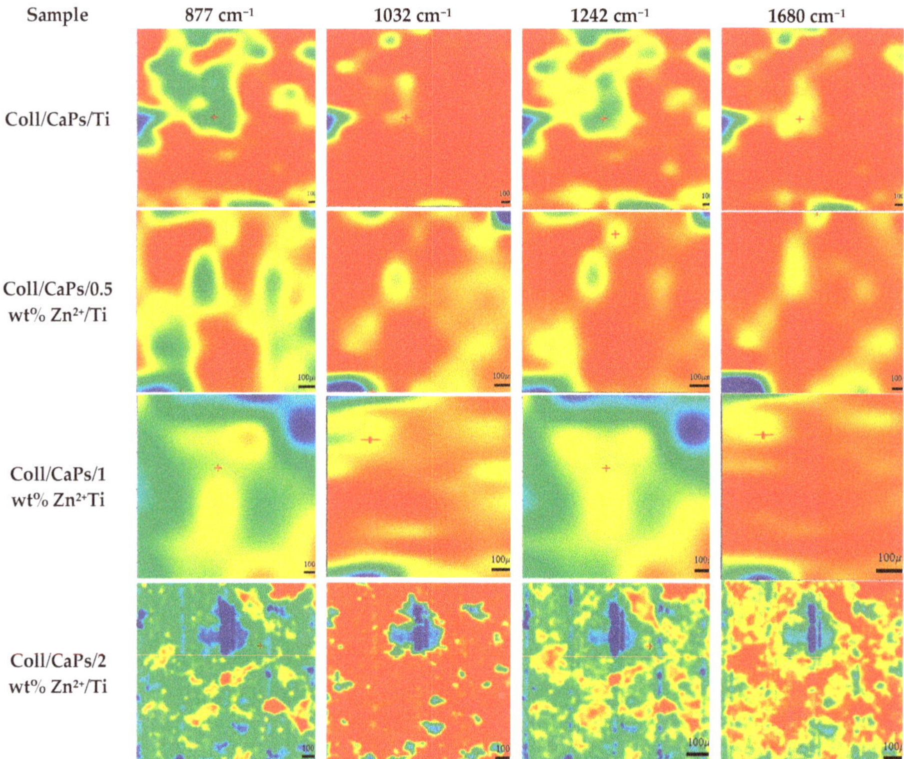

Figure 3. IR mapping of Coll/Zn(II)-CPs thin films deposited on Ti substrate by spin coating.

Direct CaPs formation in the collagen matrix can be monitored based on the peak at 1032 cm^{-1}. On the other side, the formation of carbonated apatite can be followed based on the peak at 877 cm^{-1}. The organic phase can be assigned according to the bands peaking at 1680 and 1242 cm^{-1}. In fact, the band at 1242 cm^{-1} can be also used to evaluate the collagen integrity in depositions. The decreases in intensity of the highest Zn^{2+} fraction peaking at 1032 cm^{-1}, can be accounted by cation substitutions in the apatite network. Accordingly, an excess in the Zn^{2+} concentration results in an enhanced saturation of apatite units. More Ca^{2+} ions are replaced by Zn^{2+}, promoting smooth diffused areas in both phosphate region and amide section [29].

The analysis of the maps recorded for the samples deposited by spin coating (Figure 3) showed a non-uniform layer even when using collagen. In the case of composite hydrogel depositions on Ti, large agglomerates could be observed onto the surface of the metallic substrate. These conglomerates can reach hundreds of microns (~500 μm) as clearly visible for Coll/2 wt% Zn^{2+}-CaPs sample.

An essential remark in case of MAPLE films (Figure 4) is the advanced uniformity of the composite mixture on Ti substrates.

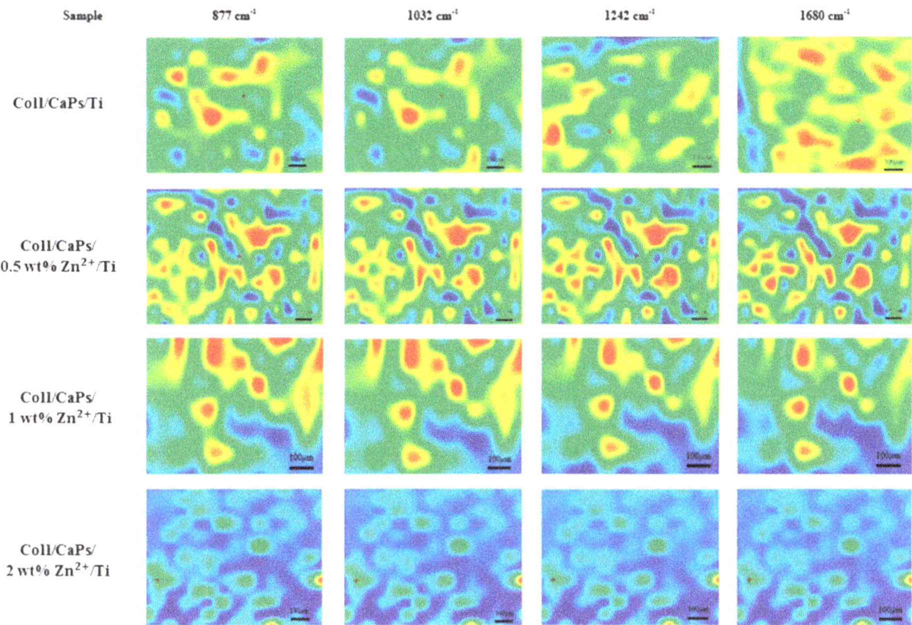

Figure 4. IR mapping of Coll/Zn(II)-CPs thin films deposited on Ti substrate by MAPLE.

A significant modification is visible in the amide region (1680–1242 cm^{-1}) of the Coll/CaPs composite that is more uniform than for the film deposited by spin coating. One may consider this feature due to the structural conservation of the collagen molecule in terms of thermal denaturation. The cleavage of peptide bonds in the collagen structure is restricted because of the low temperature during the MAPLE deposition process, while thin film quality is significantly improved. In fact, even the maps recorded for all of the four wavenumbers are quite similar, there are only marginal differences which confirm that neither compositional nor morphological (thickness) heterogeneities are present.

3.1.3. SEM

SEM images of spin coating deposited films (Figure 5) suggest that collagen is maintaining its native fibrillary structure. Although collagen and CaPs are intimately covering Ti, highly rough surfaces are obtained. A high Zn^{2+} content (2 wt%) results in a distinct morphology, characterized by small and thin particles growing inside the organic matrix (Figure 5c, yellow area). One may therefore assume that a Zn^{2+} content higher than 1 wt% could promote the crystallization process of calcium phosphates. This specific feature is significant for the development of artificial osseous regeneration [30].

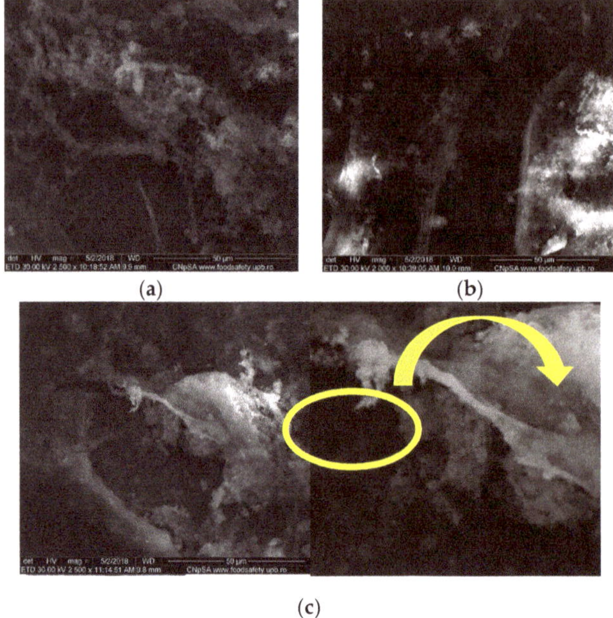

Figure 5. Scanning electron microscopy (SEM) images of Coll/CaPs thin films deposited by spin coating with different Zn(II) fractions (**a**) 0 wt%; (**b**) 1 wt%; (**c**) 2 wt%.

The preservation of the collagen structural pattern after MAPLE deposition has improved the homogeneity and morphology of the thin films (Figure 6). It should be emphasized that a highly ordered organic matrix that accommodates the inorganic particles has been formed, anticipating a potential osteo-integration behavior of the coatings. As observed, a closely-packed collagen-like architecture promotes an advanced crystallization of the substituted CaPs, proportional with the Zn^{2+} amount increase, similar to films deposited by spin coating. The difference in case of MAPLE-deposition method resides in the morphology of the particles developing inside the collagen pattern. The formation of diffused agglomerations associated to HAp was noticed for un-substituted Coll/CaPs (Figure 6a). Furthermore, different spherical nano-structures similar to CaP morphology were observed (Figure 6b,c). Small clusters were evidenced for up to 1 wt% Zn^{2+} (Figure 6c). The size decrease of droplets could be attributed to a possible Zn^{2+} super-saturation of the apatite network. The supplementary Zn-O bonds are placed outside calcium phosphate particles [31].

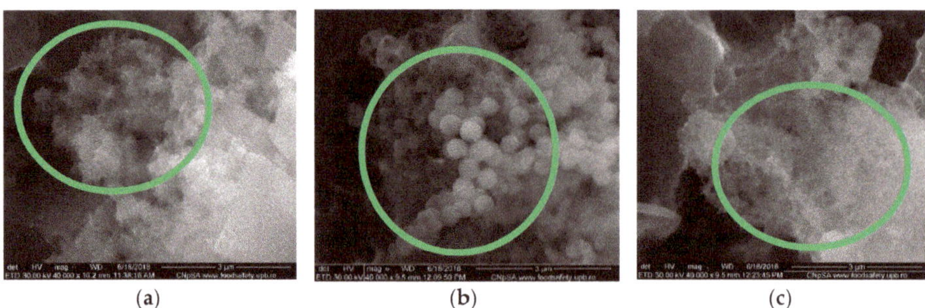

Figure 6. SEM images of the Coll/CaPs thin films deposited by MAPLE with different Zn(II) content (**a**) 0 wt%; (**b**) 1 wt%; (**c**) 2 wt%.

3.2. In-Vitro SBF Assay of Coll/Zn^{2+}-CaPs Biocomposite Coatings

3.2.1. FTIR

FTIR recorded spectra were different before (Section 3.1.1) and after immersion in SBF (Figure 7), in particular in the phosphate and amide regions, probably due to the in-vitro influence of the Zn^{2+} substitutional agent.

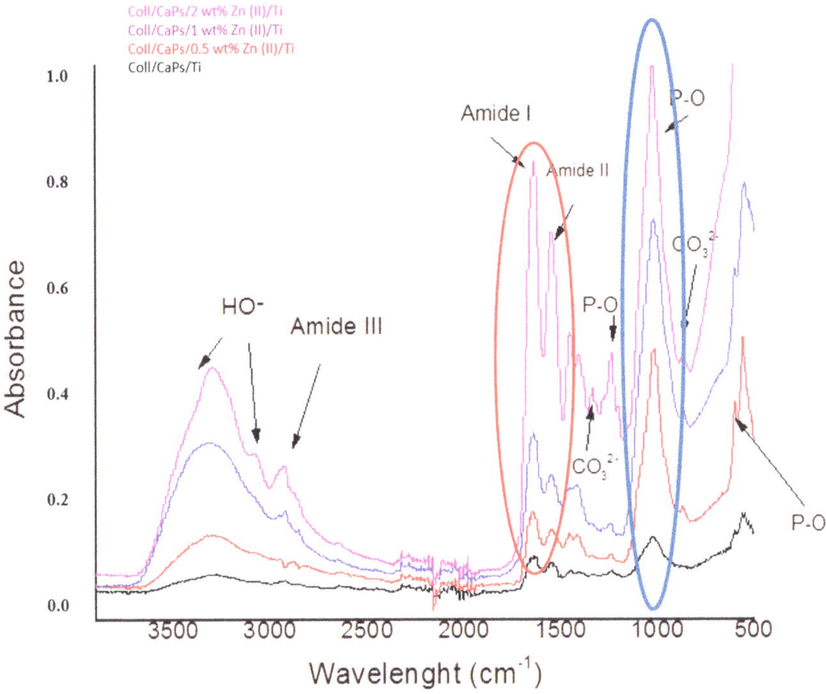

Figure 7. FTIR spectra of the Coll/Zn(II)-CPs thin films deposited by spin coating after 14 days of immersion in SBF.

The increase in Zn^{2+} induces a rise of the P-O band absorbance due to the enhanced formation of HAp, as a result of ionic exchanges during the immersion in SBF. In this context, possible cationic migrations in interstitial positions could be envisaged, followed by the generation of new phosphate functions able to bind positively charged species or interact with collagen moieties. In addition, the preservation of the inorganic template is respected to assure an adequate confinement with the collagen matrix (see the amide regions in red circle in Figure 7). The same behavior was expected for the MAPLE-deposited coatings: a decrease in the intensity of the amide and phosphate bands in respect with spin coating depositions which could be assigned to a very thin layer formation on the Ti substrate (*Data not shown*).

3.2.2. SEM and EDAX

By correlation with SEM results described in Section 3.1.3, the effect of the immersion in SBF reflected specific modifications in the case of 1wt% Zn^{2+} for both deposition methods. Figure 8 shows the formation of spherical and diffused agglomerations associated to the mineralization process inside the organic template. The further formation of CaPs as a cluster motif is observed in case of MAPLE depositions (Figure 8b, green regions). The collagen structure is not thermally affected in MAPLE, so the definite synergism between the organic network and inorganic particles promotes the HAp crystallization.

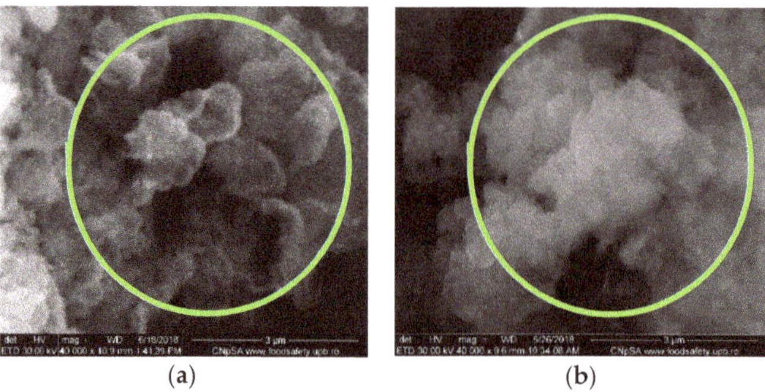

(a)   (b)

Figure 8. SEM images of the Coll/CaPs thin films with 1 wt% Zn(II)deposited by (**a**) Spin coating; (**b**) MAPLE after 14 days of immersion in SBF.

The presence of Zn along with Na, K and Mg, characteristic to the artificial mineralization process, is confirmed by EDAX spectra in Figure 9.

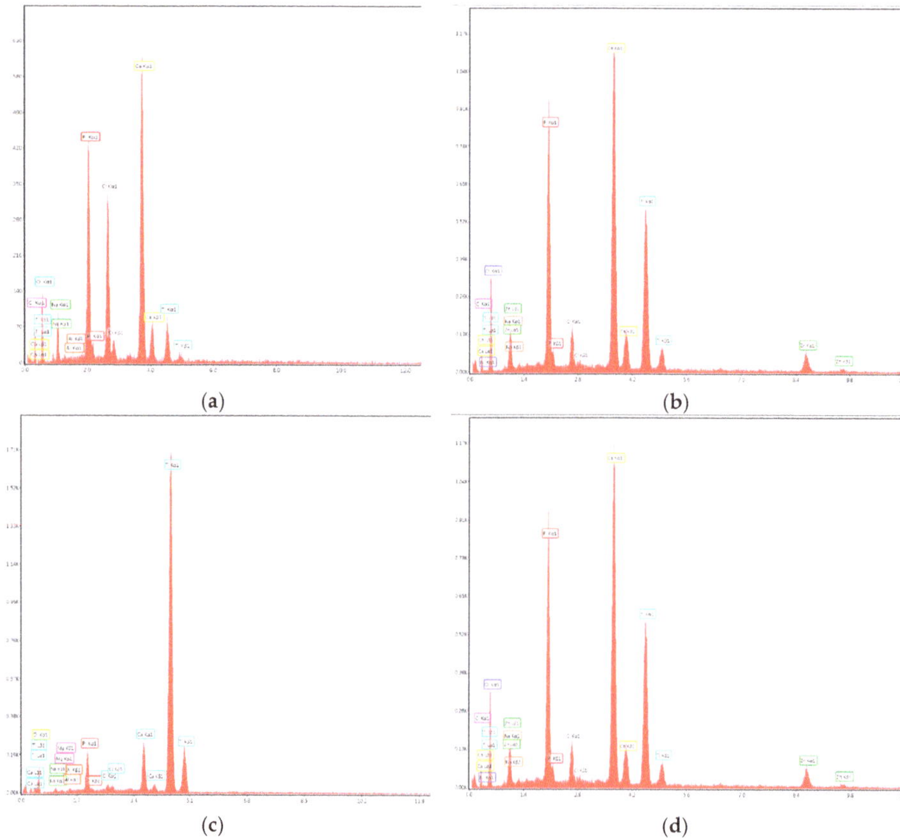

(a)   (b)
(c)   (d)

Figure 9. Energy dispersive spectrometer (EDAX), spectra of the Coll/CaPs thin films with 0 wt% and 1 wt% Zn(II) deposited by (**a**,**b**) Spin coating and (**c**,**d**) MAPLE after 14 days of immersion in SBF.

3.3. Cytological Assay of Coll/Zn^{2+}-CaPs Biocomposite Coatings

To determine the usefulness of these materials as medical devices, we conducted a pilot test for cytotoxicity using epithelial cells and a direct contact method that is able to detect weak cytotoxicity. Florescence microscopy revealed that after 24 h of incubation, epithelial cells grown on modified bioactive surfaces present a normal morphology. All tests have been made on the composites deposited by the MAPLE technique. The Coll/Zn^{2+}-CaPs biocomposite coatings favored the attachment of cells as visible from Figure 10. The density of the viable cells seems to be the same as for control and normal for this incubation time.

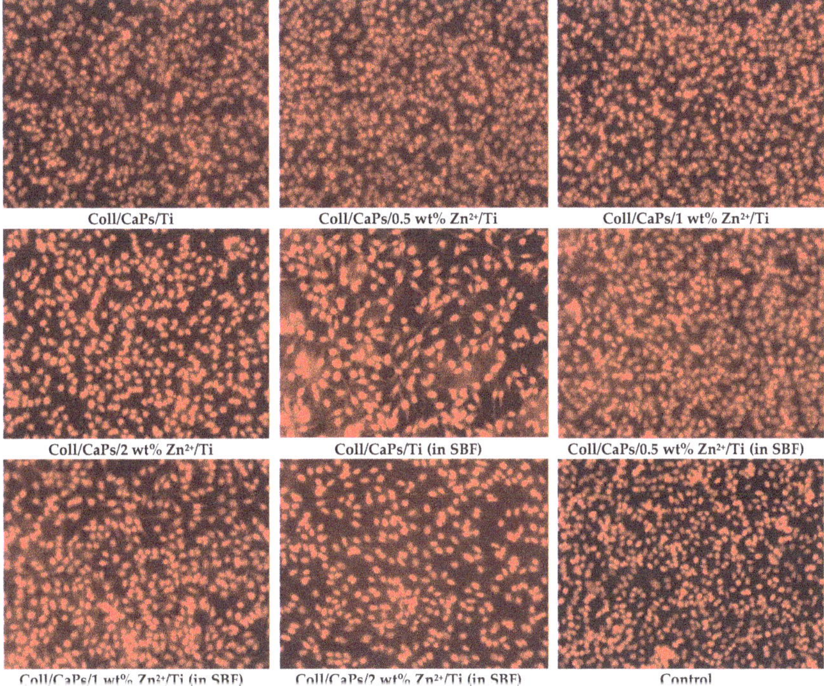

Figure 10. Fluorescent microscopy images of cells grown on Coll/Zn(II)-CPs thin films deposited on Ti by MAPLE.

Cell cycle analysis (Figure 11) revealed that the compounds immersed in SBF have induced a slight decrease in S phase (DNA replication).

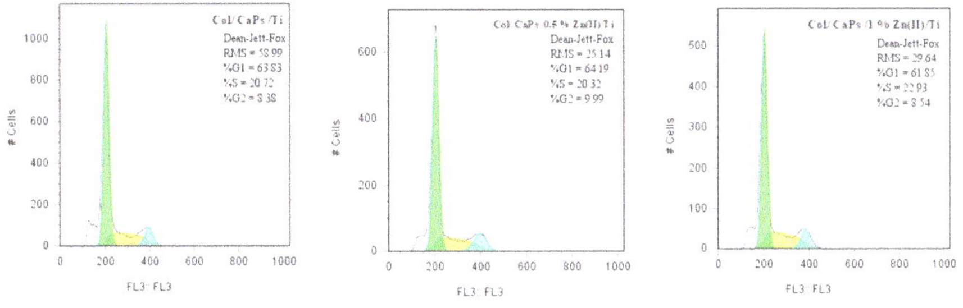

Figure 11. *Cont.*

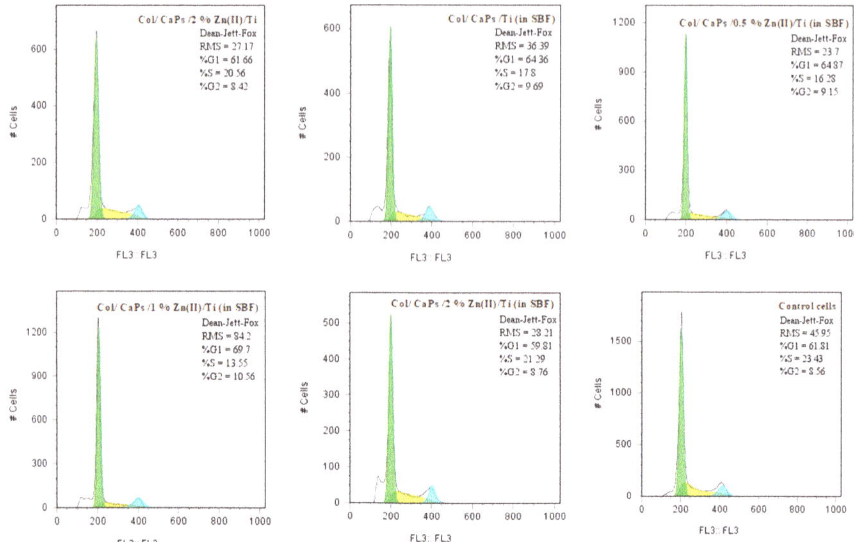

Figure 11. Cell cycle distribution of cells grown on Coll/Zn(II)-CPs thin films deposited on Ti by MAPLE.

The layers obtained via spin coating were not uniformly distributed on the metallic substrate, as demonstrated by IR microscopy. The heterogeneous deposition of the Coll/Zn^{2+}-CaPs biocomposites on Ti substrates generated differences in the biological properties of the implant and therefore no reproducible biological response was obtained and thus not presented.

4. Conclusions

A comparative study of spin coating vs. MAPLE depositions for the synthesis of Coll/Zn^{2+}-CaPs biocomposite coatings onto chemically etched Ti substrates was carried out. FTIR microscopy demonstrated that the surface homogeneity is superior in case of MAPLE coatings. In both cases, a fraction of 2 wt% Zn^{2+} induced a significant decrease of the absorbance in the P-O and amide regions. However, a further substitution of Ca^{2+} ions is conducting to a decrease of the bands intensity, most probably due to supersaturation. A similar trend in the amide area involves the confinement and synergic effects via weak interactions between the collagen matrix and CaPs particles. Zn^{2+} promoted the in-situ formation of calcium phosphate in collagen matrix, as proved by the presence of spherical CaPs. An excess of Zn^{2+} (up to 1 wt%) generated an uncontrolled crystallization of supplementary Zn–O groups attached to the Zn^{2+}-CaPs, in a cluster-like morphology. The in-vitro SBF assay confirmed the gradual formation of supplementary apatite layers in the organic matrix. High density areas were observed in the phosphate region as a result of Ca^{2+}-coordinated ions, in conjunction with an increase in the P-O band absorbance. Even in high concentrations, Zn^{2+} promoted the mineralization process, as suggested by an absorbance rise of the P–O band. The SEM analysis of coatings immersed in SBF revealed the effective mineralization of CaPs on MAPLE layers. Well-defined spherical particulates embedded into the collagen matrix proved the growth of a new apatite layer. The collagen structure was preserved under MAPLE, supporting the improved mineralization by weak interactions with numerous active sites. The MAPLE method did not produce secondary compounds that alter cell viability (morphological quantification) or normal cell cycle phases (flow cytometry assessment). One may conclude that Coll/Zn^{2+}-CaPs biocomposites synthesized by one-step MAPLE are appropriate for the development of a new generation of surface coated Ti implants with promising biological performances, and encourage us for further testing to demonstrate their capacity to sustain osteoblastic function and bone formation.

Author Contributions: L.V.A. and I.A.N.—experimental methodology of materials synthesis, Spin-Coating deposition methodology; I.L.A. and L.V.A.—IR Microscopy and FTIR; R.T.—SEM and EDAX analysis; N.M. and C.R.—MAPLE methodology; C.B.—cytological assay; L.V.A. and I.A.N.—original draft preparation.; I.A.N. and L.V.A. writing—review and editing; A.F., E.A. and I.N.M.—supervision; A.F. and E.A.—project administration and project funding.

Funding: This research was funded by 'Innovative biomaterials for treatment and diagnosis' (PN-IIIP1-1.2-PCCDI2017-0629) national project, by NATO and UEFISCDI contracts NATO G-4890 and 43 NATO/2017, Core Program 16N/08.02.2019 and UEFISCDI under the PD_52/2018 contract.

Conflicts of Interest: The authors declare no conflict of interest.

References

1. Ouerd, A.; Alemany-Dumont, C.; Berthomé, G.; Normand, B.; Szunerits, S. Reactivity of Titanium in Physiological Medium I. Electrochemical Characterization of the Metal/Protein Interface. *J. Electrochem. Soc.* **2007**, *154*, C593–C601. [CrossRef]

2. Manam, N.S.; Harun, W.S.W.; Shri, D.N.A.; Ghani, S.A.C.; Kurniawan, T.; Ismail, M.H.; Ibrahim, M.H.I. Study of corrosion in biocompatible metals for implants: A review. *J. Alloys Compd.* **2017**, *701*, 698–715. [CrossRef]

3. Haka, D.J.; Banegasa, R.; Ipaktchian, K.; Mauffrey, C. Evolution of plate design and material composition. *Int. J. Care Injured* **2018**, *49*, S8–S11. [CrossRef]

4. Young, M.F. Bone matrix proteins: Their function, regulation, and relationship to osteoporosis. *Osteoporos Int.* **2003**, *14*, S35–S42. [CrossRef] [PubMed]

5. Su, Y.; Zheng, Y.; Tang, L.; Qin, Y.X.; Zhu, D. Calcium Phosphate Coatings for Metallic Orthopedic Biomaterials. *Orthop. Biomater.* **2017**, 167–183. [CrossRef]

6. Kaczmarek, B.; Sionkowska, A.; Kozlowska, J.; Osyczka, A.M. New composite materials prepared by calcium phosphate precipitation in chitosan/collagen/hyaluronic acid sponge cross-linked by EDC/NHS. *Int. J. Biol. Macromol.* **2018**, *107*, 247–253. [CrossRef] [PubMed]

7. Ficai, A.; Andronescu, E.; Voicu, G.; Ghitulica, C.; Vasile, B.S.; Ficai, D.; Trandafir, V. Self-assembled collagen/hydroxyapatite composite materials. *Chem. Eng. J.* **2010**, *160*, 794–800. [CrossRef]

8. Jang, C.H.; Lee, H.; Kim, M.; Kim, G.H. Accelerated osteointegration of the titanium-implant coated with biocomponents, collagen/hydroxyapatite/bone morphogenetic protein-2, for bone-anchored hearing aid. *J. Ind. Eng. Chem.* **2018**, *63*, 230–236. [CrossRef]

9. Lazar, A.C.; Lacatus, R.; Pacurar, M.; Campian, R.S. Studies results regarding bisphosphonate therapy in dental medicine. *Rev. Chim.* **2018**, *69*, 1018–1022.

10. Fodor, P.; Fodor, R.; Solyom, A.; Catoi, C.; Tabaran, F.; Lacatus, R.; Trambitas, C.; Cordos, B.; Bataga, T. Autologous matrix induced chondrogenesis vs. microfracture with PRP for chondral lesions of the knee in a rabbit model. *Rev. Chim.* **2018**, *69*, 894–900.

11. Portillo-Lara, R.; Shirzaei Sani, E.; Annabi, N. Biomimetic Orthopedic Materials. *Orthop. Biomater.* **2018**, 109–139.

12. Kane, R.J.; Weiss-Bilka, H.E.; Meagher, M.J.; Liu, Y.; Gargac, J.A.; Niebur, G.L.; Wagner, D.R.; Roeder, R.K. Hydroxyapatite reinforced collagen scaffolds with improved architecture and mechanical properties. *Acta Biomater.* **2015**, *17*, 16–25. [CrossRef] [PubMed]

13. Ben-Nissan, B.; Choi, A.H. Chapter 5: Calcium Phosphate Nanocoatings: Production, Physical and Biological Properties, and Biomedical Applications. *Nanobioceram. Healthc. Appl.* **2017**, 105–149. [CrossRef]

14. Jin, G.; Qin, H.; Cao, H.; Qian, S.; Zhao, Y.; Peng, X.; Zhang, X.; Liu, X.; Chu, P.K. Synergistic effects of dual Zn/Ag ion implantation in osteogenic activity and antibacterial ability of titanium. *Biomaterials* **2014**, *35*, 7699–7713. [CrossRef] [PubMed]

15. Saxena, V.; Hasan, A.; Pandey, N. Effect of Zn/ZnO integration with hydroxyapatite: A review. *Mater. Technol.* **2017**, *33*, 79–92. [CrossRef]

16. Jankovic, A.; Erakovic, S.; Ristoscu, C.; Mihailescu Serban, N.; Duta, L.; Visan, A.; Stan, G.E.; Popa, A.; Husanu, M.A.; Luculescu, C.R.; et al. Structural and biological evaluation of lignin addition to simple and silver-doped hydroxyapatite thin films synthesized by matrix-assisted pulsed laser evaporation. *J. Mater. Sci. Mater. Med.* **2015**, *26*, 5333–5347. [CrossRef]

17. Eraković, S.; Janković, A.; Ristoscu, C.; Duta, L.; Serban, N.; Visan, A.; Mihailescu, I.N.; Stan, G.E.; Socol, M.; Iordache, O.; et al. Antifungal activity of Ag: Hydroxyapatite thin films synthesized by pulsed laser deposition on Ti and Ti modified by TiO_2 nanotubes substrates. *Appl. Surf. Sci.* **2014**, *293*, 37–45. [CrossRef]

18. Huang, Y.; Xu, Z.; Zhang, X.; Chang, X.; Zhang, X.; Chao Li, Y.; Ye, T.; Han, R.; Han, S.; Gao, Y.; et al. Nanotube-formed Ti substrates coated with silicate/silver co-doped hydroxyapatite as prospective materials for bone implants. *J. Alloys Compd.* **2017**, *697*, 182–199. [CrossRef]

19. Popa, C.L.; Bartha, C.M.; Albu, M.; Guégan, R.; Motelica-Heino, M.; Chifiriuc, C.M.; Bleotu, C.; Badea, M.L.; Antohe, S. Synthesis, characterization and cytotoxicity evaluation on zinc doped hydroxyapatite in collagen matrix. *Dig. J. Nanomater. Biostruct.* **2015**, *10*, 681–691.

20. Hall, D.B.; Underhill, P.; Torkelson, J.M. Spin coating of thin and ultrathin polymer films. *Polym. Eng. Sci.* **1998**, *38*, 2039–2045. [CrossRef]

21. Sima, F.; Davidson, P.M.; Dentzer, J.; Gadiou, R.; Pauthe, E.; Gallet, O.; Mihailescu, I.N.; Anselme, K. Inorganic-Organic Thin Implant Coatings Deposited by Lasers. *ACS Appl. Mater. Interfaces* **2015**, *7*, 911–920. [CrossRef] [PubMed]

22. Piqué, A. The Matrix-Assisted Pulsed Laser Evaporation (MAPLE) process: Origins and future directions. *Appl. Phys. A* **2011**, *105*, 517–528. [CrossRef]

23. Floroian, L.; Ristoscu, C.; Mihailescu, N.; Negut, I.; Badea, M.; Ursutiu, D.; Chifiriuc, M.C.; Urzica, I.; Dyia, H.M.; Bleotu, C.; et al. Functionalized Antimicrobial Composite Thin Films Printing for Stainless Steel Implant Coatings. *Molecules* **2016**, *21*, 740. [CrossRef]

24. Jeong, H.; Shepard, K.B.; Purdum, G.E.; Guo, Y.; Loo, Y.L.; Arnold, C.B.; Priestley, R.D. Additive Growth and Crystallization of Polymer Films. *Macromolecules* **2016**, *49*, 2860–2867. [CrossRef]

25. Chiesa, R.; Sandrini, E.; Santin, M.; Rondelli, G.; Cigada, A. Osteointegration of titanium and its alloys by anodic spark deposition and other electrochemical techniques: A review. *J. Appl. Biomater. Biomech.* **2003**, *1*, 91–107.

26. Popescu-Pelin, G.F.; Ristoscu, C.; Badiceanu, M.; Mihailescu, I.N. Protected Laser Evaporation/Ablation and Deposition of Organic/Biological Materials: Thin Films Deposition for Nanobiomedical Applications. In *Laser Ablation—From Fundamentals to Applications*; Idina, T., Ed.; InTech Open: Rijeka, Croatia, 2017; pp. 57–79.

27. Kokubo, T.; Kushitani, H.; Sakka, S.; Kitsugi, T.; Yamamuro, T. Solutions able to reproduce in vivo surface-structure changes in bioactive glass-ceramic. *J. Biomed. Mater. Res.* **1990**, *24*, 721–734. [CrossRef]

28. Narayanan, R.; Seshadri, S.K.; Kwon, T.Y.; Kim, K.H. Calcium phosphate-based coatings on titanium and its alloys. *J. Biomed. Mater. Res. Part B Appl. Biomater.* **2008**, *85 B*, 279–299. [CrossRef]

29. Walczyk, D.; Malina, D.; Krol, M.; Pluta, K.; Sobczak-Kupiec, A. Physicochemical characterization of zinc-substituted calcium phosphates. *Bull. Mater. Sci.* **2016**, *39*, 525–535. [CrossRef]

30. Nandi, S.K.; Kundu, B.; Mukherjee, J.; Mahato, A.; Datta, S.; Krishna Balla, V. Converted marine coral hydroxyapatite implants with growth factors: In vivo bone regeneration. *Mater. Sci. Eng. C.* **2015**, *49*, 816–823. [CrossRef]

31. Qiao, Y.; Zhang, W.; Tian, P.; Meng, F.; Zhu, H.; Jiang, X.; Liu, X.; Chu, P.K. Stimulation of bone growth following zinc incorporation into biomaterials. *Biomaterials* **2014**, *35*, 6882–6897. [CrossRef]

 *nanomaterials*

Article

Laser-Synthesized SERS Substrates as Sensors toward Therapeutic Drug Monitoring

Matteo Tommasini [1], Chiara Zanchi [1,2], Andrea Lucotti [1], Alessandro Bombelli [2], Nicolò S. Villa [1], Marina Casazza [3], Emilio Ciusani [4], Ugo de Grazia [4], Marco Santoro [5], Enza Fazio [5], Fortunato Neri [5], Sebastiano Trusso [6] and Paolo M. Ossi [2,*]

[1] Dipartimento di Chimica, Materiali e Ingegneria Chimica "G. Natta", Politecnico di Milano, Piazza Leonardo da Vinci 32, 20133 Milano, Italy; matteo.tommasini@polimi.it (M.T.); chiaragiuseppina.zanchi@polimi.it (C.Z.); andrea.lucotti@polimi.it (A.L.); nicolosimone.villa@mail.polimi.it (N.S.V.)

[2] Dipartimento di Energia, Politecnico di Milano, Via Ponzio 34/3, 20133 Milano, Italy; alessandro.bombelli@mail.polimi.it

[3] Division of Neurophysiopathology, Fondazione IRCCS Istituto Neurologico Carlo Besta, Via Celoria 11, 20133 Milano, Italy; marina.casazza@istituto-besta.it

[4] Laboratorio di Patologia Clinica e Genetica Medica, Fondazione IRCCS Istituto Neurologico Carlo Besta, Via Celoria 11, 20133 Milano, Italy; emilio.ciusani@istituto-besta.it (E.C.); ugo.degrazia@istituto-besta.it (U.d.G.)

[5] Dipartimento di Scienze Matematiche e Informatiche, Scienze Fisiche e Scienze della Terra, Università di Messina, Viale Ferdinando Stagno d'Alcontres 31, 98166 Messina, Italy; masantoro@unime.it (M.S.); enfazio@unime.it (E.F.); fneri@unime.it (F.N.)

[6] CNR-IPCF, Istituto per i Processi Chimico-Fisici del CNR, V.le F. S. D'Alcontres 37, 98158 Messina, Italy; trusso@ipcf.cnr.it

* Correspondence: paolo.ossi@polimi.it; Tel.: +39-02-2399-6319

Received: 12 March 2019; Accepted: 19 April 2019; Published: 1 May 2019

Abstract: The synthesis by pulsed laser ablation and the characterization of both the surface nanostructure and the optical properties of noble metal nanoparticle-based substrates used in Surface Enhanced Raman Spectroscopy are discussed with reference to application in the detection of anti-epileptic drugs. Results on two representative drugs, namely Carbamazepine and Perampanel, are critically addressed.

Keywords: noble metal nanoparticles; pulsed laser ablation; surface enhanced Raman spectroscopy; antiepileptic drugs

---

1. Introduction

Research in the field of nanomaterials opens scenarios that are anything but obvious. A particular spectroscopic technique, which aims at identifying and quantifying molecular species of biomedical interest becomes feasible when materials with specifically designed optical properties are available. In principle, it is possible to recognize a chemical species by means of the inelastic scattering of a probing laser light that discloses the features of the vibrational spectrum of the target molecule to be probed. The intrinsic limit of Raman spectroscopy is the very low scattering cross-section (about 10^{-30} cm^2 molecule^{-1}). This may be overcome when a *surface plasmon resonance* (SPR) and the associated strong increase of the scattered intensity is triggered at a nanostructured metal surface (Surface Enhanced Raman Spectroscopy, SERS). The wavelength and Full Width at Half Maximum (FWHM) of the SP, as recorded in the UV-Vis spectrum of the film, are the optical characteristics of a corrugated metal surface of specific interest when exploring its behavior as an active SERS substrate.

Near the metal surface, the electric field enhancement associated with the exciting light is most relevant with the noble metals silver and gold. In particular, large electromagnetic field enhancements [1,2] are observed at *hot spots*, corresponding to specific local surface morphologies that include sharp tips, edges and thin interparticle gaps. The surface nanostructures of artificially roughened Ag and Au thin films display many hot spots [3] making them excellent SERS substrates. Thus, the goal is to synthesize Ag and Au films with surface nanostructure engineered so as to maximize the SERS signal.

Pulsed laser ablation (PLA) makes use of two alternative methods to produce artificially corrugated nanostructured surfaces. Both methods are based on the vaporization of a solid target by high-energy laser pulses in an ambient fluid, whose role is to confine the vaporized species and to induce their mutual aggregation, obtaining clusters and nanoparticles (NPs) without chemical precursors. The fluid can be a gas or a liquid transparent to the laser radiation.

When ablation is performed using an ns-laser (nanosecond-laser) in a high pressure, neutral ambient gas, NPs are directly deposited onto appropriate inert supports where they self-arrange leading to qualitatively different surface nanostructures [4–6]. For defined experimental conditions [4], a path connects two extreme morphologies: isolated, sphere-like NPs, more or less crowded together, and a continuous metal film. The two relevant parameters to design the surface nanostructure of the growing film are the number of laser pulses (N_{LP}) and the ambient gas pressure (p_g). In fact, for a progressively increasing N_{LP} and p_g, we observe first particles more and more coalesced together whose size increases progressively while the shape departs from spherical and becomes more and more irregular. Later, islands separated by a network of randomly oriented channels with variable length and width in the few nm range develop. By increasing p_g, keeping constant N_{LP} at low values, the support is covered by increasingly smaller, spherical NPs; at high N_{LP} values, we observe islands with smaller average size and correspondingly a larger number of shorter channels (see Figure 1a,b). The N_{LP} value defines the degree of support coverage. With increasing N_{LP}, keeping constant p_g, we observe a spatial densification of the NPs on the support. The role of N_{LP} and p_g is illustrated in Figure 1 for self-assembled Au NPs on (100) Si supports. The same qualitative trend is observed for Ag NPs.

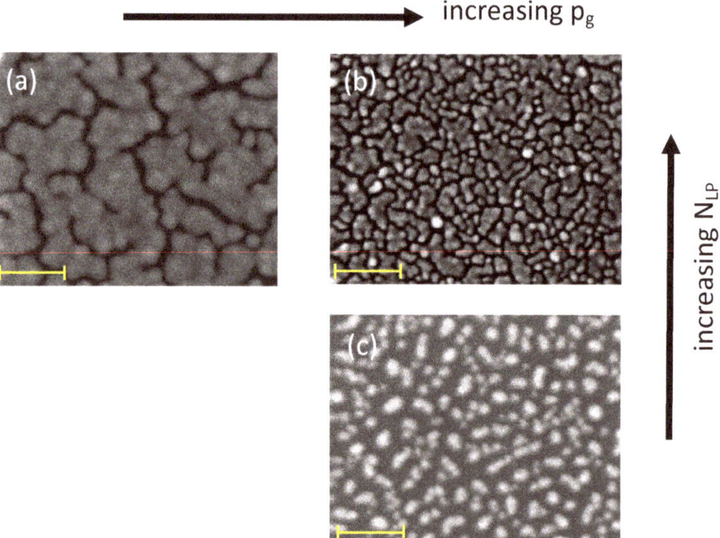

Figure 1. Scanning Electron Microscope (SEM) micrographs showing the dependence of the surface nanostructure of Au films deposited in Ar on the ambient gas pressure (p_g) and on the number of laser pulses (N_{LP}). (**a**) $N_{LP} = 3 \times 10^4$, $p_g = 10$ Pa; (**b**) $N_{LP} = 3 \times 10^4$, $p_g = 100$ Pa; (**c**) $N_{LP} = 5 \times 10^3$, $p_g = 100$ Pa. The yellow bars correspond to 100 nm. See Experimental for details.

Remarkably, for both metals, the average NP sizes observed in the Transmission Electron Microscope (TEM) agree with the predictions of a model for the plasma expansion through a high-pressure ambient gas [4,7]. Concerning the optical properties of the obtained nanostructures, for both metals, the SP wavelength decreases with increasing p_g at fixed N_{LP}, as well as, when p_g is kept fixed and N_{LP} is increased, the SP wavelength increases.

When a liquid is used as the confining medium through which the ablation plasma expands, a colloidal solution is obtained. By pulsed laser ablation in liquid (PLAL), using a nano/picosecond laser source, surfactant-free NPs are produced in a single-step approach within a time scale of a few minutes [8]. In a stationary liquid, the process consists of the production of NPs by ablation of the target, and the contemporaneous fragmentation-assembling of dispersed NPs by continuous irradiation of already synthesized particles. The laser pulse duration is a relevant parameter for NP generation [9]. With picosecond pulses (ps pulses), the relevance of melting and thermal evaporation is strongly reduced with respect to nanosecond pulses (ns pulses). With short pulses, the ablation process becomes increasingly efficient involving a nearly instantaneous vaporization with minimized heat-affected zone [10], so that the colloidal solution is produced in a shorter time. Besides this, compared to ns pulses, ps pulses allow for mitigating primary plasma shielding that is detrimental to ablation efficiency [11].

Recently, in the scientific literature that deals with drug dosage [12], there emerged a promising technique based on Raman/SERS, specifically suited to low drug concentrations [13–17] that is complementary to those in clinical use. In a potential application scenario, it would be possible to trace the concentration of a drug in blood plasma samples of clinical origin subjected to limited and rapid treatments (e.g., centrifugation, extraction with solvents). The SERS measurement exploits the interaction of the drug molecule with a nanostructured metal surface and the intensity of the signal, under controlled conditions, allows for tracing the quantity of molecules adsorbed on the metal, which is a function of the drug concentration in the tested solution. The analytical capability of SERS was shown to be comparable with High-Performance Liquid Chromatography (HPLC) technique [18], thus making possible to study the potentiality of this spectroscopic technique in the field of Therapeutic Drug Monitoring (TDM). TDM is a clinical practice that involves determining the concentration of a drug in a biological fluid, usually blood plasma. This procedure is particularly relevant for drugs characterized by a narrow therapeutic index (NTI). In such drugs, the difference between the concentration at which therapeutic effects and the (slightly larger) concentration at which side effects are observed for the patient is minimal. In therapies that use NTI drugs, it is important to know the plasma concentration of the drug. This way, the patient's clinical conditions can be associated with the required drug dose to guarantee the effectiveness of the treatment, avoiding the occurrence of side effects. TDM has been used for a long time in the clinical practice for NTI drugs. Among these drugs, we find anticancer drugs and antiepileptic drugs (AEDs). For the latter, toxicity can be induced by a small drug excess, whereas even a small reduction of the effective dosage can reduce the efficiency in controlling seizures.

Current research efforts in the SERS-TDM field point at overcoming the difficulties associated with the weakness of the signals of some drugs and/or to the background signal from much more abundant biomolecules coexisting with the drug in the fluid to be analyzed. A second relevant research direction is the development of SERS sensors (i.e., nanostructured metal surfaces) that combine a high sensitivity with spatial uniformity, control and reproducibility of the manufacturing process, not disregarding a low-cost production. Before the SERS technique can be routinely introduced into the clinical laboratory, besides the above technological issues, the treatment of the samples should also be optimized. Finally, validation procedures are required to develop quantitative SERS measurements. These imply determining drug concentration with standard reference methods (HPLC-MS or immunological assay).

In this work, we show how noble metal thin films resulting from self-assembled NPs can be produced in a controlled and reproducible way by PLA techniques. The obtained surface nanostructures are irregular and non-periodic. The peculiar morphological features of such nanostructures allow the good stability and reproducibility of the SPR of the films and lead to optimal electromagnetic enhancements in SERS [2]. Such control on plasmonic properties is required for effective application of SERS in analytics [14,19]. We have employed these films in TDM, focusing on two AEDs (Carbamazepine—CBZ, Perampanel—PER). In the case of CBZ, we prove that a SERS substrate can be re-utilized at least five times by washing it with methanol. We also show the good linear dependence of the SERS signal vs. CBZ concentration in the range 2.5×10^{-5} M to 2.1×10^{-4} M. In the case of PER, we exploit the protonation mechanism of the drug by HCl, as suggested by observed changes of the C=O stretching transition between Raman and SERS, and UV-Vis data taken on PER in acidic conditions. This paves the way to the control of the chemical enhancement pathway in SERS, which is triggered by effective chemical interaction of the analyte and the noble metal surface.

2. Experimental

2.1. Production of Au Substrates by PLA in High-Pressure Inert Gas

Au films were prepared at room temperature in a vacuum chamber, starting from a base pressure lower than 10^{-4} Pa using a KrF excimer laser (wavelength 248 nm, pulse width 25 ns, repetition rate 10 Hz, incidence angle 45°) focused onto an elemental target (Au, 99.99%) mounted on a rotating holder. The films were deposited onto pieces of glass, or (100) Si placed in front of the target at a distance of 35 mm. The target holder was rotated to avoid cratering of the target surface under repetitive ablation. Ablation was performed in Ar atmosphere at 100 Pa, with N_{LP} fixed at 2×10^4, and the laser fluence (f) kept constant at $f = 2.0$ J· cm^{-2}. Sample surface nanostructuring was observed by scanning electron microscopy (SEM) using a Zeiss Supra 40 field ion microscope (Carl Zeiss NV, Via Varesina 162, 20156 Milano, Italy). UV–vis spectroscopy measurements of SPR were performed with a PerkinElmer UV-Vis/NIR Lambda 750 spectrophotometer (PerkinElmer Italia Spa, Viale dell'Innovazione 3, 20126 Milano, Italy) over the range 190–900 nm.

2.2. Production of Ag Colloids by PLAL using Water

We carried out PLAL of an elemental target (Ag, 99.9%) in deionized water using the second harmonic (532 nm) of a laser operating at 100 kHz repetition rate with pulse width of 6–8 ps. Ablation was performed at $f = 1.5$ J· cm^{-2}, for an irradiation time of 10 min. The laser beam was focused with a galvanometric scanner to a spot of about 80 μm in diameter on the surface of the target that was scanned on a 10×10 mm^2 area with a scan speed of 800 mm· s^{-1}. The colloids were transferred on glass, or (100) Si supports by an ultrasonic spray-casting deposition method. The experimental setup consists of a deposition chamber equipped with an ultrasonic atomizer (Sonics VCX 130 W, Sonics & Materials Inc., 53 Church Hill Rd, Newtown, CT 06470, USA), a heated substrate holder and a system to remove excess vapors, thus guaranteeing standard and reproducible conditions. By ultrasonic spraying, we deposited a fraction of the produced colloids on nickel grids to perform Scanning Transmission Electron Microscopy (STEM), using an instrument operating at the primary accelerating voltage of 30 kV, at a working distance of 4 mm (Zeiss model Merlin Gemini 2).

By the same ultrasonic spraying procedure, we deposited Ag NPs onto (100) Si supports, obtaining substrates suitable for SERS measurements. In Figure 2, we show a representative STEM image (a), the average size distribution (b), and the optical absorbance spectrum (c) of Ag colloids prepared by ps-PLAL in water at the optimized laser fluence $f = 1.5$ J· cm^{-2}. Nearly spherical NPs, whose size is about 15 nm, result from the likely agglomeration and overlap of smaller NPs (see Figure 2a). The UV-Vis absorption spectrum (Figure 2c) displays a narrow SPR, as expected on the basis of the narrow size distribution of the constituent NPs. This is the outcome of the optimization of the laser fluence. In the range 0.5 J· cm$^{-2} \leq f \leq 1.5$ J· cm^{-2}, the SPR peak intensity increases and its FWHM decreases on increasing the laser fluence, keeping fixed all other deposition parameters.

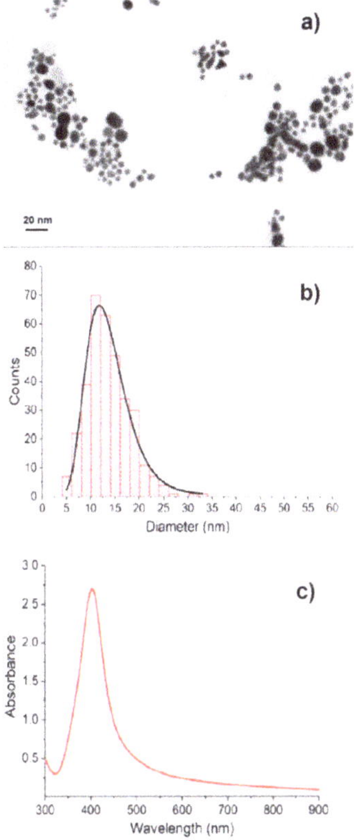

Figure 2. (**a**) Scanning Transmission Electron Microscope (STEM) images; (**b**) size distribution of nanoparticles (NPs); (**c**) optical absorbance spectrum of Ag colloids prepared by picosecond pulsed laser ablation in liquid (ps-PLAL) at the laser fluence of $1.5\,\text{J}\cdot\text{cm}^{-2}$.

2.3. Raman Spectroscopy

Raman and SERS spectra were collected by a HORIBA Jobin–Yvon LabRAM HR800 Raman Spectrometer(HORIBA France SAS, 231 rue de Lille, 59650 Villeneuve d'Ascq, France) with a solid-state laser (Laser XTRA, Toptica Photonics, TOPTICA Photonics AG, Lochhamer Schlag 19, 82166 Graefelfing (Munich), Germany) operating at 785 nm, equipped with a 600 grooves· mm^{-1} grating, a Peltier-cooled Charge-Coupled Device (CCD) detector, and notch filters to suppress Rayleigh scattering contributions. The same Raman spectrometer can be operated also with the 458 nm excitation from an Ar-ion laser.

3. Results and Discussion

The ability to produce nanostructured surfaces with a highly uniform morphology allows for designing sensors with a SERS signal of adequate reproducibility. This condition is mandatory when such sensors are used to detect analytes at low concentrations (as for drugs). In the recent past, we developed two complementary approaches to produce SERS sensors by exploiting the remarkable control in the production of nanoparticles offered by laser ablation techniques. The high control and reproducibility of these manufacturing processes allow the production of nanomaterials with optimized morphology, high sensitivity and spatial uniformity.

The first approach discussed in Section 3.1 below employs ablation in high-pressure inert gas to produce Au sensors, which we tested on the AED Carbamazepine (Figure 3a). In the second approach, which we considered at a later time, Ag colloids are produced by ablation of the target in water, and are subsequently sprayed on a support to obtain the thin film SERS sensor. We show in Section 3.2 below our early and promising results obtained by testing these PLAL sensors on a second AED, namely Perampanel (Figure 3b). The choice of Ag allows for using HCl to control pH, fostering protonation of the drug, and providing chloride ions which are known to promote SERS action on silver [20]. Carbamazepine (CBZ) is a well-established drug largely used in developing Countries, whereas Perampanel (PER) is a new generation AED.

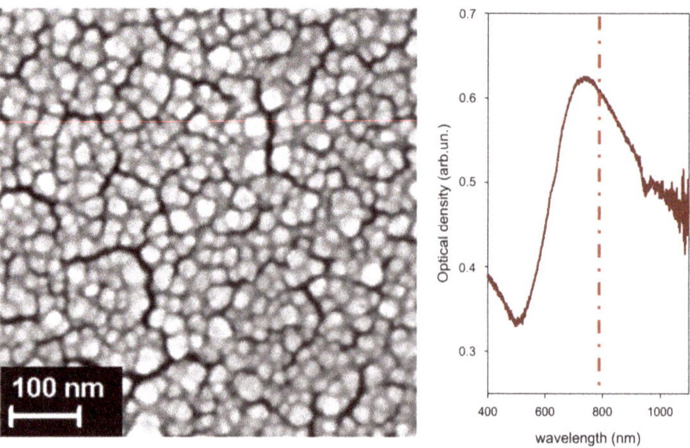

Figure 3. The chemical structure of (**a**) Carbamazepine, and (**b**) Perampanel.

3.1. Quantitative SERS Detection of Carbamazepine

Nanostructured films made of arrays of NPs produced by Pulsed Laser Deposition (PLD) of a solid Au target in high-pressure inert gas and mutually assembled on an inert (glass) support behave as SERS sensors with good performances [16].

In Figure 4, we show the surface nanostructure of an Au film deposited on (100) Si (ablation in Ar at 100 Pa; $N_{LP} = 2 \times 10^4$). As shown in Figure 4, the SPR of films produced with this set of process parameters and deposited on glass matches the popular 785 nm laser excitation found in commercial Raman instruments, including portable ones. We used substrates of this kind, either deposited on glass, or on Si, throughout all investigations on CBZ. Part of the data we discuss on CBZ was presented in [6]. Here, we complete the data analysis including additional SERS markers, and we show in detail the evolution of SERS markers upon the washing procedure of a sensor.

Figure 4. Surface nanostructure of an Au substrate deposited on (100) Si (ablation in Ar at 100 Pa, $N_{LP} = 2 \times 10^4$). The pertinent UV-Vis spectrum with the position of the exciting laser radiation (785 nm) is reported.

As reported elsewhere [16], there is agreement between the Raman and SERS features of CBZ over a wide wavenumber range. This is suggestive of a comparatively weak interaction between the Au substrate and the CBZ molecule. This fact, together with the remarkable stability of the substrates opens the way to recycle them, adopting a washing procedure with MeOH between consecutive measurements of drug concentration. The results of such washing procedure are shown in Figure 5. These results prove the following: (i) the background of the pristine sensor is blank in the spectral region of the CBZ markers; (ii) washing the sensor by immersion in MeOH for 5 min effectively removes the drug from the active surface as supported by the disappearance of the SERS markers; (iii) the successive immersion of the sensor in CBZ solution restores the initial SERS signal (i.e., the functioning of the sensor is preserved, even after five cycles of operation—which is enough to obtain the calibration curve of the sensor for four different values of concentration, see below); (iv) the interaction between CBZ and Au is weak (physisorption). Point (iv), together with the large wavelength distance between the absorption peak of CBZ (285 nm—see Figure 6) and the plasmon resonance (735 nm, very close to the laser excitation at 785 nm), indicate that the SERS of CBZ is electromagnetic in nature [21].

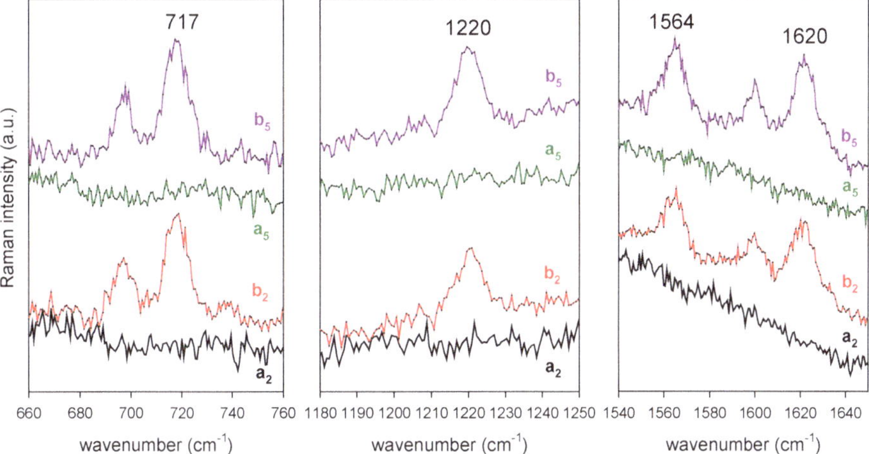

Figure 5. Effect of repeated immersions of a single Au SERS sensor (deposition on glass; ablation in Ar at 100 Pa, $N_{LP} = 2 \times 10^4$) in a concentrated CBZ solution (100 mg/L in MeOH, 60 s immersion time) and subsequent washing with MeOH. The labels indicate the spectrum recorded after washing with MeOH (**a**) and after immersion in the CBZ solution (**b**). The pedix is the step number of this repeated sequence of experiments.

After proving the re-usability of such SERS sensors, we quantified their response as a function of CBZ concentration in a range which includes the therapeutic range (2.5×10^{-5} M–5.1×10^{-5} M). Out of the seven available SERS markers [16], we selected those which show up more clearly from the background (717, 1220, 1564 and 1619 cm⁻¹—see Table 1). The measurements shown in Figure 7 were performed by using *one single substrate* at increasing CBZ concentration. Spectra were recorded by taking 3 averages of 10 s each. The laser power at the sample surface was 1 mW over a spot area of 1 um in diameter. The substrate was dipped for 5 s in a 1 mL volume of the CBZ solution, and dried before recording. The substrate was washed in MeOH for 5 min after each SERS measurement. Control SERS spectra were taken after each washing procedure showing the complete disappearance of all CBZ features. The sensitivity to drug concentration is evident in Figure 7 for all the selected CBZ markers.

Table 1. Assignment of the SERS markers of Carbamazepine (CBZ) selected in this work (after [16]).

Wavenumber (cm^{-1})	Description
717	C-H out-of-plane bending at C=C and aryl groups
1220	C-H in-plane bending at C=C and aryl groups
1564	ring stretching
1619	C=C stretching

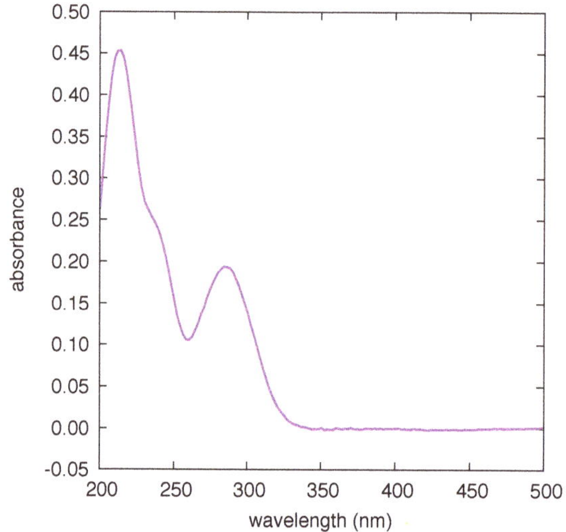

Figure 6. UV-Vis absorption spectra of a methanol solution of Carbamazepine (CBZ) (2.6×10^{-5} M, 1 cm pathlength quartz cuvette, Jasco V-570 spectrometer (Jasco Europe S.R.L., Via Luigi Cadorna 1, 23894 Cremella (LC), Italy)).

In Figure 8, we report the SERS spectrum of CBZ at the total concentration of 5.0×10^{-5} M in blood serum from an epileptic patient. We used an Au substrate deposited on (100)Si to record the SERS spectrum adopting a higher laser power (10 mW) and a longer integration time (100 s). We observe four SERS signatures coincident with CBZ signatures, notwithstanding the large fraction (about 70%) of CBZ bound to albumin, as well as the complexity of a biological matrix such as the blood serum has. It is interesting to notice that the peculiar nanostructure of our SERS sensors produced by PLD in gas results in plasmonic hot spots localized at the narrow (few nm wide) channels that separate from each other Au islands made by agglomerated NPs on the support [2]. When samples are composed by molecules characterized by different molecular weights, and thus different diffusivities, this unique substrate morphology may enhance the probability that small molecules (such as drugs) reach the hot spots with respect to more bulky species.

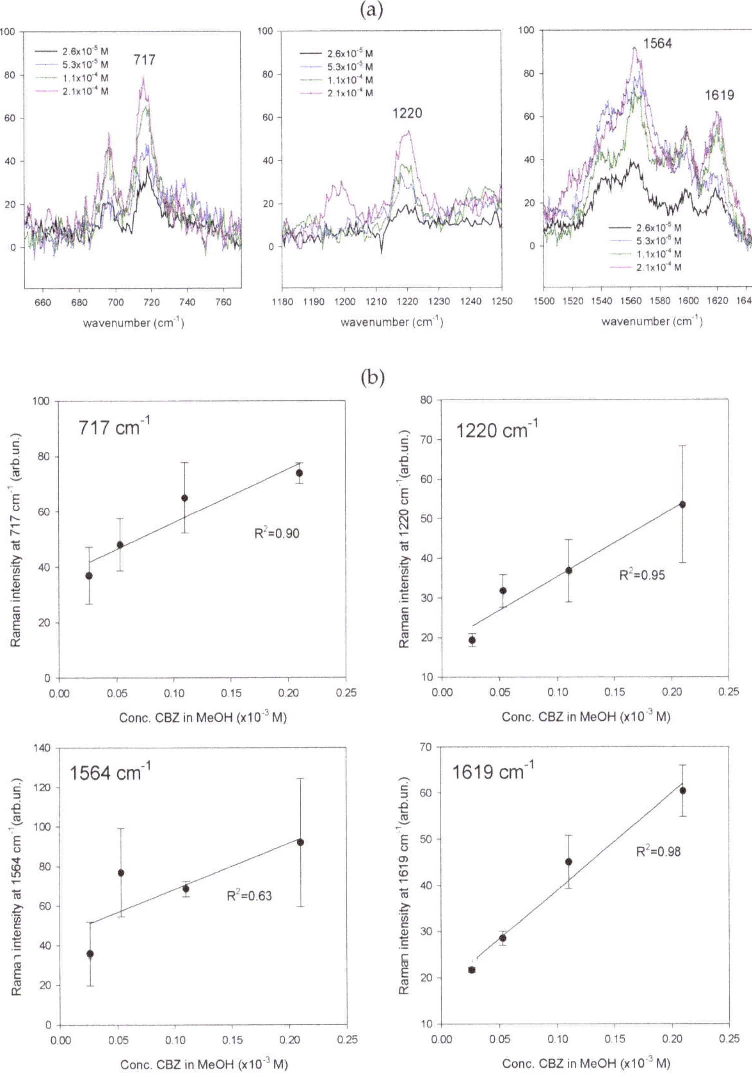

Figure 7. (**a**) average SERS spectra of CBZ in MeOH at different concentrations including the therapeutic range, in selected regions where the four CBZ markers are located (717, 1220, 1564 and 1619 cm^{-1}); (**b**) calibration curves obtained by averaging the SERS intensities recorded at each peak vs. the corresponding CBZ concentration (range 2.5×10^{-5} to 2.1×10^{-4} M). Averages were done after baseline subtraction over a minimum of 3 to a maximum of 5 spectra, depending on concentration; vertical bars are the corresponding standard deviations.

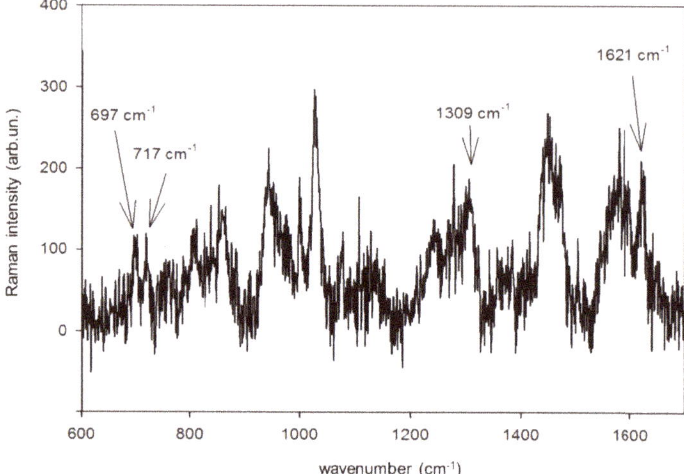

Figure 8. SERS spectrum of CBZ at the total concentration of 5.0×10^{-5} M in blood serum, recorded on an Au substrate deposited on (100)Si (Ar at 100 Pa, $N_{LP} = 2 \times 10^4$).

3.2. SERS of Perampanel in Acidic Water Solutions

For SERS measurements of PER, we used an Ag substrate ultrasonically sprayed on a (100)Si support starting from colloids synthesized by ps-PLAL (see Experimental). The surface nanostructure (Figure 9a) consists of nearly spherical NPs in part coalesced to give spheroidal shapes and more complex agglomerates. The UV-Vis absorption spectrum from a companion film sprayed on glass is shown in Figure 9b.

Moving from our previous observation that an acidic environment (using HCl) fosters SERS on such Ag substrates through the protonation of PER, extracted from Fycompa® tablets [22], we prepared aqueous solutions of PER (Cayman Chemical Item No. 23003; CAS 380917-97-5) at the concentration of 5×10^{-5} M at different pH values, starting from a concentrated methanol solution of PER in water acidified with HCl. The pH was checked each time before adding the drug. Our preparation procedure of PER solutions suggests that the charge state of PER plays a relevant role to SERS measurements, thus making evident the importance of the chemical enhancement mechanism [23].

In Figure 10, we show that the absorbance of PER displays a systematic dependence on the pH of the solution. The protonation process appears to start at pH 3, and is achieved at pH 2 (and lower). We tested the SERS performance of the Ag substrate on the aqueous solution of protonated PER at pH 2. We prepared a 3×10^{-4} M aqueous solution of PER by mixing a suitable amount of a concentrated methanol solution of PER in water acidified with a mixture of HCl and H_2SO_4 in a 1:9 molar ratio. This expedient allows for limiting the quantity of Cl^- in the solution. Indeed, we verified that excessive amounts of Cl^- lead to the preferential formation of Ag-Cl surface bonds. The pH was checked before adding PER. The SERS spectrum was recorded with 458 nm excitation wavelength through a 50× microscope objective (NA = 0.75) at the solid–liquid interface formed by the solution droplet on the substrate, with 20 s exposure time (2 averages) and a laser power at the sample of 10 mW, on a spot area of 1 µm in diameter.

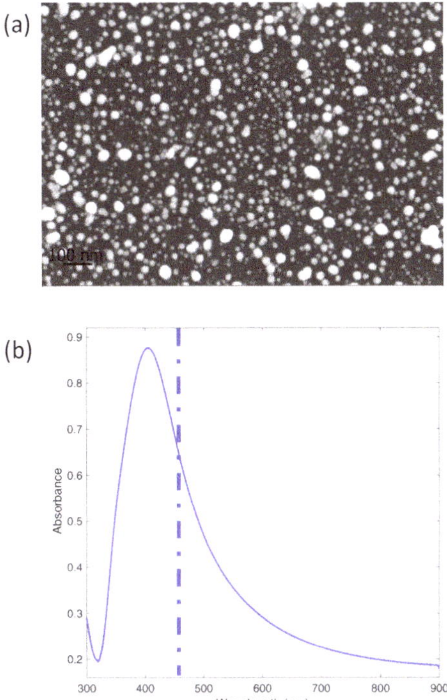

Figure 9. (a) surface nanostructure of an Ag substrate prepared by ps-PLAL at f = 1.5 J $\cdot$ cm^{-2} on a (100)Si support; (b) UV-vis spectrum taken on an analogous film sprayed on a glass support; the position of the exciting laser radiation is reported.

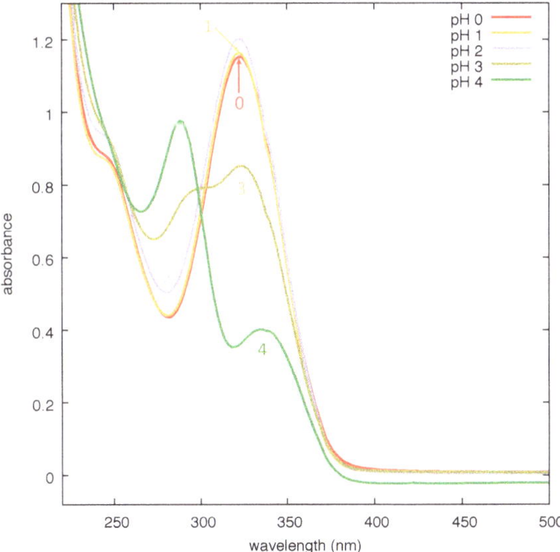

Figure 10. UV-Vis absorption spectra of aqueous solutions of Perampanel (PER) (5 $\times$ 10^{-5} M) prepared at various pH (1 cm pathlength quartz cuvette, Jasco V-570 spectrometer). The label on each curve corresponds to the pH value.

In Figure 11, the SERS spectrum is compared with the Raman spectrum of solid PER, as received (laser excitation 785 nm). We observe that our early attempts to take SERS from *neutral* PER solutions failed. Remarkably, many of the major Raman features of solid PER appear in the SERS spectrum of an *acidic* PER solution (Figure 11). In particular, the following SERS lines of PER can be observed with a good signal to noise ratio: 666, 830, 877, 1000, 1018, 1158 (broad), 1225, 1279, 1394, 1447, 1483, 1514, 1599, 2231 cm^{-1}. Most of these lines match with those recently measured for protonated PER on Au substrates produced by PLA [24], namely 670 (666, this work), 875 (877), 1000 (1000), 1135 (1158) cm^{-1}.

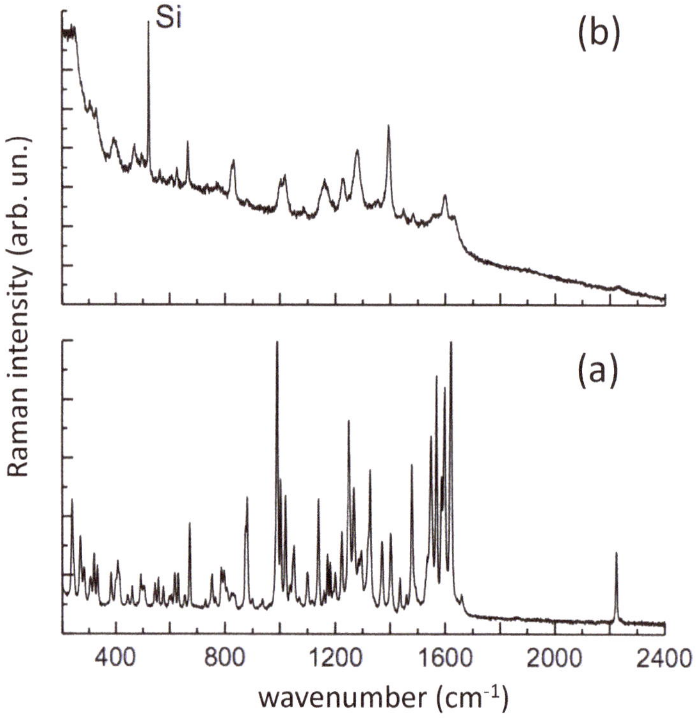

Figure 11. (**a**) Raman spectrum of solid Perampanel (PER) as received (785 nm excitation); (**b**) SERS spectrum of 3×10^{-4} M PER aqueous solution prepared at pH 2 with a mixture of HCl:H$_2$SO$_4$ in a 1:9 molar ratio. The SERS spectrum (458 nm excitation) was recorded at the solid–liquid interface of an Ag substrate prepared by ps-PLAL at $f = 1.5$ J · cm^{-2} on a (100)Si support.

Based on Density Functional Theory calculations (DFT B3LYP/6-31G(d,p)) [24], the principal Raman lines of PER can be assigned as in Table 2. The additional SERS lines of PER which can be detected with the Ag substrate produced by PLAL in this work are the following: 830, 1018, 1225, 1279, 1394, 1447, 1483, 1514, 1599, 2231 cm^{-1} (see Table 2 for details). Among these SERS signals, those assigned to a collective in-plane C-H bending, coupled with collective ring deformations, can be considered as a fingerprint of PER. These SERS signals find their correspondence in the Raman spectrum of the solid (1436, 1478, 1549, 1569, 1598, 1619 cm^{-1}). The changes in relative intensity and the slight wavenumber shifts can be explained by the fact that SERS features belong to the protonated PER, whereas the Raman spectrum was recorded on the neutral species.

Table 2. Assignment of the markers of Perampanel (PER) discussed in this work (after [24]).

Wavenumber (cm^{-1})	Description
666	collective in-plane C-H bending
830	out-of-plane C-H bending
877	collective out-of-plane C-H bending of the three outer rings of PER
1000	trigonal ring deformation of the three outer rings of PER
1018	ring deformation
1158	collective C-H in-plane bending of the ring carrying the C≡N group
1225	in-plane C-H bending, ring deformation
1279	in-plane C-H bending, ring deformation of heterocycles
1394	in-plane C-H bending, central ring deformation
1447	in-plane C-H bending, pyridine ring deformation
1483	collective in-plane C-H bending coupled with collectve ring deformations
1514	ring deformation of the only phenyl group of PER, in-plane C-H bending
1599	ring stretching
2231	C≡N stretching

Notably, the weak C=O stretching peak of solid PER (1658 cm^{-1}) disappears in the SERS spectrum of protonated PER. We deduce that the protonation of PER most likely occurs on the carbonyl.

4. Conclusions

We exploited the production by laser ablation of noble metal substrates engineered at the nanometer level with specific optical properties to perform SERS of selected AEDs toward a clinical application of the technique. An optimization of the substrate performance and the use of portable, possibly miniaturized Raman spectrometers are the next steps to address a point-of-care perspective.

Author Contributions: Conceptualization, M.T., F.N., P.M.O., S.T., E.F., M.C., E.C. and U.d.G.; methodology, C.Z., A.L., E.F. and S.T.; investigation, A.B., N.S.V., M.S., A.L., C.Z. and S.T.; resources, M.C., E.C. and U.d.G.; data curation, C.Z., E.F. and S.T.; writing, P.M.O., M.T. and C.Z.; supervision, M.T., P.M.O., C.Z., A.L., E.F. and F.N.

Funding: This research received no external funding.

Conflicts of Interest: The authors declare no conflict of interest.

Abbreviations

The following abbreviations are used in this manuscript:

AEDs	Antiepileptic Drugs
CBZ	Carbamazepine
FWHM	Full Width at Half Maximum
HPLC	High-Performance Liquid Chromatography
N_{LP}	number of laser pulses
NPs	nanoparticles
NTI	narrow therapeutic index
p_g	ambient gas pressure
PER	Perampanel
PLA	Pulsed Laser Ablation
PLAL	Pulsed Laser Ablation in Liquid
PLD	Pulsed Laser Deposition
SEM	Scanning Electron Microscopy
SERS	Surface Enhanced Raman Spectroscopy
SPR	Surface Plasmon Resonance
STEM	Scanning Transmission Electron Microscopy
TDM	Therapeutic Drug Monitoring
TEM	Transmission Electron Microscopy

References

1. Solís, D.M.; Taboada, J.M.; Obelleiro, F.; Liz-Marzán, L.M.; García de Abajo, F.J. Optimization of Nanoparticle-Based SERS Substrates through Large-Scale Realistic Simulations. *ACS Photonics* **2017**, *4*, 329–337. [CrossRef]
2. Agarwal, N.R.; Ossi, P.M.; Trusso, S. Driving electromagnetic field enhancements in tailored gold surface nanostructures: Optical properties and macroscale simulations. *Appl. Surf. Sci.* **2019**, *466*, 19–27. [CrossRef]
3. Halas, N.J.; Moskovits, M. Surface-enhanced Raman spectroscopy: Substrates and materials for research and applications. *MRS Bull.* **2013**, *38*, 607–611. [CrossRef]
4. Ossi, P.M.; Bailini, A. Cluster growth in an ablation plume propagating through a buffer gas. *Appl. Phys. A* **2008**, *93*, 645–650. [CrossRef]
5. Neri, F.; Ossi, P.M.; Trusso, S. Cluster Synthesis and Assembling in a laser-generated plasmas. *Riv. Nuovo Cim.* **2011**, *34*, 103–149.
6. Tommasini, M.; Zanchi, C.; Lucotti, A.; Fazio, E.; Santoro, M.; Spadaro, S.; Neri, F.; Trusso, S.; Ciusani, E.; de Grazia, U.; et al. Laser Synthesized Nanoparticles for Therapeutic Drug Monitoring. In *Advances in the Application of Lasers in Materials Science*; Ossi, P.M., Ed.; Springer Nature Switzerland: Berlin, Germany, 2018; Chapter 12. [CrossRef]
7. Bailini, A.; Ossi, P.M. Expansion of an ablation plume in a buffer gas and cluster growth. *EPL* **2007**, *79*, 35002. [CrossRef]
8. Acacia, N.; Barreca, F.; Barletta, E.; Spadaro, D.; Currò, G.; Neri, F Laser ablation synthesis of indium oxide nanoparticles in water. *Appl. Surf. Sci.* **2010**, *256*, 6918–6922. [CrossRef]
9. Schwenke, A.; Wagener, P.; Nolte, S.; Barcikowski, S. Influence of processing time on nanoparticle generation during picosecond-pulsed fundamental and second harmonic laser ablation of metals in tetrahydrofuran. *Appl. Phys. A* **2011**, *104*, 77–82. [CrossRef]
10. Miotello, R.K.A. Comments on explosive mechanisms of laser sputtering. *Appl. Surf. Sci.* **1996**, *96*, 205–215. [CrossRef]
11. Pathak, K.; Povitsky, A. Plume dynamics and shielding characteristics of nanosecond scale multiple pulse in carbon ablation. *J. Appl. Phys.* **2008**, *104*, 113108. [CrossRef]
12. Zheng, Y.B.; Kiraly, B.; Weiss, P.S.; Huang, T.J. Molecular plasmonics for biology and nanomedicine. *Nanomedicine* **2012**, *7*, 751–770. [CrossRef]
13. Hidi, I.J.; Mühlig, A.; Jahn, M.; Liebold, F.; Cialla, D.; Weber, K.; Popp, J. LOC-SERS: Towards point-of-care diagnostic of methotrexate. *Anal. Methods* **2014**, *6*, 3943–3947. [CrossRef]
14. Jaworska, A.; Fornasaro, S.; Sergo, V.; Bonifacio, A. Potential of Surface Enhanced Raman Spectroscopy (SERS) in Therapeutic Drug Monitoring (TDM). A Critical Review. *Biosensors* **2016**, *6*, 47. [CrossRef]
15. Litti, L.; Amendola, V.; Toffoli, G.; Meneghetti, M. Detection of low-quantity anticancer drugs by surface-enhanced Raman scattering. *Anal. Bioanal. Chem.* **2016**, *408*, 2123–2131. [CrossRef]
16. Zanchi, C.; Lucotti, A.; Tommasini, M.; Trusso, S.; de Grazia, U.; Ciusani, E.; Ossi, P.M. Laser tailored nanoparticle arrays to detect molecules at dilute concentration. *Appl. Surf. Sci.* **2017**, *396*, 1866–1874. [CrossRef]
17. Litti, L.; Ramundo, A.; Biscaglia, F.; Toffoli, G.; Gobbo, M.; Meneghetti, M. A surface enhanced Raman scattering based colloid nanosensor for developing therapeutic drug monitoring. *J. Colloid Interface Sci.* **2019**, *533*, 621–626. [CrossRef]
18. McLaughlin, C.; MacMillan, D.; McCardle, C.; Smith, W.E. Quantitative Analysis of Mitoxantrone by Surface-Enhanced Resonance Raman Scattering. *Anal. Chem.* **2002**, *74*, 3160–3167. [CrossRef]
19. Kuttner, C. Plasmonics in Sensing: From Colorimetry to SERS Analytics. In *Plasmonics*; Gric, T., Ed.; IntechOpen: Rijeka, Croatia, 2018; Chapter 9.
20. Muniz-Miranda, M.; Sbrana, G. Quantitative Determination of the Surface Concentration of Phenazine Adsorbed on Silver Colloidal Particles and Relationship with the SERS Enhancement Factor. *J. Phys. Chem. B* **1999**, *103*, 10639–10643. [CrossRef]
21. Yamamoto, Y.S.; Itoh, T. Why and how do the shapes of surface-enhanced Raman scattering spectra change? Recent progress from mechanistic studies. *J. Raman Spectrosc.* **2016**, *47*, 78–88. [CrossRef]

22. Santoro, M.; Fazio, E.; Trusso, S.; Tommasini, M.; Lucotti, A.; Saija, R.; Casazza, M.; Neri, F.; Ossi, P.M. SERS sensing of perampanel with nanostructured arrays of gold particles produced by pulsed laser ablation in water. *Med Devices Sens.* **2018**, *1*, e10003. [CrossRef]
23. Morton, S.M.; Silverstein, D.W.; Jensen, L. Theoretical Studies of Plasmonics using Electronic Structure Methods. *Chem. Rev.* **2011**, *111*, 3962–3994. [CrossRef] [PubMed]
24. Zanchi, C.; Lucotti, A.; Tommasini, M.; Pistaffa, M.; Giuliani, L.; Trusso, S.; Ossi, P.M. Pulsed laser deposition of gold thin films with long-range spatial uniform SERS activity. *Appl. Phys. A* **2019**, *125*, 311. [CrossRef]

MDPI

St. Alban-Anlage 66

4052 Basel

Switzerland

Tel. +41 61 683 77 34

Fax +41 61 302 89 18

www.mdpi.com

Nanomaterials Editorial Office

E-mail: nanomaterials@mdpi.com

www.mdpi.com/journal/nanomaterials

9 783039 284542